HUMAN FACTORS
IN
INTELLIGENT
TRANSPORTATION
SYSTEMS

HUMAN FACTORS IN TRANSPORTATION
A Series of Volumes Edited by
Barry H. Kantowitz

Bainbridge • *Complex Cognition and the Implications for Design*

Barfield/Dingus • *Human Factors in Intelligent Transportation Systems*

Billings • *Aviation Automation: The Search for a Human-Centered Approach*

Garland/Wise/Hopkin • *Aviation Human Factors*

Noy • *Ergonomics and Safety of Intelligent Driver Interfaces*

Parasuraman/Mouloua • *Automation and Human Performance: Theory and Application*

HUMAN FACTORS IN INTELLIGENT TRANSPORTATION SYSTEMS

Edited by

WOODROW BARFIELD
THOMAS A. DINGUS
Virginia Polytechnic Institute and State University

LEA   **LAWRENCE ERLBAUM ASSOCIATES, PUBLISHERS**
1998   **Mahwah, New Jersey**   **London**

Lawrence Erlbaum Associates, Inc., Publishers
10 Industrial Avenue
Mahwah, New Jersey 07430

Library of Congress Cataloging-in-Publication Data

Human factors in intelligent transportation systems / edited by
  Woodrow Barfield, Thomas A. Dingus.
      p.   cm. — (Human factors in transportation series)
    Includes bibliographical references and indexes.
    ISBN 0-8058-1433-7 (cloth : alk. paper). — ISBN 0-8058-1434-5
  (pbk. : alk. paper)
    1. Roads—Social aspects—United States. 2. Intelligent Vehicle
  Highway Systems. I. Barfield, Woodrow. II. Dingus, Thomas A.
  III. Series: Human factors in transportation.
  HE355.H94 1997
  388.1'0973—dc21   97-14434
                                                        CIP

Printed in the United States of America
10 9 8 7 6 5 4 3 2 1

CONTENTS

v

SERIES FOREWORD

Barry H. Kantowitz
Battelle Human Factors Transportation Center

The domain of transportation is important for both practical and theoretical reasons. All of us are users of transportation systems as operators, passengers, and consumers. From a scientific viewpoint, the transportation domain offers an opportunity to create and test sophisticated models of human behavior and cognition. This series covers both practical and theoretical aspects of human factors in transportation, with an emphasis on their interaction.

The series is intended as a forum for researchers and engineers interested in how people function within transportation systems. All modes of transportation are relevant, and all human factors and ergonomic efforts that have explicit implications for transportation systems fall within the series purview. Analytic efforts are important to link theory and data. The level of analysis can be as small as one person, or international in scope. Empirical data can be from a broad range of methodologies, including laboratory research, simulator studies, test tracks, operational tests, fieldwork, design reviews, or surveys. This broad scope is intended to maximize the utility of the series for readers with diverse backgrounds.

I expect the series to be useful for professionals in the disciplines of human factors, ergonomics, transportation engineering, experimental psychology, cognitive science, sociology, and safety engineering. It is intended to appeal to the transportation specialist in industry, government, or academic, as well as the researcher in need of a testbed for new ideas about the interface between people and complex systems.

This volume focuses on intelligent transportation systems, an exciting new research area for human factors. It is far too expensive to increase the capacity of the U.S. highway system by building more and more new roads. Instead, the Federal Highway Administration is funding research to demonstrate that adding intelligence to the system is a cost-effective solution. Because the human driver is the key to successful implementation of this solution, human factors plays a prominent role in intelligent transportation systems. The chapters in this book represent the cutting edge of this human factors research. Both practical and theoretical aspects of transportation human factors are illustrated in these chapters, thus achieving a major goal of this book series.

CONTRIBUTORS

Woodrow Barfield
Industrial and Systems Engineering
302 Whittemore Hall
Virginia Tech
Blacksburg, VA 24061

John R. Bloomfield
Principal Research Scientist
Center for Computer Aided Design
  and Simulation
208 Engineering Research Facility
University of Iowa
Iowa City, IA 52242

John L. Campbell
Battelle Research Corporation
P.O. Box 5395
40000 NE 41st Street
Seattle, WA 98105

Thomas A. Dingus
Center for Transportation Research
1700 Kraft Drive, Suite 2000
Blacksburg, VA 24061

Rebecca N. Fleischman
1906 S.W. Edgewood Road
Portland, OR 97201

Dennis J. Folds
Georgia Tech Research Institute
Georgia Institute of Technology
Atlanta, GA 30332

Martha N. Hanley
Hughes Aircraft Company
P.O. Box 92426
RE-RO-R512
Los Angeles, CA 90009

Cheryl M. Hein
Hughes Research Laboratories
3011 Malibu Canyon Road
MA-254-RL96
Malibu, CA 90265

Avraham D. Horowitz
Staff Research Scientist
General Motors R&D Center
Warren, MI 48090

Melissa C. Hulse
Performance and Safety Sciences, Inc.
1610 Luster's Gate Road
Blacksburg, VA 24060

Steven K. Jahns
Project Engineer
PACCAR Technical Center
1261 Farm to Market Road
Mount Vernon, WA 98273

Barry H. Kantowitz
Battelle Research Corporation
P.O. Box 5395
40000 NE 41st Street
Seattle, WA 98105

Michael J. Kelly
Head, Human Factors Branch
Georgia Tech Research Institute
Georgia Institute of Technology
Atlanta, GA 30332

Rhonda A. Kinghorn
Battelle Research Corporation
P.O. Box 5395
40000 NE 41st Street
Seattle, WA 98105

Ronald Knipling
Chief, Research Division
U.S. Department of Transportation
NHTSA, 400 7th Street SW
FHWA OMC HCS30
Washington, DC 20590

Francine H. Landau
Hughes Electronics Corporation, retired
2813 Moraga Drive
Los Angeles, CA 90077

John D. Lee
Battelle Research Corporation
P.O. Box 5395
40000 NE 41st Street
Seattle, WA 98105

Lee Levitan
Senior Principal Research Scientist
Honeywell Inc.
HTC, 3660 Technology Drive
MN 65-2600
Minneapolis, MN 55418

Fred Mannering
Department of Civil Engineering
University of Washington
Seattle, WA 98195

Truman Mast
Office of Safety and Traffic Operations
  Research and Development
Federal Highway Administration
Turner Fairbanks Highway Research
  Center
6300 Georgetownpike
McLain, VA 22101

Ann E. Miller
2747 S.E. 35th Ave.
Portland, OR 97202

Linda Ng
Department of Civil Engineering
University of Washington
Seattle, WA 98195

Jan H. Spyridakis
Department of Technical Communication
Box 352195
University of Washington
Seattle, WA 98195

William A. Wheeler
Senior Research Fellow
Accident Research Centre
Monash University
Clayton, VIC 3168
Australia

PREFACE

Techniques to improve driving safety or to reduce urban congestion have often focused on capital intensive strategies such as the development of new roads, new road designs, or light rail to increase the system capacity. An alternative strategy that is currently being explored, and which is less capital intensive, is to design an "intelligent transportation system" (ITS). Such a system will provide the driver with a vast variety of information—information about alternative routes or roadside services to information about crash avoidance.

However, an ITS is only one of many possible approaches to solving the critical and difficult problems of reducing traffic congestion or vehicle crashes, and it is highly unlikely that any single solution will produce a quick fix, given the complexity of the problems associated with the movement of people and material from one location to another. For example, we all know that urban travel patterns are intrinsically related to developments in city structure, the location of workplaces and drivers' residences, and the characteristics and activities of drivers.

However, information-based solutions to transportation problems do present a number of attractive features. They are relatively inexpensive compared to traditional solutions to transportation problems (e.g., building new roads and bridges are very expensive); they have low social and environmental impact; and in addition to their primary goal of improving the safety and efficiency of roadway use, they can produce a number of secondary benefits by carrying public relations and educational messages.

Further, the importance of an ITS is clear when one considers the specific benefits that are expected to occur from its usage. For example, an ITS is expected to reduce traffic congestion, improve navigational performance, decrease the likelihood of accidents, reduce fuel costs and air pollution, and improve driver efficiency. The safety aspects of an ITS are expected to occur because an ITS will provide timely and accessible information delivered in the car on traffic regulations, alternative routes to take, the avoidance of hazardous situations, and safety advisory and warning messages. This information will be especially beneficial in decreased visibility situations due to poor weather or congestion.

This book is concerned with human factors issues that relate to the design and use of the ITS. Intelligent transportation systems emcompass a wide range of cutting-edge technologies being applied to the driver–vehicle–highway infrastructure system in order to improve throughput, safety, and efficiency. Generally, these technologies include advanced sensor, computer, communication (radio/optical), and control technologies which are being used to avoid crashes and regulate the flow of vehicles along roads and highways. The potential of the ITS to make a significant contribution to the problem of transporting people and goods, and the amount of funding being directed toward ITS development, has attracted attention from researchers in several fields—these include behavioral scientists, policy analysts, and others, in addition to engineers.

In this volume, a major effort is made to integrate information from diverse academic fields, including behavioral science, human factors engineering, and civil and transportation engineering, on the topic of the design and use of ITS. It is our belief that ITS technology should be designed from the user's perspective, that is, considering their needs for information, and their cognitive, motor, and visual information-processing abilities. Thus, for ITS to be successful, from a human factors perspective, it is important to determine what information should be provided to users, and in what form. The design of ITS from a user's perspective will involve many complex design issues; we expect that these issues will not be adequately addressed in a few studies, or in a relatively short time period. Rather, we advocate a continuing research program focusing on field and laboratory studies that is adequately funded by the U.S. government and appropriate industries so that our highway infrastructure remains second-to-none.

Based on the thoroughness in which a variety of topics related to ITS development and use are covered, the book can be used as a textbook on the topic of human factors in ITS for courses in civil engineering, industrial engineering, and psychology. Further, the volume will also provide a valuable technical reference source for those involved in the development of ITS technologies, or highways and transportation systems and automotive industries in general. Finally, we hope the book serves as a stimulant for more thorough research on the design and use of ITS, and that the resulting systems that comprise the ITS are "usable" from the driver and traffic engineer–manager points-of-view.

We acknowledge the help we received from Ray O'Connell and his staff at Lawrence Erlbaum Associates. Further, Sondra Guideman provided technical support in the preparation of the manuscript. Finally, without the oustanding technical contribution of the chapter authors this book would not be completed.

Woodrow Barfield
Tom Dingus

INTRODUCTION TO ITS

Truman Mast
Federal Highway Administration

WHAT IS ITS?

The Intelligent Transportation System (ITS) Program is a cooperative effort by government, private industry, and academia to apply advanced technology to resolving the problems of surface transportation. The objective is to improve travel efficiency and mobility, enhance safety, conserve energy, provide economic benefits, and protect the environment. The current demand for mobility has exceeded the available capacity of the roadway system. Because the highway system cannot be expanded, except in minor ways, the available capacity must be used more efficiently to handle the increased demand. Traffic congestion in urban areas and on heavily traveled intercity corridors continues to rise rapidly with the annual cost to the nation in lost productivity alone being about $100 billion. The enormous costs of wasted fuel and environmental damage are not included in this estimate. Moreover, traffic accidents kill more than 40,000 people and injure another five million every year (DOT IVHS Strategic Plan, 1992).

ITS will apply advanced information processing, communications, sensing, and computer control technologies to the problems of surface transportation. Considerable research and development efforts will be required to produce these new technologies and to convert technologies developed in the defense and space programs to solve surface transportation problems.

IMPORTANT ROLE OF HUMAN FACTORS

Human factors considerations are critical to the eventual success of solving surface transportation problems. ITS will substantially change the driving task and the way traffic is managed. Human factors considerations play a vital role in ITS development and application, from the design of in-vehicle navigation displays and controls, to the provision of safe and efficient transition from manual to automatic control in highly automated systems.

The extensive use of operational tests in the ITS program makes development of substantive human factors guidelines particularly important. Early and continued acceptance of ITS by the public and policy makers relates directly to the human engineering that goes into the early design prototypes. The public often believes that the system they see in operational tests will be the final product, making quality human factors imperative to the success of the program.

ITS TECHNOLOGY AREAS

ITS has been subdivided into six interlocking technology areas. This book addresses human factors concerns for four of these areas.

ATIS. Advanced Traveler Information Systems include a variety of systems that provide real-time, in-vehicle information to drivers regarding navigation and route guidance, motorist services, roadway signing, and hazard warnings.

AVCS. Advanced Vehicle Control Systems refer to systems that aid drivers in controlling their vehicle particularly in emergency situations and ultimately taking over some or all of the driving tasks.

CVO. Commercial Vehicle Operations address the application of ITS technologies to the special needs of commercial roadway vehicles including automated vehicle identification, location, weigh-in-motion, clearance sensing, and record keeping.

ATMS. Advanced Traffic Management Systems monitor, control, and manage traffic on streets and highways to reduce congestion using vehicle route diversion, automated signal timing, changeable message signs, and priority control systems.

Two technical areas are not specifically addressed here in individual chapters, but many aspects of them are covered in associated chapters. These technical areas are:

ARTS. Advanced Rural Transportation Systems include systems that apply ITS technologies to the special needs of rural systems and include emergency notification and response, vehicle location, and traveler information.

APTS. Advanced Public Transportation Systems enhance the effectiveness, attractiveness, and economics of public transportation and include fleet management, automated fare collection, and real-time information systems.

ADVANCED TRAVELER INFORMATION SYSTEMS (ATIS)

The major human factors problem area in ATIS is the design of in-vehicle displays and controls. The wide range of equipment designs and the extensive diversity of the driving population make this task particularly challenging. The four ATIS functional subareas are referred to as In-Vehicle Information Systems (IVIS) and are discussed next.

IRANS (In-Vehicle Routing and Navigation Systems)

IRANS provide individual motorists with real-time congestion information, depict the best route to their destination and provide explicit turning directions at each choice point. Use of such systems will reduce congestion because drivers will either know what alternate routes to take when a major route is congested, or can postpone discretionary trips. Safety will be increased because drivers can focus on the primary driving task. Designing safe and effective interfaces for in-vehicle navigation and routing advisory displays is an important human factors challenge for ATIS. Driver needs must be accounted for with regard to the amount and coding of information on maps and other in-vehicle displays.

IMSIS (In-Vehicle Motorist Services Information Systems)

Virtually any consumer service listed in the telephone yellow pages can be incorporated into IMSIS. In conjunction with IRANS, motorists will derive significant mobility benefits because they can easily make efficient en route trip changes and combine multiple purpose trips. Motorists will always have ready access to information on hospitals and emergency medical services, increasing safety and comfort. Driver interaction with the IMSIS is an important human factors issue. Making the systems safely accessible to drivers while traveling on high-speed expressways is highly desirable, but it definitely poses a formidable human factors challenge.

ISIS (In-Vehicle Signing Information Systems)

Highway signing deficiencies are among the most frequent complaints received by automobile clubs and traffic engineers, ranging from "the signs don't have the information I need," to "there is too much information." ISIS will supplement and augment information presented on existing on-road signs. The major human factors challenge is to design ISIS so they can be easily tailored to individual driver requirements in terms of information processing capabilities as well as individual trip needs.

IVSAWS (In-Vehicle Safety Advisory Warning Systems)

Using radio transmission devices, IVSAWS provide safety warnings for moving or fixed hazards that are beyond the drivers' sight distance. A major safety benefit will be experienced by motorists traveling in rural areas where the highest percentage of accidents occur. Human factors studies are required to determine the signal strength and sensory mode for the warning display to sufficiently attract and divert the driver's attention toward the potential hazard. To retain credibility, false alarms must be minimized, and the messages must be reserved for highly dangerous situations where the violation of driver expectancy is an important factor.

ADVANCED TRAFFIC MANAGEMENT SYSTEMS (ATMS)

ATMS integrate vehicle detection, communication, and control functions to be responsive to dynamic traffic conditions to increase the efficiency of roadway networks. The central ATMS component, the Traffic Management Center (TMC), is the element with the most human factors concerns. The operations of the TMC include receiving incoming information from a variety of sources including pavement sensors, video cameras, mobile phones, probe vehicles, police cruisers, and fleet managers. This information is fused and processed to determine the location and severity of problems and ascertain optimal re-routing strategies and appropriate changes to traffic control devices. Finally, the TMC must output the information to a variety of receivers, ranging from individual vehicles to fleet managers to traffic control devices.

Control Room Ergonomics

Optimal operator-machine interfaces must be incorporated into the TMC workstation designs. Some TMC response outputs will be determined automatically via computer simulation models and will require operator monitoring tasks. Other outputs will be carried out manually by the operator and will often require utilization of various decision aids, particularly when different sources provide conflicting incoming information.

Operator Team Configuration and Training

The number of personnel, their skill levels, and training requirements must be determined to guarantee efficient and reliable TMC operation during normal, emergency, and system failure operations. To maintain credibility, the TMC failure modes must be designed such that most breakdowns are transparent to the motoring public, making team configuration, coordination and communication among operators crucial to TMC system design.

This introduction provides a broad overview of ITS and a general discussion of some of the important human factors issues. A detailed examination of these issues and related research is presented in the following chapters.

ADVANCED VEHICLE CONTROL SYSTEMS (AVCS)

AVCS will enhance the human capacity to perceive or respond to the driving environment and will significantly change the driving task. Appropriate incorporation of human factors in the design of the AVCS will be imperative to assure that the potential safety benefits are realized and that the systems achieve public acceptance. AVCS consist of the three functional subareas discussed next.

SES (Sensory Enhancement Systems)

Sensory enhancement systems will expand human sensory capabilities beyond normal limits by providing an extended line of sight in darkness, fog, and other reduced visibility conditions. The major human factors challenges in the design of SES focus on determining what visual cues are most essential to safe operation of the vehicle and how these augmented cues should be displayed. Human factors investigations of the amount of information, location of displays, and required mapping to the external world are required.

ODAS (Obstacle Detection and Avoidance Systems)

ODAS consist of two related subsystems, obstacle detection and obstacle avoidance. Detection systems will alert drivers when their vehicle's trajectory puts them in danger of colliding with a fixed object or another vehicle.

Avoidance systems will actually take control of the vehicle and maneuver it out of harm's way in situations where human responses are not adequate to avoid a collision. Human factors investigation in this area will focus on methods of alerting drivers to hazards.

ACS (Automated Control Systems)

The technology for automatic control of vehicles exists, to some extent, today. Systems such as adaptive cruise control, in which the accelerator and brake are automatically regulated to maintain a pre-set distance from a lead car, have been demonstrated. The full realization of ACS is the Automated Highway System (AHS) in which coordinated groups of vehicles will be under automatic, centralized control. The benefits of these systems include enhanced safety and increases in throughput, as more vehicles are moved faster and more efficiently in a highly controlled environment. The most critical human factors issues in the ACS and AHS relate to methods of transferring control between the driver and the system and user acceptance. The system designs must make sure drivers do not lose situational aware-ness of the driving task, to avoid problems when control is transferred back to the driver and to assure that negative carryover effects do not jeopardize safety. With the high speeds and close headways projected in the AHS, user acceptance issues will be paramount.

COMMERCIAL VEHICLE OPERATIONS (CVO)

The term CVO includes a wide variety of commercial vehicles. It represents a unique application of ITS because of the characteristics of the subpopu-lation of drivers, vehicle design, commercial aspects, pertinent roadway features, and the nature of long-haul trucking operations. Due to the demand for efficient and economically competitive operations, CVO will be the van-guard for implementing many ITS systems.

MAJOR AREAS OF CONCERN

Regulatory Demands

Commercial vehicles are regulated by federal, state, and local governments, and enforcement of these regulations can cause delays and congestion on the road. A variety of ITS technologies are under development to automate many of these activities such as computerized record-keeping systems to file licenses and permits, report registration information, calculate fuel taxes, maintain log books, and collect tolls. The variety of regulations and reporting needs of CVO requires increased information processing by the operators. Human factors investigations addressing the combination of cognitive and physical workloads involved, as well as possible ITS-related solutions, are required.

Fleet Management

Fleet management is concerned with the timeliness and reliability of commercial carriers. ITS technologies in this area are likely to include automatic vehicle tracking and in-vehicle routing displays. Navigation and route guidance systems for CVO must accommodate specific routing regulations governing the weight, size, and cargo of the vehicle. In-vehicle displays for CVO will present more information than similar displays in personal vehicles. Furthermore, the physical environment of a truck cab including the windshield design, and the excessive noise and vibrations, present unique human factors challenges for the design of in-vehicle displays.

Safety

Many ITS technologies are designed to improve CVO safety, including ODAS, warning systems for load shifts, and impending vehicle rollover messages on freeway ramps. Driver impairment issues due to fatigue and other factors are critical in the CVO area. Human factors research will determine the optimal methods for providing safety information to CVO drivers. Furthermore, methods of detecting possible safety problems before they become emergencies, such as driver fatigue or excessive speed, will require significant human factors investments.

REFERENCE

Intelligent Vehicle Highway Society of America. (1992). *Strategic plan for intelligent vehicle-highway systems in the United States*. Washington, DC: Author.

1

DESCRIPTION AND APPLICATIONS OF ADVANCED TRAVELER INFORMATION SYSTEMS

Melissa C. Hulse
Performance and Safety Sciences, Inc.

Thomas A. Dingus
Woodrow Barfield
Virginia Polytechnic Institute and State University

To alleviate traffic congestion and to more effectively use existing transportation resources, a major national and international effort is currently made to integrate knowledge on driver behavior and decision making into the design of Advanced Traveler Information systems (ATISs). In-vehicle ATISs are often viewed as one of the cornerstones of an Intelligent Transportation System (ITS), also known as an Intelligent Vehicle Highway System (IVHS), which involves the use of sensor, computer, communication (radio-optical), and control technologies for regulating the flow of vehicles along roads and highways. In this chapter, we provide an overview of several existing ITSs, including a broad overview of the technologies associated with the design of an ITS. These systems are described in more detail in subsequent chapters along with the human factors issues associated with their design and use.

Techniques to reduce urban congestion have typically focused on capital-intensive strategies such as the development of new roads or light rail to increase the system capacity. However, an alternative strategy, which is less capital-intensive, calls for designing an information system based specifically on the traffic information needs of private and commercial drivers. Along these lines, a number of major new efforts to alleviate traffic congestion have centered around the development of transportation information systems. For example, foreign development efforts in the areas of motorist navigation and information systems are already well underway in West Germany, Great Britain, France, and Japan; in West Germany, Great Britain, and Japan, these efforts are already to the point of public testing. In the United States, recent initiatives in this area have been spurred by an announcement from the Federal Highway Administration (FHWA) of a High Priority National Program Area in Advanced Motorist Information Systems for Improved Traffic Operations. The single largest effort in the United States at this time is a cooperative project between the FHWA, the California Department of Transportation, and General Motors known as Pathfinder. Pathfinder, as its name implies, focuses on the assessment of communications technology for route guidance and in-car navigation in response to incidents and traffic congestion.

An ATIS is only one of many possible approaches to reducing traffic congestion, and it is highly unlikely that any single solution will produce a quick fix given the tremendous increase in the amount of traffic on our major roadways and the complexity of the problems associated with the movement of people and material from one location to another. For example, it is well known that urban travel patterns are intrinsically related to developments in city structure, the locations of workplaces, the locations of drivers and their residences, and the characteristics and activities of drivers (Barfield, Haselkorn, Spyridakis, & Conquest, 1989). However, information-based solutions to transportation problems do present a number of attractive features: They are relatively inexpensive compared to other solutions (e.g., the building of new roads); they have low social and environmental impact; and in addition to their primary goal of improving the efficiency of roadway use, they can produce a number of secondary benefits by carrying public relations and educational messages. Furthermore, the importance of an ATIS is clear when one considers the specific benefits expected from its usage. For example, an ATIS is expected to reduce traffic congestion, improve navigational performance, decrease the likelihood of accidents, reduce fuel costs and air pollution, and improve driver efficiency. Another benefit associated with an ATIS is that this system will provide safety advisory and warning messages to the motorist. The safety aspects of an ATIS are expected to occur because an ATIS will provide additional, more timely, and more accessible information on traffic regulations, guidance, and hazardous situations. This information will be especially beneficial in decreased visibility situations because of poor weather or congestion (Mobility 2000, 1989).

In the context of designing an in-vehicle transportation system, several researchers have stated that for an ATIS to be effective, its design must be based on a comprehensive understanding of the traffic information needs of drivers (Barfield, Haselkorn, Spyridakis, & Conquest, 1989; Barfield, Spyridakis, Conquest, & Haselkorn, 1989). Specifically, an ATIS is expected to affect four aspects of driver behavior: departure time, means of transportation (buses, train, car pools, etc.), pretrip route choice, and on-road route modification. However, to positively affect driver behavior, we must first understand the decision-making processes of drivers and driving behavior (Barfield, Haselkorn, Spyridakis, & Conquest, 1991; Conquest, Spyridakis, Haselkorn, & Barfield, 1993; Ng, Barfield, & Mannering, 1995; Wenger, Spyridakis, Haselkorn, Barfield, & Conquest, 1990). This idea represents the human factors approach to the design of transportation systems. The application of human factors knowledge and guidelines has already proved beneficial in the design of airplane cockpits, automobiles, computer software, and input devices, and is expected to have a major impact on the design of the ATIS. The following sections discuss the ATIS in detail, including the benefits expected from the use of the ATIS, a description of the basic ATIS components, and a description of ATISs currently in use or being developed in the United States and worldwide.

EXPECTED ATIS BENEFITS

Congestion

It has been suggested that by the year 2020, highway traffic in many areas of the United States will be reduced to 11 miles per hours (mph). In fact, the average highway speeds on some southern California freeways are already down to 31 mph, while the average speeds during the morning and evening rush hours are lower yet. Congestion is a serious problem in urban and suburban areas and will continue to be so until major steps are made to alleviate it. An ATIS is designed to give drivers real-time traffic information in their cars, allowing them to avoid areas of high congestion, select alternative modes of travel that decrease the amount of traffic on congested roads, or delay departure times, thereby further decreasing the amount of traffic during times of peak congestion. Reduced levels of congestion will improve air quality, decrease personal stress, and likely improve worker health, attitude, and job performance (Mobility 2000, 1989, p. 8).

Environment

It has been argued that implementing the ATIS will have a positive effect on the environment. With better routing information given directly to drivers through in-vehicle displays, less time will be spent driving, thus curtailing car emissions and other pollutants and also reducing the need for fuels. As noted by Mobility 2000, an "IVHS will improve energy efficiently by reducing congestion and improving travel planning and routing. . . . IVHS has environmental benefits through fuel savings, reduced vehicle emissions, and reduced noise levels" (p. 4).

Mobility

Another objective for the ATIS is to increase the mobility of vehicles, which can be accomplished by reaching other objectives, including the reduction of congestion and improving routing efficiency via the implementation of a system such as the In-Vehicle Routing and Navigation System (IRANS), an In-Vehicle Signing Information System (ISIS), the In-Vehicle Safety and Warning System (IVSAWS), and the In-Vehicle Motorist Service Information System (IMSIS). As a result, urban areas will more efficiently manage their existing streets and freeways through improved traveler information and traffic control systems, and both rural and urban area travelers will benefit from improved security, comfort, and convenience. Measured, quantified improvements to mobility will include reduced congestion, accommodation of increased travel and higher trip speeds, less motorist confusion and aggravation, augmented and enhanced driver capabilities, lower cost in the transportation element of producing goods and services, and reduced driver fatigue and frustration (Mobility 2000, 1989, p. 4).

Productivity

Individuals who drive for business-related purposes, especially carriers, can use ITS technologies as key tools to reduce costs and improve productivity. New ITS technologies allow faster dispatching, fuel-efficient routing, and more timely pickups and deliveries.

Safety

Implementing various systems into the driving scenario requires consideration of safety issues as well. Mobility 2000 notes that "many believe that IVHS technologies, such as traveler information systems providing in-vehicle advisory and warning messages, plus future control assist systems, will usher in a new, substantially increased level of motoring safety. . . . Safety benefits will be substantial; they will include reduced fatalities, injuries, and property damage. Further, reducing accidents will keep lanes open and minimize the frustration that can contribute to further accidents" (p. 4). However, if human factors knowledge is not considered in the design of the ITS, these safety benefits may not occur; in fact, the resulting systems may be highly dangerous!

OVERVIEW OF THE ATIS

To accomplish the overall ATIS goal of safer and more efficient travel, several classes of systems have been identified within the ATIS umbrella: the IRANS, IMSIS, ISIS, and IVSAWS (Perez & Mast, 1992). Lee, Morgan, Wheeler, Hulse, and Dingus (1993) outlined the following proposed functional capabilities for each of these systems.

In-vehicle Routing and Navigation System (IRANS)

An IRANS provides drivers with information about how to get from one place to another, as well as information on traffic operations and recurrent and nonrecurrent urban traffic congestion. At this time, seven functional components have been identified: (a) trip planning, (b) multimode travel coordination and planning, (c) predrive route and destination selection, (d) dynamic route selection, (e) route guidance, (f) route navigation, and (g) automated toll collection.

Trip Planning. This component involves route planning for long or multiple-destination journeys. It may involve identifying scenic routes and historical sites, as well as coordinating hotel accommodations, restaurants, and vehicle service information.

Multimode Travel Coordination and Planning. This feature provides the driver with information for coordinating different modes of transportation (such as buses, trains, and subways) in conjunction with driving a vehicle. Such information might include real-time updates of actual bus arrival times and anticipated travel times.

Predrive Route and Destination Selection. This function allows the driver to select any destination or route while the vehicle is in PARK. These predrive selections include entering and selecting the destination, a departure time, and a route to the destination. System information might include real-time or historical congestion information, estimated travel time, and routes that optimize a variety of parameters.

Dynamic Route Selection. This component encompasses any route selection system while the vehicle is not in PARK and includes presenting updated traffic and incident information that might affect the driver's route selection. In addition, the system would alert the driver if he or she makes an incorrect turn and leaves the planned route. Dynamic route selection can generate a new route that accommodates the driver's new position.

Route Guidance. This capability includes turn-by-turn and directional information and can be in the form of a highlighted route on an electronic map, icons indicating turn directions on a headup display (HUD), or a voice commanding turns.

Route Navigation. This function provides information to help the driver arrive at a selected destination, but does not include route guidance. It supplies information typically found on paper maps, which might include an in-vehicle electronic map with streets, direction orientation, current location of vehicle, destination location, and location services or attractions.

Automated Toll Collection. This system would allow a vehicle to travel along a toll roadway without stopping to pay tolls, which would be deducted automatically from the driver's account as the vehicle is driven past toll collection areas.

In-vehicle Motorist Service Information System (IMSIS)

An IMSIS provides (a) broadcast information on services or attractions, (b) access to a directory of services and attractions, (c) coordination destinations, and (d) message transfer capability.

Broadcast Information on Services–Attractions. This information is similar to what might otherwise be found on roadside signs. It may be very similar to the directory of services and attractions information, but the driver does not need to look for this broadcast information; it is presented as the vehicle travels down the road.

Directory of Services–Attractions. This directory provides information about motels; hotels; automobile fuel and service stations; and emergency medical, entertainment, and recreational services.

Coordination with Destination. This function enables the driver to communicate and make arrangements with various destinations. This may include restaurant and hotel reservations.

Message Transfer. This capability enables drivers to communicate with others while driving. Currently, cellular telephones and CB radios provide this capability. In the future, the ATIS may improve upon this technology by automatically generating preset messages at the touch of a button and by receiving messages for future use. Message transfer might involve both text and voice messages.

In-vehicle Signing Information Systems (ISIS)

An ISIS provides noncommercial routing, warning, regulatory, and advisory information.

Roadway Sign Guidance Information. Guidance information includes street signs, interchange graphics, route markers, and mile posts.

Roadway Sign Notification Information. Notification information alerts drivers to potential hazards or changes in the roadway. This information includes merge signs, advisory speed limits, chevrons, and curve arrows.

Roadway Sign Regulatory Information. Regulatory information includes speed limits, stop signs, yield signs, and turn prohibitions.

In-Vehicle Safety and Warning System (IVSAWS)

An IVSAWS provides warnings on immediate hazards and road conditions affecting the roadway ahead of the driver. It provides sufficient advanced warning to permit the driver to take remedial action such as slowing down and offers the capability of both automatic and manual aid requests. This system does not encompass in-vehicle safety warning devices for imminent danger requiring immediate action such as lane-change/blind-spot warning devices or imminent collision warning devices.

Immediate Hazard Warning. An IVSAWS may provide proximate hazard information to the driver by indicating the relative location of a hazard, the type of hazard, and the status of emergency vehicles in the area. Specifically, this might include notifying the driver of an approaching emergency vehicle or warning the driver of an accident immediately ahead.

Road Condition Information. This function provides information on traction, congestion, construction, and so forth within some predefined proximity to the vehicle or the driver's route.

Automatic Aid Request. This feature provides a Mayday signal in circumstances requiring an emergency response where a manual aid request is not feasible and where immediate response is essential, such as in the case of severe accidents. The signal will provide location information and potential severity information to the emergency response personnel.

Manual Aid Request. This component encompasses those services needed in an emergency such as those of police, ambulance, wrecker, or fire department. It will allow the driver to request emergency service from a vehicle without needing to locate a phone, know the appropriate phone number, or even know the current location. This function might also include feedback to notify the driver of the status of the response, such as the expected arrival time of the service.

Thus far in the evolution of the ATIS, the vast majority of developed systems and empirical research has centered around IRANS applications. IMSIS functions are represented in a few instances, and the ISIS and IVSAWS are greatly underrepresented in early system development. A summary of the functions of recent ATIS projects provided by Rillings and Betsold (1991) illustrates the emphasis on the IRANS up to the present time. These projects include the Pathfinder, TravTek, AMTICS, RACS, AUTOGUIDE, ATLAS, CARMINAT, CARIN, HAR, ARI, and RDS, which are described in greater detail in later sections of this chapter.

A major reason for the development activity centered around IRANS applications apparently is the number of potential benefits and the perceived marketability of such systems. Navigating to an unknown destination is a difficult task that in most cases is performed inefficiently or unsuccessfully. Outram and Thompson (1977) as cited in Lunenfeld (1990) estimate that between 6% and 15% of all highway mileage is wasted due to inadequate navigation techniques. This results in a monetary loss of at least 45 billion dollars per year (King, 1986). An additional cost that potentially can be reduced by widespread use of navigation systems is traffic delay. Several systems under development are designed to interface with advanced traffic management centers that eventually will be based in metropolitan areas. Once such systems are in place, real-time information on traffic delays can be broadcast to in-vehicle systems. These systems then can continuously calculate the fastest route to a destination during travel. Such a capability, if widely used, has the potential for increasing efficiency for an entire traffic infrastructure.

Although most effort to date has been expended on IRANS development, the other ATIS subsystems, namely the IMSIS, ISIS, and IVSAWS, hold promise for improving driving efficiency and safety. A paper by Green, Serafin, Williams, and Palke (1991) rated the relative costs and benefits of ATIS features. Based on ratings by four IVHS human factors experts regarding the costs and benefits associated with accidents, traffic operations, and driver needs and wants, several IVSAWS features were found to be most desirable in future systems. Several in-car signing features were also highly ranked. In contrast, some IRANS features were ranked relatively low, primarily because of the potential safety cost of using such systems. (Green et al., 1991).

TECHNOLOGY ISSUES

The following sections describe some of the technology components associated with the design of an ATIS. This material is presented to give the reader a broad overview of the emerging technologies being used to design ATISs.

Communication Technology

Media currently used for data and voice communications include infrared systems, FM sideband, mobile satellite services, cellular systems, radio frequency (RF) data networks, inductive loop systems, and the Shared Trunked Radio System (STRS) (Kirson, 1991; Weld, 1989).

Infrared Systems

Infrared systems use roadside beacons to transmit and receive information, respectively, to and from equipped automobiles. Relative to other technologies these systems provide an excellent rate of data transfer and have a low cost. However, they must be in proximity to the car, and environmental conditions can disrupt the signal. Infrared beacons could be used to support Automatic Vehicle Identification (AVI) systems with either one-way or two-way communications. Beacons can also be used for navigation aids. The beacons can update an automobile's position on the map database as the car passes by and provide information about upcoming intersections (Weld, 1989).

FM Sideband

FM sideband technology takes advantage of sideband radio and TV frequencies and broadcast information. This format is inexpensive and requires no additions to the automobile. The United States has used a highway advisory system on the AM dial since the 1970s. Several European companies have developed more complex systems that broadcast a code at the start of the message so that it will be received only by the cars that will be affected by the information. Other advances allow the drivers to listen to noncritical information at their leisure and have critical information mute their radio or tape player (Davies, Hill, & Klein, 1989). These more advanced systems require a device to decode and present the information. Possible display formats include in-dash information displays and speech synthesis. Usability of this system is limited because it provides only one-way communication and because some areas of the country are not suitable for receiving FM transmissions.

One possible short-term use of sideband technology involves providing up-to-date traffic information to all drivers in a local area. For example, units could be sold with varying degrees of complexity. The low-end model would intercept all information and display it on a small monitor. More expensive models would use coded signals to present information that is relevant only to the driver in the area. These systems would be ideal for the traveler who does not need route guidance or trip planning, but needs to know current traffic and road conditions.

Mobile Satellite Services

Mobile satellite services are advantageous because they can transmit and receive information directly to and from an automobile regardless of geographic location. Relative costs for satellite systems are low, and transmission speeds are relatively high (about 2,400 bits/sec). A disadvantage of satellites is their large footprint requiring several cities to use the same channel, which limits the total usage.

Cellular Systems

Cellular technologies use land-based centers, each with several cells or transmission channels capable of transmitting information. A mobile unit uses the strongest cell to communicate and can be handed off to another cell when a stronger one comes into range. Newer digital cellular links being formed will improve the reliability and transfer rate of information. At the time of this writing, approximately 2% of Americans use cellular technology, and cells are already becoming overloaded. This problem, plus the fact that many areas are not cellular equipped, suggests that cellular technology as it exists today will not be beneficial to large-scale ATIS uses. For cellular communications to be a useful source for communications in the future, cells will have to accept more users at one time and will need a greater range.

Radio Frequency (RF) Data Networks

RF data networks may prove to be an expandable resource for ITS applications. The system operates in a manner similar to the cellular communication links but does so at a much lower cost. This cost is lessened both on the users' end and on the transmitter construction end. Each new location of coverage requires its own antenna. RF technology is a mature area and has proven to be successful. An application of radio transmission called Packet Access Radio (PAR) uses short spikes of data either sent through normal RF nodes or bounced off satellites or meteor scatter (Williams, 1989). Williams claimed that meteor bounce could be an excellent low-cost data communication medium. The biggest drawback of PAR is its speed of transmission. Some transfers can take several minutes.

Inductive Loop Systems

Inductive loop systems are mounted under the roadway surface and used mainly to detect and track vehicles. An alternate use would allow communications between the loop and the automobile. The major drawbacks of this system include a low data rate and a range limited to the length of the loop. Also, cost of installation is high because each loop must be buried under the road surface. ATIS use of this technology includes AVI. For example, commercial drivers entering an area may need to follow a specific path to avoid dangerous areas. Loop systems could be used to guide the vehicles in the right direction and inform a control center when a vehicle has entered a dangerous area.

Shared Trunked Radio System

The Shared Trunked Radio System (STRS) operates in the same way as cellular radio but uses a 300 MHz band. Mobile units either lock onto a control channel or scan available channels for transmission. The major difference between a cellular unit and STRS is that a cellular unit needs enough dedicated cells to cover all users, whereas STRS users share cells that are not being used. STRS covers only about 30% of the land area of the United States, and adding more transmission units would be very expensive.

DISPLAY TECHNOLOGY

As automotive-compatible technologies to display information increase, the need to present more information quickly and saliently has become more important. For the ITS, the major categories of current, near-term, and future displays are, respectively, vacuum fluorescent displays (VFDs), cathode ray tubes (CRTs), liquid crystal displays (LCDs), and head-up displays (HUDs).

Vacuum Fluorescent Displays (VFDs)

VFDs are currently the most common type of displays found in automobiles. They are used in clocks, digital speedometers, message centers, audio systems, and temperature control systems. VFDs are produced by exciting a phosphor-coated anode. The current colors available are blue, green, yellowish-green, greenish-yellow, yellowish-orange, orange, and reddish-yellow. With filters, the color combinations increase. VFDs provide high luminance at low-voltage cost, are highly readable, and have a life span of more than 10,000 hours. Recent advances in VFD technology have increased the size of the display area, created a greater range of colors, and reduced the voltage use to half-duty cycle. Future advances will give full-color displays larger than the current maximum of five-by-seven inches, utilize graphics, increase luminance to 24,000 ml for HUD technology, and run at a much lower power consumption rate. (See Akiba et al., 1991; Iwasa, Kikuchi, Yamaguchi, Minato, & Ohtsuka, 1991, for more detailed information on VFDs.)

Cathode Ray Tube Displays (CRTs)

CRTs are now being used in some automotive applications, but are more commonly seen in computer displays and TV screens. CRTs have high resolution, can present many colors, and because of their maturity are currently the least expensive of the major display technologies. The major disadvantages of CRTs for automotive uses are their large size, weight, and power consumption. Another disadvantage is that as the screen size gets larger, the image gets dimmer, a critical factor in the glare-ridden vehicle environment.

Liquid Crystal Displays (LCDs)

LCDs are the type of display currently receiving the most attention by far from researchers in display development. According to Nordwall (1989), "LCDs generally offer savings of about 60% in volume, 70% in weight, and 80% in power compared with cathode ray tubes" (p. 56). LCDs can also display color using built-in filters. However, LCDs are nonemissive with a narrow viewing angle, and some types have difficulty operating under high and low temperatures (Erskine, Troxell, & Harrington, 1988).

There are several different types of LCD displays available. The type currently used in automobiles is the twisted nematic (TN) configuration. In its simplest form, an LCD works by applying voltage to a "sandwich" cell consisting of a polarizer, a glass substrate, a transparent conductor, and an alignment layer on either side of the twisted nematic liquid crystals. When voltage is applied to the cell, the conductors cause the alignment layers to "untwist" the liquid crystals. When this happens, the polarizers line up and let light through. A good general description of LCD technology is presented in Firester (1988). A more recent type of LCD called a double-layered super-twisted nematic liquid crystal display (DL-STN LCD) also has been developed. Its major benefits are a wider range of temperature operation, lower voltage usage, and increased contrast ratio (Matsumoto, Nakagawa, & Muraji, 1991).

Although the technology exists to create large flat-panel displays, they are expensive. Nearly half of this cost is in the fabrication process. Much of the current research is focused on creating cheaper and larger panels to house displays (for an example, see Takeda et al., 1989). A large flat-panel LCD would allow one large panel to replace the several smaller gauges and dials in most current automobiles. These displays could be made to fit the sizes of each manufacturer's models and could be programmed to have different appearances. The displays could also be user-definable so that they include only the information that the users want to see.

Because LCDs are nonemissive, they need an external source to light them. The two most common types of backlighting lamps for LCDs are the cold-cathode fluorescent lamp (CFL) and the hot-cathode fluorescent lamp (HFL). The HFL provides a higher luminance than the CFL, but does so at the cost of a higher operating temperature and a shorter life (approximately 3,000 hours). A newly developed type is the warm-cathode fluorescent lamp (WFL). The WFL offers twice the intensity of the CFL and 10,000 hours of operation, equal to that of the CFL. The WFL also operates at a lower voltage level and can be constructed with a thin film heater to help the liquid crystals operate better at temperatures below 20°C (Itoh, Yoshida, Terada, Horil, & Kuniyasu, 1991).

Head-Up Displays (HUDs)

A HUD can use one of several projection sources to project an image such as a speedometer display or warning indicator onto the windshield. This information appears to be floating in space in front of the vehicle. A HUD has the advantage of allowing the drivers to keep their eyes forward and observe information without having to accommodate them to the dashboard. HUDs have been successfully used in aircraft by giving the pilot a "window" through which to fly. An automotive HUD is different from an avionics HUD in that the scenery behind the display is often more complex for the driver than for the pilot. A second difference is that automotive HUDs are displayed, not at optical infinity as in an aircraft, but at a closer distance, somewhere between 6 and 24 feet (Stokes, Wickens, & Kyte, 1991).

HUDs are commonly produced either by reflecting an LCD off the windshield with a half mirror or direct reflection, or by using a light source to illuminate a holographic element on the windshield. Proponents of both systems claim success with each, and much research is currently being conducted to create the best optical picture at the lowest cost. (For detailed information, see Patterson, Farrer, & Sargent, 1988; Wood & Thomas, 1988.) A high-luminance VFD has also been proposed as an alternate HUD projector because it needs no backlighting and is resistant to shock and temperature conditions.

NAVIGATION TECHNOLOGY

According to French (1988), there are three basic types of navigation systems that include autonomous navigation systems, radio navigation systems, and proximity beacon systems. Autonomous navigation systems are capable of operating without the need for external sources such as satellites or road beacons. To track the distance and angles traveled most autonomous systems use dead reckoning, which employs wheel rotations and directions turned to estimate the current position of the car. The Japanese Multi-AV system uses a new optical fiber gyroscope to more accurately sense vehicle turns (Harrell, 1991; Oshishi & Suzuki, 1992). Dead reckoning systems have historically been relatively reliable, but periodically get out of calibration. The result is that the driver is given incorrect information about location or route status. Therefore, provisions for out-of-position information and simple location adjustment are necessary. These features necessarily utilize route-map displays. (See Fig. 1.1B for an example of a route-map display.)

Radio navigation systems use satellites to keep track of an automobile's position. Each vehicle to be tracked has a unique code that can be "seen" by the satellite and reported to a base station. The most commonly used satellite system is the Global Positioning Satellite (GPS) system, but there are many companies around the world competing for the market. Most systems that use radio navigation also use dead reckoning to account for areas where signals may be blocked. Currently the GPS system of satellites is incomplete over the United States. In addition, satellite signals are often blocked by tall structures in major cities. Therefore, it appears likely that many IVHSs will always require backup navigation systems if a GPS is used.

Proximity beacon systems use short-range transmitters to send signals to passing cars with receivers. These beacons are also usually combined with dead reckoning and satellite signals. The beacons allow a correction factor to update the position indicated by the other methods. A drawback to beacon technology is its cost of construction and maintenance.

Navigation Interfaces

The most common method of presenting information to drivers with navigation systems is through video monitors that display a map of the area and the automobile's current position (see previous sections). Some systems also use voice synthesis to convey messages to the driver. Much research must yet be conducted to determine which properties of voice are most salient in the driving situation. The Back Seat Driver, an MIT-developed navigation system, uses only speech to guide drivers around the Boston area. (See Davis & Schmandt, 1989, for a detailed discussion of some potential problems with speech-based directions.)

Synthesized speech is not as intelligible as digitized speech (Marics & Williges, 1988). Intelligibility is particularly important in vehicle environments because vehicle interiors are often noisy. It may be some time before intelligibility is of sufficient quality, particularly in low-cost speech synthesis systems required for in-vehicle use. However, despite its lower intelligibility, synthesized speech is desirable for many IVHS applications because large databases are sometimes required for communication of certain information (e.g., next street name in a large city; Dingus & Hulse, 1993).

**DESCRIPTION OF U.S.-BASED ATIS
PROJECTS/SYSTEMS**

Most available reports of U.S.-based ATIS systems and projects are of a descriptive nature. Field or laboratory evaluations are lacking with the exception of the TravTek system and Pathfinder. Pathfinder had a limited scope (25 cars) and was the first domestic IVHS project. It was primarily undertaken to demonstrate the feasibility of IVHS and promote further study. The TravTek operational testing phase was completed in March of 1993. A description of each planned or completed U.S. ATIS project is presented below.

TravTek

Travel Technology (TravTek) a demonstration system developed by General Motors and involving the City of Orlando, the Florida Department of Transportation (DOT), the Federal Highway Administration (FHWA), and the American Automobile Association (AAA), is nearing completion. The TravTek system was a complete IVHS infrastructure, including a traffic management center (TMC), traffic monitoring and sensing, and route-guidance information. The goal of TravTek was to reduce congestion and provide information on geographic attractions and services. The TravTek interface linked the drivers of 100 test vehicles to real-time information via digital data broadcasts. Avis rental car customers and solicited subjects participated in the testing.

The majority of ATIS reports discuss the TravTek system. To those who understand the constructs involved in TravTek, the constellation of in-vehicle navigation systems is roughly represented. Using the latest technology, the driver is aided in various tasks of navigation, route selection, route guidance, local information, and system interface. Human factors design considerations have been utilized since the inception of the system. The driver accesses information in three vehicle modes: predrive (PARK), drive (vehicle in motion), and zero speed. Both visual and auditory sensory channels are used to inform the driver of navigation, route selection, route guidance, local information, and system interface information. Extensive research into the needs and functions of both driver interface modalities was accomplished prior to the start of data collection. In addition, two visual display formats were available to the driver: a turn-by-turn graphic guidance screen (see Fig. 1.1A) and a color route map (see Fig. 1.1B; Fleischman et al., 1991). Several reports and publications are available describing the system. For system architecture, see the report by Rillings and Lewis (1991). Information on task analysis can be found in an article by Krage (1991). Human factors design aspects are described by Carpenter et al. (1991) and by Fleischman et al. (1991). Finally, the design of the auditory interface is described in Means et al. (1992).

Results reported by Fleischman, Thelen, and Dennard (1993) found that drivers familiar with the area (local drivers) using TravTek used the route guidance feature for approximately half of their trips. The same was true for drivers unfamiliar with the area (rental car drivers). Rental drivers used

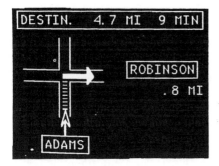 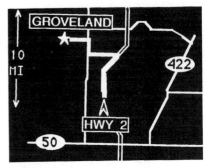

FIG. 1.1. (a) TravTec turn-by-turn guidance map display example; (b) route map display example.

the yellow pages feature most often in entering a destination, whereas local drivers more often used exact street addresses. Both groups chose "fastest" for the type of route they preferred to use in getting to the destination, as opposed to the alternative choices of "no interstates" or "no toll roads." The results also showed that all drivers used the "turn-by-turn" map type most often when driving in comparison to the route map. Furthermore, drivers used an optional supplementary voice guidance feature the majority of the time when driving to a destination.

Pathfinder

As the first in-vehicle navigation system project in the United States, Pathfinder involved CalTrans, General Motors, and the FHWA. This project focused on a 13-mile stretch of the Santa Monica Freeway, where 25 vehicles were equipped with Etak modified displays. Information on accidents, congestion, highway construction, and route diversion was presented to the driver either on the map display or through digital voice. The final phase of evaluation took place in the spring of 1992.

Pathfinder is one of the few projects with a publication describing human factors aspects of system design. Mammano and Sumner (1989) made the following general design observations: (a) By digitizing common words and synthesizing less common ones, voice messaging intelligibility was improved and space was saved in the computer memory; (b) data was filtered to avoid overload of messages during peak traffic times.

ADVANCE

The largest operational test of IVHS will be based in Chicago and its northwest suburbs. The Advanced Driver and Vehicle Advisory Navigation ConcEpt (ADVANCE) brings together the efforts of major IVHS manufacturers:

Ford, Toyota, Nissan, Saab, Volvo, Peugeot, Etak, Navigation Technologies, DonTech, Motorola, and Sun Microsystems. Institutional bodies of Illinois are also involved, including the Illinois Universities Transportation Research Consortium, the City of Chicago, and the Illinois DOT. This project is still in the data collection stage, but it is nearing completion. The ADVANCE operational test is very similar to TravTek, yet also focuses on both arterial roadways and highways.

Travelpilot

A joint project of Bosch and Etak, Travelpilot is a second-generation navigation system. This device forms the core of the Pathfinder system. It is also used in over 400 emergency vehicles in Los Angeles. The system consists of wheel sensors, compass, microcomputer with CD-ROM map data base, and a 4.5-in. vector-drawn monochrome display. It uses dead reckoning to update the navigation display. In addition, Travelpilot can be linked to communication systems for real-time data display.

DriverGuide

Pretrip out-of-vehicle route guidance is conducted using this system. Users enter origin-destination pairs and receive a printed set of instructions. The system was tested on French air travelers visiting San Francisco.

Oldsmobile Guidestar/Zexel Navigation System

The Oldsmobile Guidestar/Zexel System is a navigation system available on selected 1995 Oldsmobile models and via several rental car agencies in Florida and California. The system utilizes scroll lists for destination entry and has a choice of visual displays including a route map and a turn-by-turn display presented via a color LCD measuring approximately 4 inches diagonally. A voice guidance feature is also available. The system control and display unit is depicted in Fig. 1.2.

Delco Electronics Telepath 100

The Telepath 100 provides a simple form of navigation guidance by displaying dynamic heading and distance information to a selected destination (Wu & Welk, 1994). As shown in Fig. 1.3, this system utilizes a low-cost LCD-based display to show approximate direction to the destination and alphanumeric instructions via a seven-segment display. As a driver approaches a turn, the heading indicator points in the direction of the turn. When the vehicle approaches the destination, a text message alerts the driver that the destination is near.

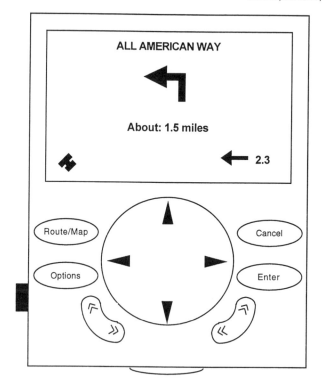

FIG. 1.2.   Depiction of the Oldsmobile Guidestar control and display unit.

FAST-TRAC

The Ali-Scout system developed as a joint project of the Federal Republic of Germany, Siemens, Volkswagen, Blaupunkt, and others is used as part of the FAST-TRAC project in Oakland County, Michigan. The display is a simple LCD readout that shows driving instructions with arrows at appropriate intersections. Infrared communication occurs at beacons located at key intersections to update the vehicle information systems. FAST-TRAC is a relatively low-cost system on a per vehicle basis, but it requires intersections equipped with transmitting beacons.

ROGUE

Navigation Technologies Corporation has developed the ROute GUidance Expert system (ROGUE) for daily in-vehicle navigation. The ROGUE software draws on the NavTech digital streetmap databases. Embedded in the CD-ROM database is information that simulates human intuition about routing, such as time of day (e.g., rush hour). The system can run on a stand-alone

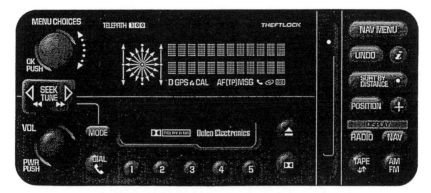

FIG. 1.3. Delco Electronic's Telepath 100.

basis or with an infrastructure updating its information. The stand-alone option is used as a selling point because global position satellites (GPS) and communication infrastructures can be cost prohibitive. Points of interest are also coded into the database. ROGUE uses an in-dash CRT display.

The expert system for the ROGUE in-vehicle route guidance is described in a report by Silverman (1988). Six design concepts were used in developing ROGUE: (1) provide route planning expertise; (2) provide effective and efficient directions; (3) provide navigation guidance during travel; (4) detect driving errors; (5) operate without external equipment; and (6) maximize driver comfort and safety.

The driver interface is also described in the Silverman report. A VDT mounted in the instrument cluster delivers requested navigation information. The display is monochrome, but provides line graphics as well as text capability. Also, a speech synthesis unit aurally provides directions. The system can be toggled to give spoken directions or a chime when directions are updated on the screen, alerting the driver to glance at them. Driver input is provided via an alphanumeric keyboard.

CARIN

Philips Corporation's CAR Information and Navigation system is an early implementation of the CD-interactive format for storing digital maps (Thoone, Dreissen, Hermus, & van der Valk, 1987). Vehicle location is accomplished by dead reckoning and map matching. Included in the system design is a radio datalink for traffic information. The driver is guided with synthesized speech in conjunction with a pictogram display (similar to that of the ALI-Scout).

The system requires a keyboard for driver input, whereas a supplemental color touchscreen is optional. A flat-panel display is used in the basic configuration, which shows stylized map graphics to supplement the audio. Maps are presented "heading up."

SmartRoutes

Liebesny (1992) discussed the SmartRoutes system that will service the Boston metropolitan area. This system will use real-time data from a traffic information center. Drivers will be able to access this information through the use of land line or cellular phone, e-mail, cable television, direct fax, or computer modem. Various automated mechanisms such as interactive audiotext and video graphics have been developed to disseminate the information. Liebesny (1992) recommended that the information be kept current and the system design updated continuously with a maximum acceptable aging period of 15 minutes. He also suggested developing a coordinated public-private partnership to handle the full aspects of incident management.

TRIPS

The Transportation Resources Information Processing System (TRIPS) includes dispatching of single-trip car pooling or parataxi systems, enabling drivers and riders to use Touch-Tone telephones, personal computers, and videotext terminals to obtain information on local traffic information and alternative route information (Ratcliff & Behnke, 1991).

TRAFFIC REPORTER

Traffic Reporter (TR) is a PC- and graphic-based ATIS developed by researchers from the University of Washington. The system seeks to improve traffic flow by influencing commuter behavior and decisions concerning alternative routes, departure times, and transportation modes. The system provides up-to-the minute information about freeway speeds and congestion by using information generated from on-freeway sensors. The system is PC-based and designed to be installed and used in locations such as malls, home, work, and lobbies of office buildings.

OVERVIEW OF ATIS SYSTEMS-PROJECTS OUTSIDE THE UNITED STATES

Other countries seem to be ahead of the United States in IVHS technology development. This advantage stems partly from pure need because the traffic congestion in areas such as Europe and Japan is considerably worse than it is in this country. We would do well, however, to begin implementing a structured system of traffic management before our problems grow to the levels observed in other parts of the world. The systems in other countries were all formed as joint operations between government, industry, and research institutions. Without such collaborative effort, projects of this magnitude will have little chance of success.

Generally, the goal of organizing groups within other countries has been to establish an entire system, complete to the last detail and designed to be country- or citywide, not just to limit a system to individual units within cars. This form of organization has not been adopted in the United States, at least not on a scale equivalent to that of the European and Japanese systems. In addition to traffic flow and route navigation information, developments from other countries include driving aids such as collision avoidance and driving condition monitors.

EUROPEAN ATIS PROJECT-SYSTEMS

Europe has several large-scale programs in progress under the umbrella of Road Transport Informantics (RTI), which is the equivalent of the U.S. ITS thrust. The main programs are DRIVE and PROMETHEUS. These two programs are separated by the organizations that formed them, but their goals are largely the same. DRIVE is under the control of the Commission of European Communities (CEC), whereas PROMETHEUS is part of the EUREKA platform, an industrial research initiative involving 19 countries and European vehicle manufacturers. Although the projects are in fact separate, close cooperation exists between the two in order to reach a common goal. Actual system development is the primary goal of the PROMETHEUS project, whereas DRIVE tends to focus on issues of human behavior and implementation of systems into the entire European community. Material describing these programs in more detail can be found in McQueen and Catling (1991), Hellaker (1990), Kemeny (1990), and Transport Canada (1992).

DRIVE

The intention of Dedicated Road Infrastructure for Vehicle Safety in Europe (DRIVE) is to move Europe toward an Integrated Road Transport Environment (IRTE) by improving traffic efficiency and safety and by reducing the adverse environmental effects of the motor vehicle. DRIVE focuses on the infrastructure requirements, traffic operations, and technologies of interest to public agencies responsible for the European road transport systems. Another focus of the DRIVE project is the human user and related issues that will necessarily be addressed in the implementation of in-vehicle systems.

The first phase of the project, DRIVE I, started in 1989. Its goal was to establish the overall working plan from which a European IRT environment could be developed. In the beginning, the DRIVE program was seen only as a feasibility study. However, as DRIVE progressed, it became apparent that it offered a realistic opportunity for system development. This resulted in DRIVE II, which emphasized the implementation of pilot projects developed as a result of DRIVE I. DRIVE II was scheduled to end in 1995, and products were released into the marketplace at that time. The DRIVE II workplan identified seven pilot project areas: (1) demand management, (2) traffic and travel information, (3) integrated urban traffic management, (4) integrated interurban traffic management, (5) driver assistance and cooperative driving, (6) truck fleet management, and (7) public transit management.

Groeger (1991) discussed the task of learning to drive and proposed a intelligent vehicle system that could continue to teach drivers as their skills mature or deteriorate. The Personalized Support And Learning Module (PSALM) is a concept derived as part of the DRIVE project. It will use the sensors likely to be in place in many IVHS-equipped automobiles to monitor driver performance. Individualized and task-specific instruction will be possible with PSALM. (For detailed individual project descriptions of DRIVE, see Keen & Murphy, 1992.)

PROMETHEUS

The PROgraM for European Traffic with Highest Efficiency and Unprecedented Safety (PROMETHEUS) project was initiated in 1986 as part of the EUREKA program, a pan-European initiative aimed at improving the competitive strength of Europe by stimulating development in such areas as information technology, telecommunications, robotics, and transport technology. The project is slated to last 7 years. PROMETHEUS is a precompetitive research project whose output will be a common technological platform to be used in turn by the participating companies once the product development phase begins. The overall aims of PROMETHEUS fall into four categories:

1. *Improved driver information:* To provide the driver with information from new sources of technology that were not previously available. Currently the lack of information or the inability to assess a hazard is often the primary cause of accidents.
2. *Active driver support:* To aid the driver in an informative way or by active intervention when the driver fails in some way at the driving task.
3. *Cooperative driving:* To establish a network of communication between vehicles in order to provide drivers with relevant information for areas en route to their destination.
4. *Traffic and fleet management:* To establish systems for the efficient use of the road network, ranging from highway flow control to fleet operations.

The emphasis of PROMETHEUS is on systems with a large in-vehicle component in their design. The ultimate aim is for every vehicle to have an on-board computer to monitor vehicle operation, provide the driver with information, and assist with the actual driving task. A centralized communications network will also be a component of the system to provide two-way communications between each vehicle and a control center.

The research phase, covering the past 4 years, has largely been completed. The current thrust is a move into the definition phase, in which the emphasis has shifted to field tests and demonstrations. To evaluate systems in each of the following areas, 10 common European demonstrations have been identified: vision enhancement, emergency systems, proper vehicle operation, commercial fleet management, collision avoidance, test sites for traffic management, cooperative driving, dual mode route guidance, autonomous intelligent systems, travel information systems, and cruise control.

These demonstrations were largely competed in 1994. The second phase will be somewhat modified to reflect the near-market status of some products under development, thus moving away from the program's noncompetitive origins.

To bring products to market more quickly in Europe, European Road Transport Telematics Implementation Coordination Organization (ERTICO) was created in November 1991. Its objectives are to pool the information from the many individual projects and identify strategies for exploiting the results of DRIVE, PROMETHEUS, and other individual programs. The goal is for ERTICO to create a climate for market-driven investments to ensure European dominance in advanced vehicle technologies.

Descriptions of Individual RTI/IVHS Systems

Many individual RTI/IVHS systems are now being tested throughout Europe. A short description of some individual systems is presented to enhance the reader's understanding of the development currently taking place in Europe. Descriptions largely describe the driver interface rather than the actual system hardware and communication network information.

Autoguide and the Ali-Scout are dynamic in-vehicle route guidance systems. That is, the system gives routing recommendations to drivers who are dependent on real-time traffic conditions. The display unit is mounted on the dashboard of the car and controlled with a hand-held remote control similar to a television remote. At the start of a journey, the driver can enter a grid reference or a preprogrammed destination. The system uses dead reckoning and roadside infrared transmitter-receiver beacons to guide the driver to the selected destination. The beacons serve the system by correcting cumulative errors and updating traffic information. The navigation information is presented to the driver through the use of icons and arrows to indicate directions. There is also a digitized speech unit that supplements the visual directions. The Autoguide system has undergone extensive testing in London, whereas Ali-Scout systems are being tested in Berlin. (For more information, refer to one of the following articles: Catling & Belcher, 1989; Jeffery, Russam, & Robertson, 1987; Jurgen, 1991; Morans, Kamal, & Okamoto, 1991; von Tomkewitsch, 1989.)

Trafficmaster from the United Kingdom was the first commercially available in-vehicle system to provide dynamic traffic information to the driver. It is a map-based system that provides only information on traffic flow; it does not actively suggest routes. The display screen is a 101 × 82 mm LCD. "Hard" push buttons for control of guidance functions are mounted next to the display (Jurgen, 1991).

TRAVELPILOT is a German autonomous navigation system based on the American ETAK Navigator sold by Blaupunkt Bosch Telecom. This system displays vehicle location on a dashboard-mounted CRT map stored on CD-ROM. The maps move relative to the vehicle's position, which is determined through the use of dead reckoning and map matching. The display is a small CRT that can show maps with highlighted routes or driving instructions with intersection maps and street names. Hard buttons mounted on either side of the CRT are changeable function controls. The system has reportedly sold over 1,000 units in its first year on the market and will be available soon in the United States for certain areas. (For more information, refer to Suchowerskyj, 1990; Morans et al., 1991.)

JAPANESE PROJECTS AND SYSTEMS

Japan is far ahead of any other country in implementing a large-scale traffic control system that utilizes in-vehicle technology. The reason for this accelerated pace is largely because the Japanese have the greatest need for such a system. Over vast portions of metropolitan areas in Japan, the average traffic speed is below 10 mph during much of the day. The small geographic area and large population has led the government to install traffic control systems in all large cities and on most urban and interurban freeways in Japan. These systems employ the latest technology, such as fiber optics communications and LED changeable message signs displaying both text and graphics in color. Japan has also made substantial investments in the development of driver information systems. Over 50 corporations have collaborated to develop in-vehicle systems marketed as units to be purchased by individuals who use the governmental road network system. The main IVHS initiatives currently are RACS, AMTICS, and VICS. Within RACS, the Ministry of Construction (MC) promoted and funded the Digital Road Map Association. This group was given the task of preparing and maintaining a national digitized roadmap database. The results of this work are available on compact disc in a standard format. This format is used by both RACS and AMTICS, as well as by the various manufacturers of autonomous vehicle navigation systems (Ervin, 1991).

RACS

The Road/Automobile Communication System (RACS) is sponsored by MC, the Highway Industry Development Organization (HIDO), and 25 private companies. The system consists of vehicles equipped with dead reckoning navigation systems, roadside communication units (beacons) that are distributed about 2 km apart throughout the road network, and a control center. There are three types of roadside beacons: Type 1 transmits location to the vehicle to zero-out cumulative navigation errors; type 2 transmits congestion and other traffic information in addition to location; and type 3 provides two-way communications with the vehicle so that information about the vehicle, such as vehicle locations and automatic debiting of tolls as well as emergency calls, can be transmitted to the control center. The MC recently announced a major beacon installation program consisting mostly of type 1. At present there are about 1,000 beacons around Tokyo. Beacon installation was scheduled to proceed throughout Japan at a rate of about 10,000 per year until 1994, with a gradual increase in the number of types 2 and 3 beacons. Travel-time savings of 3% to 5% are expected, representing a significant reduction in fuel consumption and air pollution.

AMTICS

The Advanced Mobile Traffic Information and Communications System (AMTICS) is sponsored by the National Police Agency (NPA), the Ministry of Posts and Telecommunications (MPT), the Japan Traffic Management and Technology Association (JSK), and 59 private companies. It employs in-vehicle equipment very similar to that of RACS, with the exception of the communication interface. The AMTICS datalink is essentially a one-way means of broadcasting traffic data from a cellular system of terminals. It is intended to convey a wide variety of information, including congestion information, travel-time predictions, traffic regulations, railway timetables, and advice on special events. This information is available from static terminals at railway stations, hotel lobbies, and other public locations, as well as in the vehicle. A large-scale test of AMTICS was held in Osaka in 1990. The results suggest that an individual travel-time reduction of about 7% could be achieved with in-vehicle navigation systems that provide congestion information to the driver. This would amount to individual travel-time savings of about $300 million in the Osaka area if all cars were equipped, with similar savings to the community because of reduced congestion. (For more information on the AMTICS system, see Okamoto, 1988; Okamoto & Nakahara, 1988; Okamoto & Hase, 1990.)

VICS

The Vehicle Information and Control System (VICS) is a new program formed under the combined direction of the MPT, MC, and NPA, with the goals of attempting to resolve the competition between RACS and AMTICS and to define a common system that would use the best features of both. This venture is meeting with some opposition, however, by those who feel that the competition between the two systems was improving them both. A digital microcellular radio system has been proposed to provide two-way road–vehicle communications and location information, essentially combining the tools used by each respective system. Although VICS may have a long-term future as part of an integrated driver information system for Japan, it will take some years to implement it. In the meantime, a common RACS–AMTICS system using RACS type 1 beacons and the broadcast of information to drivers via their FM car radios (such as RDS-TMC) is the likely direction for further development.

A Nissan System

A digital map-based system is sold with the Nissan Cedric, Gloria, and Cima models in Japan. It is very similar in design to the ROGUE and Travelpilot systems discussed in the previous section. Three scales of map display are available: 1:25,000 (street grid by blocks), 1:100,000 (default arterial roads), and 1:4000,000 (macro). The heading of the map can be toggled to either "north up" or "direction of travel" at the top. Vehicle location is always positioned in the center of the scrolling map display. For safety, minor roads are not displayed while the vehicle is in motion, thus reducing eye-glance time. Also, while the car is being driven, the system inhibits all switch operation, except for "changing of scale" and "display rotation mode." It is not clear what the "display rotation mode" feature entails from the research described (Tanaka, Hirano, Nobuta, Itoh, & Tsunoda, 1990).

SUMMARY AND CONCLUSIONS

To summarize, this chapter describes the potential benefits that could arise from using an ATIS. These include a reduction in traffic congestion, improved navigational performance, a decrease in the likelihood of an accident, reduced fuel costs and air pollution, and improved driver efficiency. In addition to the description of system benefits, the chapter lays a foundation for further discussion of human factors ATIS issues by providing a description of current ATIS technology, systems in production or under development, and projects underway or completed. As discussed in several subsequent chapters, for ATIS benefits to be realized, designers of traffic information systems must implement human factors knowledge and design principles into the design of these systems.

REFERENCES

Akiba, H., Davis, R. J., Kato, S., Tatiyoshi, S., Torikai, M., & Tsunesumi, S. (1991). Technological improvements of vacuum fluorescent displays for automotive applications. In Vehicle Information Systems and Electronic Display Technology, *SAE Paper Series* (SAE No. 910349), 59–67. Warrendale, PA: Society of Automotive Engineers.

Barfield, W., Haselkorn, M., Spyridakis, J., & Conquest, L. (1989). Commuter behavior and decision making: Designing motorist information system. *Proceedings of the Human Factors Society 33rd Annual Meeting* (pp. 611–614). Santa Monica, CA: Human Factors Society.

Barfield, W., Haselkorn, M. P., Spyridakis, J., & Conquest, L. (1991). Integrating commuter information needs in the design of motorist information system, *Transportation Research, 25*(A), 71–78.

Barfield, W., Spyridakis, J., Conquest, L., & Haselkorn, M. (1989). Information requirements for real-time motorist information systems. *Vehicle Navigation and Information Systems (VNIS '89) Conference* (pp. 101–103). New York: IEEE.

Carpenter, J. T., Fleischman, R. N., Dingus, T. A., Szczublewski, F. E., Krage, M. K., & Means, L. G. (1991). Human factors engineering the TravTek driver interface. *Vehicle Navigation and Information Systems Conference Proceedings* (pp. 749–756). Warrendale, PA: Society of Automotive Engineers.

Catling, I., & Belcher, P. (1989). Autoguide: Route guidance in the United Kingdom. *Vehicle Navigation and Information Systems Conference Proceedings* (pp. 467–473). Warrendale, PA: Society of Automotive Engineers.

Conquest, L., Spyridakis, J., Haselkorn, M. P., & Barfield, W. (1993). The effect of motorist information on commuter behavior: Classification of drivers into commuter groups, *Transportation Research: B, 2*, 183–201.

Davis, J. R., & Schmandt, C. M. (1989). The back seat driver: Real time spoken driving instructions. *Vehicle Navigation and Information Systems Conference Proceedings* (No. CH2789, pp. 146–150). Toronto, Ontario, Canada.

Davies, P., Hill, C., & Klein, G. (1989). Standards for the radio data system-traffic message channel. In *SAE Technical Paper Series* (SAE No. 891684, pp. 105–115). Warrendale, PA: Society of Automotive Engineers.

Dingus, T. A., & Hulse, M. C. (1993). Some human factors design issues and recommendations for automobile navigation information systems. *Transportation Research, 1C*(2), 119–131.

Erskine, J. C., Troxell, J. R., & Harrington, M. I. (1988). Systems and cost issues of flat panel displays for automotive application. *Automotive Displays and Industrial Illumination, 958*, 49–58.

Ervin, R. D. (1991). *An American observation of IVHS in Japan.* Ann Arbor, MI: The University of Michigan Press.

Firester, A. H. (1988). Active matrix liquid crystal display technologies for automotive technologies. *Automotive Displays and Industrial Illumination, 985*, 80–85.

Fleischman, R. N., Carpenter, J. T., Dingus, T. A., Szczublewski, F. E., Krage, M. K., & Means, L. G. (1991). Human factors in the TravTek demonstration IVHS project: Getting information to the driver. *Proceedings of the Human Factors Society 35th Annual Meeting* (pp. 1115–1119). Santa Monica, CA: Human Factors Society.

Fleischman, R. N., Thelen, L. A., & Dennard, D. (1993). A preliminary account of TravTek route guidance use by renter and local drivers. *Vehicle Navigation and Information Systems Conference Proceedings* (pp. 120–125). Warrendale, PA: Society of Automotive Engineers.

French, R. L. (1988). Road transport informatics: The next 20 years. In *SAE Technical Paper Series* (SAE No. 881175). Warrendale, PA: Society of Automotive Engineers.

Green, P., Serafin, C., Williams, M., & Paelke, G. (1991). What functions and features should be in the driver information systems of the year 2000? Vehicle Navigation and Information Systems Conference Proceedings (SAE No. 912792, pp. 483–498). Warrendale, PA: Society of Automotive Engineers.

Groeger, J. A. (1991). Supporting training drivers and the prospects for later learning. In *Advanced Telemetrics in Road Transport, Proceedings of the DRIVE Conference* (pp. 314–330). Brussels.

Harrell, B. (1991). Nissan refines its vehicle navigation system. *Nissan Technology Newsline.* October.

Hellaker, J. (1990). PROMETHEUS: Strategy. In *SAE Technical Paper Series* (SAE No. 901139, pp. 195–199). Warrendale, PA: Society of Automotive Engineers.

Itoh, M., Yoshida, M., Terada, T., Horil, M., & Kuniyasu, S. (1991). Development of color STN LCD for automotive information display. In *SAE Technical Paper Series* (SAE No. 910064, pp. 41–47). Warrendale, PA: Society of Automotive Engineers.

Iwasa, T., Kikuchi, Y., Yamaguchi, H., Minato, T., & Ohtsuka, I. (1991). Large scale message center vacuum fluorescent display for automotive applications. In *SAE Technical Paper Series* (SAE No. 910350, pp. 69–73). Warrendale, PA: Society of Automotive Engineers.

Jeffery, D., Russam, K., & Robertson, D. I. (1987). Electronic route guidance by AUTOGUIDE: The research background. *Traffic Engineering and Control, 28*(10), 525–529.

Jurgen, R. (1991). Smart cars and highways go global. *IEEE,* May, 26–36.

Keen, K., & Murphy, E. (Eds.). (1992). *DRIVE 92: Research and technology development in advanced road transport telemetrics* (Rep. No. DRI203). Brussels: Commission of the European Communities.

Kemeny, A. (1990). PROMETHEUS: Design technics. In *Proceedings of the International Congress on Transportation Electronics.* No. 901140 (pp. 201–207). Warrendale, PA: Society of Automotive Engineers.

King, G. F. (1986). Driver attitudes concerning aspects of highway navigation. *Transportation Research Record, 1093,* 11–21.

Kirson, A. M. (1991). RF data communications considerations in advanced driver information systems. *IEEE Transactions on Vehicular Technology, 40*(1), 51–55.

Krage, M. K. (1991). The TravTek Driver information system. In *SAE Technical Paper Series.* SAE No. 912820 (pp. 739–747). Warrendale, PA: Society of Automotive Engineers.

Lee, J. D., Morgan, J., Wheeler, W. A., Hulse, M. C., & Dingus, T. A. (in press). *Development of human factors guidelines for ATIS and CVO: Description of ATIS/CVO functions.* Federal Highway Administration Report.

Liebesny, J. P. (1992). SmartRoute systems: The nation's first private, area-wide ATIS. *ITE Compendium of Technical Papers,* 49–51.

Lunenfeld, H. (1990). Human factor considerations of motorist navigation and information systems. *Vehicle Navigation and Information Systems Conference Proceedings* (pp. 35–42). Warrendale, PA: Society of Automotive Engineers.

Mammano, F., & Sumner, R. (1989). PATHFINDER system design. *Vehicle Navigation and Information Systems Conference Proceedings* (No. CH2789-6/89/0000-0484). Warrendale, PA: Society of Automotive Engineers.

Marics, M. A., & Williges, B. H. (1988). The intelligibility of synthesized speech in data inquiry systems. *Human Factors, 30*(6), 719–732.

Matsumoto, T., Nakagawa, Y., & Muraji, H. (1991). Double-layered super-twisted nematic liquid crystal display for automotive applications. *Vehicle Information Systems and Electronic Display Technologies* (SAE No. 910351, pp. 75–80). Warrendale, PA: Society of Automotive Engineers.

McQueen, B., & Catling, I. (1991). The development of IVHS in Europe. In *SAE Technical Paper Series* (SAE No. 911675, pp. 31–42). Warrendale, PA: Society of Automotive Engineers.

Means, L. G., Carpenter, J. T., Szczublewski, F. E., Fleishman, R. N., Dingus, T. A., & Krage, M. K. (1992). Design of the TravTek auditory interface (Tech. Rep. GMR-7664). Warren, MI: General Motors Research and Environmental Staff.

Mobility 2000. (1989). Proceedings of the workshop on intelligent vehicle highway systems. *Texas Transportation Institute*, Texas A&M.

Morans, R., Kamal, M., & Okamoto, H. (1991). IVHS. *Automotive Engineering, 99*(3), 13–20.

Ng, L., Barfield, W., & Mannering, F. (1995). A survey-based methodology to determine information requirements for advanced traveler information systems. *Transportation Research: C.*

Nordwall, B. D. (1989). Navy chooses LCD technology for new A-12 color displays. *Aviation Week and Space Technology, 131*, 56–57.

Okamoto, H. (1988). Advanced mobile information and communication system (AMTICS). In *SAE Technical Paper Series* (SAE No. 881176, pp. 2–10). Warrendale, PA: Society of Automotive Engineers.

Okamoto, H., & Hase, M. (1990). The progress of AMTICS: Advanced mobile traffic information and communication system. In *SAE Technical Paper Series* (SAE No. 901142, pp. 217–224). Warrendale, PA: Society of Automotive Engineers.

Okamoto, H., & Nakahara, T. (1988). An overview of AMTICS. International Congress on Transportation Electronics Proceedings (pp. 219–228).

Oshishi, K., & Suzuki, H. (1992). The development of a new navigation system with beacon receiver (Tech. Rep., pp. 1–11). Ann Arbor, MI: Nissan Research and Development Center.

Outram, V. E., & Thompson, E. (1977). Driver route choice. *Proceedings of the PTRC Summer Annual Meeting*, University of Warwick, UK. Cited in Lunenfeld, H. (1990). Human factor considerations of motorist navigation and information systems. *Vehicle Navigation and Information Systems Conference Proceedings* (pp. 35–42). Warrendale, PA: Society of Automotive Engineers, 35–42.

Patterson, S., Farrer, J., & Sargent, R. (1988). Automotive head-up display. *Automotive Displays and Industrial Illumination, 958*, 114–123.

Perez, W. A., & Mast, T. M. (1992). Human factors and advanced traveler information systems (ATIS). *Proceedings of the Human Factors Society 36th Annual Meeting* (pp. 1073–1077). Santa Monica, CA: Human Factors Society.

Ratcliff, R., & Behnke, R. W. (1991). Transportation Resources Information Processing System (TRIPS). In *Proceedings of Future Transportation Technology Conference and Exposition* (SAE No. 911683, pp. 109–113). Warrendale, PA: Society of Automotive Engineers.

Rillings, J., & Betsold, R. J. (1991). Advanced driver information systems. *IEEE Transactions on Vehicular Technology, 40*(1), 31–40.

Rillings, J., & Lewis, J. (1991). TravTek. In *SAE Technical Paper Series* (SAE No. 912819, pp. 729–737). Warrendale, PA: Society of Automotive Engineers.

Silverman, A. (1988). An expert system for in-vehicle route guidance. In *SAE Technical Paper Series* (SAE No. 881177, pp. 1–13). Warrendale, PA: Society of Automotive Engineers.

Stokes, A., Wickens, C., & Kyte, K. (1990). *Display technology: Human factors concepts*. Warrendale, PA: Society of Automotive Engineers.

Suchowerskyj, W. E. (1990). Vehicle navigation and information systems in Europe: An overview. In *SAE Technical Paper Series* (SAE No. 901141, pp. 209–215). Warrendale, PA: Society of Automotive Engineers.

Takeda, S., Ezawa, H., Kuromaru, A., Kawade, K., Takagi, Y., & Suzuki, Y. (1989). Fine pitch tab technology with straight side wall bump structure for LCD panel. In *IEEE* (pp. 343–351). Kawasaki-City, Japan: Toshiba Corporation, Tamagawa Works.

Tanaka, J., Hirano, K., Nobuta, H., Itoh, T., & Tsunoda, S. (1990). Navigation system with map-matching method. In *SAE Technical Paper Series* (SAE No. 900471, pp. 45–50). Warrendale, PA: Society of Automotive Engineers.

Thoone, M., Driessen, L., Hermus, C., & van der Valk, K. (1987). The car information and navigation system CARIN and the use of compact disc interactive. In *SAE Technical Paper Series* (SAE No. 870139, pp. 1–7). Warrendale, PA: Society of Automotive Engineers.

Transport Canada. (1992). *Intelligent Vehicle Highway Systems: A synopsis*. Transport Canada Report Number 11145E.

von Tomkewitsch, R. (1989). The LISB field trial, forerunner of AUTOGUIDE. In *Proceedings of the Institution of Mechanical Engineers* (No. C391/060, pp. 320–326).

Weld, R. B. (1989). Communication flow considerations in vehicle navigation and information systems. *IEEE* (No. CH2789-6/89/0000, pp. 373–375). New York: The Institute of Electrical and Electronics Engineers.

Wenger, M., Spyridakis, J., Haselkorn, M. P., Barfield, W., & Conquest, L. (1990). An in-depth examination of motorist behavior: Considerations with regards to the design of motorist information systems, *Journal of the Transportation Research Board, 1282*, 159–167.

Williams, R. E. (1989). Packet access radio: A fast, shared data service for mobile applications. *IEEE* (No. CH2789-6, pp. 389–391). New York: The Institute of Electrical and Electronics Engineers.

Wood, R., & Thomas, M. (1988). A holographic head-up display for automotive applications. *Automotive Displays and Industrial Illumination, 958*, 30–48.

Wu, E. Y., & Welk, D. L. (1994). An alternative approach to automobile navigation. *Proceedings of the IVHS America Annual Meeting* (pp. 50–54). Washington, DC: IVHS America.

2

PERCEPTUAL AND COGNITIVE ASPECTS OF INTELLIGENT TRANSPORTATION SYSTEMS

John D. Lee
Barry H. Kantowitz
Battelle Memorial Institute

This chapter describes some cognitive and perceptual characteristics of drivers that may influence the success of Intelligent Transportation Systems (ITS). Cognitive characteristics imply constraints on what information drivers require and how that information can be best displayed. Thus, the cognitive characteristics of drivers help to define information requirements and formats for display and control. Although cognitive characteristics help to define design requirements, they are not the only factors involved. In fact, driver behavior and information needs also may be understood by a close examination of specific driver tasks and by a functional description of the driver's interaction with the ITS. In other words, multiple factors such as ITS functional capabilities, environmental factors, and driver characteristics provide the context for driver interaction with ITS and play an important role in determining information that should be presented to drivers. Figure 2.1 shows that an ITS design depends on considering each of these elements.

Driver behavior and the associated design implications depend then on understanding both the context in which the driver operates and driver cognitive characteristics. For example, the cognitive characteristics of private and commercial drivers may not be dramatically different, but their information requirements and interaction with ITS components will certainly differ. This chapter examines driver cognitive characteristics to identify driver information requirements and the form that this information should take. This review does not attempt to completely specify driver information require-

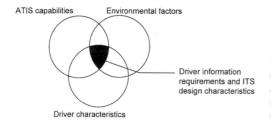

ATIS capabilities   Environmental factors

Driver information
requirements and ITS
design characteristics

Driver characteristics

FIG. 2.1. The joint influence of ITS capabilities, environmental factors, and driver characteristics on information requirements and design characteristics.

ments and ITS design parameters, but simply draws upon the human factors and experimental psychology literature to identify and describe cognitive constraints that may be particularly important in the design of ITS.

Some research findings in this chapter have been drawn from the domains of aviation, process control, and human computer interaction. Although many of these findings may apply to driving in general, substantial differences do exist between the average driver and the subjects of studies in these more specialized domains. For example, military pilots used in some studies represent a very homogeneous population of young, well-trained, highly skilled people who are perceptually and mentally acute. Other studies have used college students as subjects and therefore have sampled a narrow range of age and expertise. Drivers belong to a much more diverse group, with a wide range of educational, physical, perceptual, mental, and attitudinal differences. Furthermore, drivers, even private drivers, have developed a high level of proficiency that is seldom replicated in most laboratory research paradigms.

In addition to differences in experimental subjects from various domains, the tasks in these domains also may differ considerably from that of driving. For example, even though flying and driving share many of the same perceptual–motor demands, driving can be a more attention-demanding task. Specifically, pilots may be able to tolerate distractions lasting for seconds if not minutes, whereas distractions lasting only a few seconds can be disastrous to a driver. Likewise, the constant demand of visual attention provides a clear distinction between driving and the domains of process control and human computer interaction. Therefore, caution should be taken before using empirical results from other domains to describe the cognitive characteristics of drivers. However, some of the cognitive characteristics identified in other domains do represent fundamental characteristics of humans that will apply to drivers.

CHAPTER ORGANIZATION

This review of driver cognitive characteristics has been organized into three areas: perception and action, information assimilation and decision making, and driver knowledge and attitudes. These areas share similarities with the distinctions of control, maneuvering, and navigation (Janssen, Alm, Michon, & Smiley, 1993); however, they are somewhat broader and place a greater emphasis on the factors related to interaction with ITS components than on driving in particular. These three areas also share many characteristics that differentiate response tendencies, rules, and frames (Kantowitz, Triggs, & Barnes, 1990) as well as skill-, rule-, and knowledge-based behaviors (Rasmussen, 1983).

The boundaries among these areas can be fuzzy, yet skill-, rule-, and knowledge-based behaviors represent a well-accepted taxonomy or human behavior. Skill-based behavior is sensory–motor activity that proceeds without conscious control and is typical of the continuous, moment-to-moment control of a vehicle. Rule-based behavior is guided by stored rules that have been developed through trial-and-error experience, prior problem solving, or written or verbal instructions. Knowledge-based behavior occurs in unfamiliar situations for which no predefined rules exist, such as navigating through an unfamiliar city using a map. Thus, knowledge-based behavior depends on knowledge about general laws and relationships that governs the interaction between objects and events. Each of these types of behavior has different limits and suggests different factors that must be considered in ATIS design.

Similar distinctions have been made to explain stimulus–response compatibility in terms of a nested hierarchy of response tendencies, rules, and frames (Kantowitz et al., 1990). These categories refer to the internalized representation that guides the interpretation and response to information. Frames govern the organization of knowledge, rules refer to simple condition-action pairs, and response tendencies refer to overlearned associations. As an example, speed and lane control depend on response tendencies and therefore require little conscious effort for experienced drivers. In driving, a rule might include response to a stop light. If the light is red, then the driver slows to a stop. Drivers depend on frames to cope with complex and unusual circumstances that require deliberate decision making. For instance, frames govern interpretation of maps and diagnosis of vehicle malfunctions. Because these categories mediate interpretation, a single stimulus may generate multiple interpretations, depending on whether it is interpreted in terms of a response tendency, rule, or frame.

The distinctions that have been made to describe fundamentally different modes of behavior (Kantowitz et al., 1990; Rasmussen, 1986) are reflected in the three categories that organize this chapter: (a) perception and action, (b) information assimilation and decision making, and (c) driver knowledge and attitudes. Like the distinction among response tendencies, rules and frames, the three areas discussed in this chapter, exemplify distinctly different representations of human behavior, with each area defining a different way that individuals internalize and respond to information. Although these areas are discussed as discrete categories, driving behavior depends on a complex interplay involving activity from several areas, so perception is often intimately linked to decision making as well as to knowledge and attitudes. No area can be ignored, and ITS design must accommodate drivers' limits and tendencies embodied in each area. The categories used to organize in this chapter do not represent a rigorous psychological theory, but they provide a very useful framework for considering the complexities of driver behavior. As suggested by Kantowitz et al. (1990) and Rasmussen (1983), such distinctions provide a useful guide for empirical research as well as for system design. Specifically, these distinctions represent a parsimonious taxonomy that promotes a comprehensive approach and avoids a piecemeal listing of human limitations.

Figure 2.2 overviews the three areas of cognitive characteristics discussed in this chapter. Based on the nested hierarchical model developed by Kantowitz et al. (1990), this figure shows how the three areas of driver characteristics influence driver behavior. The first area addresses perceptual and motor limitations affecting information access and driver response capabilities. These limitations govern how well drivers can acquire information and respond. The second area describes the cognitive characteristics that might influence how well drivers integrate and understand information. These characteristics reflect factors that influence how drivers interpret information and select from multiple options. The third area represents characteristics that define how drivers develop and organize knowledge and the factors that influence driver attitudes. The cognitive characteristics in this area explain how people develop and adapt knowledge to resolve unusual problems, and how attitudes develop and influence acceptance and use of technology. This framework shows that information from the environment

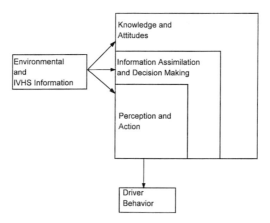

FIG. 2.2. An overview of the three areas of cognitive characteristics and their relationship to driving behavior.

and ITS components may be mediated by characteristics of the perceptual system, by decision making limits, and by driver attitudes. It also shows that driver attitudes may influence decision making, which in turn may influence individual actions. Thus, even though cognitive characteristics will be described as discrete classes of behavior, actual driving behavior involves a continuous interplay of all three categories. This chapter contains three examples of how each class of cognitive characteristics (perception and action, information assimilation and decision making, and driver knowledge and attitudes) might influence ITS design. The first section describes how driver characteristics might be used to specify the design of an in-vehicle signing system. The second section discusses how parameters of a collision avoidance system could be tuned to augment drivers decision making capabilities. The final section explains how navigation knowledge and attitudes toward technology may influence drivers' use of routing and navigation systems.

PERCEPTION AND ACTION

Drivers' perceptual and motor characteristics affect their interaction with ITS technology. These characteristics refer to the limits associated with the process of receiving and acting on well-learned stimulus–response pairings. As such, these cognitive characteristics pertain to drivers' perceptual and motor limits rather than cognitive limits associated with understanding and interpreting information. Thus, these limits refer to drivers' abilities to see or hear signals and execute responses to them. Inaccurate perception and response will inhibit the more complex cognitive characteristics of information processing, decision making, attitudes, and knowledge. Specifically, perceptual and action limits influence drivers' abilities to identify the presence of warning icons, react appropriately to direction arrows given by a navigation aid, and accurately depress buttons to control a multifunction display.

An analysis of driver response to warning sirens provides a good example of how perceptual and action limits can influence driver performance and safety. Perceptual limits may turn the warning sirens of emergency vehicles into confusing noise rather than useful information. In many instances emergency sirens are not easily localized and the driver is left wondering if action is needed and how it should be taken. To aid the driver the siren should specify the distance and direction of an approaching emergency vehicle. In the best circumstances it may be a difficult for drivers to interpret this information. Furthermore, this is often complicated by car windows that are closed and car radios playing at loud volumes. These factors mask and distort the information needed to detect and estimate the distance and direction of approaching emergency vehicles.

Interviewing ambulance drivers, Caelli and Porter (1980) found that experienced ambulance drivers believe their sirens are poorly heard and localized by car drivers, and report that car drivers do not respond to sirens until they are within 100 meters from the rear and 200 meters from the front of a car. Investigating drivers' perception of sirens empirically, Caelli and Porter (1980) examined several factors that mediate drivers' ability to localize and respond to sirens and found that sirens often fail to provide drivers with adequate information for accurately estimating the distance and direction of approaching ambulances. They found that drivers often made reversal errors, identifying a siren from the front when it actually was emanating from the rear. These errors were accented when one window was open and reduced when all the windows were closed. However, with their windows closed drivers had a much greater difficulty detecting the siren. Drivers consistently overestimated distances to the approaching siren, with their errors increasing as the ambulance approached more directly from the rear. Thus, sirens appeared to provide inadequate information regarding the distance of the ambulance, but rarely enabled a driver to localize an ambulance accurately.

Even though sirens failed to support drivers' localization of emergency vehicles, the drivers consistently expressed high levels of confidence in their judgments, even when their direction and distance estimates were dramatically different from reality. They remained highly confident of their direction estimates when these differed by more than 20°. For distance estimates, drivers remained confident of their judgments even when they believed ambulances to be twice as far away as they actually were (Caelli & Porter, 1980).

ITS technology may enable drivers to identify and localize emergency vehicles better through in-vehicle warning systems. Such systems could signal the approach of an ambulance with an array of speakers whose output corresponds to the location and direction of the approaching vehicle. The system could bypass several problems that interfere with drivers' ability to localize sirens. Because they are inside the vehicle, these systems are not susceptible to the perceptual confusion that can result from partially open windows. They also avert the problem of masking associated with highway noise and closed windows. Furthermore, the in-vehicle signals can be configured to avoid other sources of masking, such as signals that override car radios. These warning systems could also provide information tailored to the particular vehicle based on its location relative to the emergency vehicle. Specifically, for an emergency vehicle approaching from a considerable distance, the driver could be warned: "Emergency vehicle approaching from ahead." If the emergency vehicle is closer, then the message might be more urgent: "Pull to the right and stop immediately." The success of in-vehicle systems used to warn drivers of approaching emergency vehicles depends

TABLE 2.1
A Summary of Perception and Action Cognitive Characteristics and Related Information
Requirement

Cognitive Characteristics	Design Considerations
Limited range of color perception	Minimum illumination, redundant coding
Loss of visual acuity with low illumination	Minimum illumination, size limits
Color blindness	Redundant coding
Reduced visual accommodation	Information location
Visual attention and limited glance time	Limited information transmission
Impaired hearing ability	Choice of frequency range and level
Age-related increase in response time	Preview information
Motor control accuracy and speed	Size and distance of controls
Perceptual interaction	Coordination of information flows
Response interaction	Coordination of response requirements, preview information
Startle effect of warnings	Adjustment to ambient noise level
Population stereotypes	Actions suggested/required by interface should match expectancies

on the ability of these systems to provide accurate distance and direction
cues. Without careful calibration, in-vehicle warning systems could produce
direction and distance cues that are even more confusing that those received from actual sirens. Parameters important in calibrating these systems
include ambient noise, signal intensity, and signal frequency.

Beyond this specific example, many other important driver characteristics
fall into the category of perception and action constraints. Table 2.1 summarizes some of these characteristics.

INFORMATION ASSIMILATION
AND DECISION MAKING

Certain driver characteristics are associated with the integration and understanding of information. Unlike drivers' perceptual and motor characteristics,
these characteristics do not describe cognitive characteristics that directly
influence the drivers' ability to access and act on stimuli in the environment.
Instead, these characteristics govern drivers' abilities to interpret and understand the meaning of displays and the decisions they imply. Thus, these
characteristics go beyond whether drivers can accurately perceive and act
on signals from the environment. They describe whether drivers can access
the semantics contained in displays and messages. More specifically, these
characteristics determine how drivers integrate information from the environment and ITS components to select alternate routes that avoid congestion

and roadway hazards, use in-vehicle databases to find points of interest, and integrate warning messages to avoid impending dangers.

In the future, technology may augment drivers' decision making in a number of ways. For example, a collision avoidance system may monitor the position of a car relative to others on the roadway and alert drivers when distance between the cars becomes unsafe. This system provides a good example of an applied technology whose success depends on how well drivers are able to make decisions that must integrate information from an ITS component with information from the surrounding roadway. Although such technology offers great potential for improving highway safety, its beneficial effects depend not on the capability of the technology alone, but on the joint performance of the collision avoidance system and the driver. Many technologically sophisticated systems fail, not through any technological shortcoming, but because designers failed to consider the role of the human (Wiener, 1985; Woods, Potter, Johannesen, & Holloway, 1991). Therefore, the success of a collision avoidance system will depend on the joint capabilities and limits of the technology and the driver.

Traditional warning systems have been binary (e.g., either on or off). Such binary systems present several design parameters that must be estimated in evaluating when and how they should be actuated. A vital issue with traditional warning displays is the problem of false alarms. In extreme cases, operators have dealt with false alarms by disabling warning systems, which led to severe accidents in transportation and nuclear power systems. The false alarm problem is quite likely for a collision avoidance system, which will tend to have liberal activation criteria. Vehicles traveling in adjacent lanes can close to within several feet of each other without inherent danger. Furthermore, normal lane drift behavior produces occasional instances of high closure rates that alone are not necessarily unsafe. A miss (failure to warn the driver of an approaching vehicle) would be a costly collision avoidance system error. Therefore, collision avoidance systems must be tuned to err on the side of avoiding misses. The price for this caution is always a high false alarm rate. In aviation, a first-generation collision avoidance system was rejected by pilots because of high false alarm rates in congested traffic areas. It is imperative then that a collision avoidance system device benefit from prior understanding of how people behave in systems with high false alarm rates.

The theory of signal detection is well-suited to address the effects of collision avoidance system parameters. In particular, an extension of signal-detection theory called the alerted monitor system has been developed to explain the decision-making behavior of people using devices (e.g., a collision avoidance system) that are prone to high false alarm rates (Sorkin & Woods, 1985). This model has led to the development of a new type of warning display, called the Likelihood Alarm Display (LAD), that is an effective decision aid for systems with high false alarm rates (Sorkin, Kantowitz, & Kantowitz, 1988). A LAD alerts the operator about event likelihood as computed by an automated monitoring system. Using a mathematical algorithm, the LAD prompts the operator to monitor a display when a signal is likely. This prompt is graded according to the calculated likelihood ratio. Hence, LADs produce graded output as opposed to the binary output of the normal alarm system. In addition to collision avoidance system parameters such as activation criteria and warning signal mode, it is important to study likelihood alarms as an additional parameter because they illuminate the driver decision-making process. Understanding how likelihood alarms are effective in a collision avoidance system gives us a better appreciation of how other collision avoidance system parameters affect lane-change decisions.

This theoretical rationale for LADs does not necessarily imply that likelihood information can be used effectively in practical operational settings. As a first test of the theory's practical utility, Sorkin, Kantowitz, and Kantowitz (1988) used LADs in a dual-task laboratory simulation. A dual-task was selected to impose stress and attentional demands more similar to those in operational environments than could be induced by a single display monitoring task. The LAD used in this experiment generated continuous likelihood information to prompt the operator in a graded fashion about the possible need to attend to a warning display. Operators performed a primary task of one-dimensional tracking combined with a secondary task that required monitoring an array of four three-digit numbers. The experiment asked several questions: Can operators use likelihood information to improve monitoring performance? Does the added mental workload of attending to LADs diminish overall system performance? How do LADs affect allocation of attention across tasks?

Results were positive and strongly suggested that LADs could improve operator performance. The LADs not only improved monitoring performance, but they also improved primary task tracking. This positive outcome was attributed to LADs' improvement of attention allocation among tasks by providing information that became integrated into operator decisions. LADs did not necessarily add to the operator's attentional load, although the generality of this result beyond the 2- and 4-state LAD formats tested in the experiment remains to be established. Triggs and Drummond (1993) further tested LADs in a driving simulator by comparing drivers' unaided decision making with their decision making aided by a LAD. Their results suggested that LAD enhances decision making performance for inexperienced and experienced drivers equally.

A LAD can be likened to an expert giving warnings coded according to urgency. A low-urgency signal would resemble the expert saying, "You might consider looking to see if there is another vehicle in your blind spot." Low-urgency signals have high false alarm rates, but these do not upset the operator because he has already been told that the warning is not urgent. Busy operators will delay looking in response to low-urgency signals until it is convenient for them. A high-urgency signal would resemble the driving expert shouting "LOOK OUT! YOU'RE GOING TO COLLIDE INTO THAT CAR BESIDE YOU!" Chances for a false alarm are lower, and drivers learn to give immediate attention to high-urgency warnings.

Because LADs provide continuous output, the warning signal could use any dimension that has a wide range. For example, auditory frequency might be a viable coding dimension with low frequencies conveying low urgency and higher frequencies conveying increasing urgency. In practice, it is often simpler to divide the continuum into a few discrete levels. Short verbal messages are coded for each level (Sorkin et al., 1988). For example, a driving collision avoidance system for a three-level LAD might signal (a) check mirror, (b) be careful, or (c) LOOK OUT!

Using the LAD concept to help drivers interpret collision avoidance information provides one example of how ITS technology should be designed to accommodate driver cognitive characteristics. The example shows how ITS information should be integrated with driver decision making processes. As with other ITS components, collision avoidance systems provide information that augments information the driver currently perceives directly from the environment. Thus, design of these systems must consider concepts such as LAD that facilitate driver decision making by enabling drivers to combine their expertise with information from the ITS component. More generally, this example shows the need to consider how characteristics of in-vehicle technology influence information assimilation and decision making. It is not enough to ensure that warning tones are heard. They must also be effective in guiding drivers to make better decisions. Some other driver characteristics associated with information assimilation and decision making are shown in Table 2.2. This table also identifies design considerations for each cognitive characteristic.

KNOWLEDGE AND ATTITUDES

Drivers' characteristics concerning knowledge and attitudes influence their long-term interaction with ITS components. These characteristics govern whether and how drivers use ITS capabilities. Driver attitudes and knowledge are important because even if drivers correctly perceive and understand information generated by an ITS component, their attitudes and knowledge may inhibit their acting on this information. For example, drivers' knowledge affects how well they integrate the information provided by navigation aids to determine their location and desirable routes. In addition, driver attitudes will likely have a powerful influence on drivers' compliance with route

TABLE 2.2
A Summary of Information Assimilation and Decision-Making Cognitive Characteristics
and Related Information Requirements

Cognitive Characteristics	Design Considerations
Limited divided attention	Collocate similar information.
Limited focused attention	Recognize the potential to distract drivers from their primary task with excessive information.
Age-related decrements of divided attention	Provide redundant information sources.
Mental overload	Minimize noncritical information in times of high workload.
Faulty task prioritization	Include priority cues in messages to drivers.
Time misestimation	Display estimated time for routine and alternate routes to show travel time savings associated with deviation.
Limited literacy (written)	Use icons, make system use possible without reading a manual.
Limited literacy (math)	Minimize need to calculate values.
Language differences	Use icons.
Lack of computer familiarity	Do not rely on computer metaphors.
Icon interpretation	Identify with text coupled with standardization.

guidance and warning notifications. This section examines driver navigation knowledge (the relationship between paths and obstacles that make up the road network) and the factors that shape attitudes and response to ITS information. These two factors are likely to combine and influence drivers' acceptance of alternate routes that bypass traffic congestion and roadway hazards. This example shows that human factors considerations in ITS design must go beyond ensuring that drivers can perceive and understand information from ITS components. It also illustrates the need to consider the forces that shape knowledge and attitude development. Frequently, this will require a consideration of social pressures that shape driver behavior (Wilde, 1976).

Navigational Knowledge

Several characteristics that define how people develop, organize, and use navigation knowledge have implications for ITS. For example, if navigation depends on a mental representation or map of the environment, then the characteristics of such representations may have important consequences for the way drivers interpret navigation and traffic congestion information. Thus, understanding and predicting how drivers will use this type of information requires an understanding of how drivers organize topographic knowledge and mental models of congestion patterns. Although Tolman (1948) showed that mental maps may be crucial for guiding behavior, Siegel and White (1975) suggested that the knowledge of the environment may only vaguely correspond to a map. Instead of the uniform, detailed, and veridical representation of spatial relationships associated with electronic or paper maps, cognitive maps are likely to be fragmented, distorted, and in many instances may contain only a series of landmarks with little detail to connect them. Understanding the content, organization, and development of mental maps and navigation knowledge should facilitate the development of useful navigation aids. For instance, navigational information that is incompatible with the driver's internal representation may be ignored or misinterpreted. Thus, the characteristics of navigational knowledge are important for aiding navigation in familiar and unfamiliar locations.

In examining the way people organized information regarding locations within cities, Lynch (1960) found that their knowledge fell into five categories: paths, edges, districts, nodes, and landmarks. *Paths* are routes on which people move. Because people observe the city while moving through it on these paths, the paths often form the predominant elements of their image of the city. As such, the paths also strongly influence how other elements of the city are integrated into the image. *Edges* form boundaries between other elements of the image. They may be obstacles or barriers that separate regions (rivers, bays, and railroad tracks), or they may be indefinite lines that join two adjoining regions (state lines). *Districts* consist of large homogeneous areas of cities made up of many city blocks. The area within a district shares some common, identifying characteristics that define the part of the city that belongs to the district. *Nodes* represent strategic points within the city and are defined by the confluence of paths, concentrations of physical characteristics, or the center of districts. *Landmarks* are external points of reference defined by a physical object such as a sign, building, park, store, lake, or mountain. Some landmarks are quite large and visible for many miles (mountains), and some are small and visible only on close examination (store fronts, trees, and doorways).

The importance of each of these attributes depends on both the person and the city. For instance, Lynch (1960) found that some cities had more distinct districts, so these were a more predominant feature in people's images. Likewise, images differed among people with some people emphasizing paths, while others emphasized nodes or districts. In some cases identical physical features might be considered differently. For example, a highway might represent an edge to a pedestrian, but a path to a driver (Lynch, 1960). Because the city image depends on the geographical characteristics of the city and on the behavior patterns of the person, conventional maps may not match drivers' concepts of the city. By considering both the geographical characteristics of the city and the behavior patterns of the person, designers can anticipate drivers' information requirements and accommodate their preferred perspectives.

The disassociation between the organization of information in conventional maps and drivers' spatial knowledge is further amplified by Chase (1982). He claimed that people's conception of cities may be far removed from that represented in maps. With taxi drivers, Chase (1982) showed that mental representations of a city are better described as a hierarchy than as a map. Instead of an accurate spatial representation of the geography, drivers have a hierarchical organization of locations within neighborhoods, neighborhoods within districts, and districts within larger geographic regions. Thus, observations of Lynch (1960) and Chase (1982) showed that drivers' knowledge of the city may not correspond to the conventions used in most maps. These results suggest that the effectiveness of electronic maps may be enhanced by organizing their content to reflect the mental representations of drivers.

Expertise and Navigation

It has been found that expertise determines how different environmental attributes contribute to navigation. In a study by Lynch (1960), distant landmarks were recalled by most people, but only inexperienced people used them for navigation. Specifically, people with little knowledge of the city tended to consider it in terms of large regions with general characteristics and broad directional relationships. More experienced people relied on paths and the relationships between paths. Like the novice, experienced people relied on landmarks. However, experienced people focused on small landmarks that provided specific cues to aid the navigation process.

Siegel and White (1975) reported similar conclusions. They also suggested that the development of navigational knowledge begins with landmark knowledge. Route knowledge develops next through paired association of bearing and landmark changes. The final stage of navigational knowledge development integrates routes into a complex network, which has been termed survey knowledge. The integration of individual routes into survey knowledge facilitates judgments of straight-line distance and the relative locations of many objects, even if a route between them has never been traveled.

With experience, navigational knowledge often develops from landmark to survey knowledge. Evidence suggests, however, that even highly experienced travelers may not possess complete survey knowledge. Chase (1982) found that seasoned taxi drivers rarely develop deep knowledge of the spatial layout of cities, relying instead on base and secondary routes. They use the base routes to arrive at the general location, then use the secondary routes to arrive the exact location. This enables expert drivers to traverse a complex city without detailed survey knowledge. Similarly, although longtime city residents had often developed images and procedures for successful navigating, many reported that once they left their well-traveled routes they might not be able to navigate successfully (Lynch, 1960). These residents also drew maps of the city that were often incomplete, missing important detail, and including severe distortions. The maps that people generate are seldom precise descriptions of a city. Lynch (1960) described it thus:

> The image itself was not a precise, miniaturized model of reality, reduced in scale and consistently abstracted. As a purposive simplification, it was made by reducing, eliminating, or even adding elements to reality, by fusion and distortion, by relating and structuring the parts. (p. 87)

Commonly, features of the city were related using a flexible structure that preserved the sequential relationship between features on a path, but distorted the relationship between features on different paths. This distorted city image facilitated travel between locations on paths, but it made travel between locations off these paths confusing. These findings suggest that ITS guidance must be carefully integrated with drivers' sometimes distorted and limited knowledge of city topography.

Some of these distortions in drivers' survey knowledge take predictable forms. Chase and Chi (1979) found that when people reconstruct street maps, they tend to force streets into a rectilinear grid, distorting the true directions of the streets so that major streets are drawn to run north and south and perpendicular to each other. Another systematic distortion is the tendency for distance estimates to reflect the time taken to scan an imagined map, rather then the true distance. Because scan time can depend on items that fill the map, such as the number of cities, number of intersections, and other geographical features, distance estimates can be overestimated for complex areas of the map (Thorndyke, 1980). These biases suggest that in-vehicle maps should be carefully designed to avoid enhancing these biases by minimizing superfluous detail and preserving the true orientation of the roadway network.

Mental Models and Navigation Knowledge

Mental models may also be a useful concept to consider in the context of navigation knowledge. Research in the domain of process control and human–computer interaction has shown that people develop mental models of complex systems to guide their observations, predict future performance, and plan potential courses of action (Moray, 1988). In addition, mental models aid people in transferring knowledge about a well-known system to a new, but related, system (Gentner & Gentner, 1988). Thus, mental models significantly influence interactions with complex systems. Just as mental models aid operators in organizing their knowledge of complex industrial plants, mental models may be an important factor in drivers' organization of complex street and traffic information. The way a driver develops and uses mental models may be crucial to the driver's use of route guidance systems, particularly systems that provide routes based on criteria supplied by the driver. If users have an accurate model of traffic patterns, then they will be more likely to know when the system supplies inaccurate or misleading information.

Mental models developed to facilitate operator interaction with complex systems are often simplified representations of the whole system. That is, a mental model will contain only a subset of the relationships embodied in a complex system. Thus, there is a many-to-one mapping from the original system to the person's mental model of it. Ashby (1956) and Moray (1988) termed this reduced or simplified model a homomorph of the original system. A homomorphic model of the original system has several advantages to a complete, or isomorphic, model of the system. Its advantages lie in its relative simplicity and the corresponding reduction in mental workload. Its disadvantages, however, lie in the fact that the true state of the actual system cannot be unambiguously deduced by knowing the state of the homomorph (Moray, 1988). Furthermore, Moray (1988) suggested that humans have the natural tendency to create homomorphs that express the essential elements of the system in the simplest way possible. Because people have the tendency to create and then rely upon such simple models, there is the risk that the models they create will be useful only in limited and routine circumstances. When unusual situations occur, operators may not be able to generate a more detailed and complete model of the system, resulting in poor performance. In the context of navigation, these results imply that navigation knowledge organized as a mental model may not contain the detail required to cope with novel situations. In addition, the process of developing mental models may introduce relationships that are inconsistent with the actual system. Sanderson and Murtagh (1989) showed how these "buggy" mental models lead to predictable errors. In the driving domain, "buggy" mental models of a street network might lead drivers to use inefficient routes. Information provided by ITS might be able to rectify this situation by providing the driver with an accurate description of the street network.

Although mental models can enhance performance and speed learning, while decreasing task difficulty, they can also lead to predictable errors. The errors that result from applying erroneous analogies to new systems are particularly important because people spontaneously generate these analogies, based on surface similarities that may not always correspond to structural relationships. For example, assumptions about rush-hour traffic in a large city such as Dallas may not hold for Seattle where topography and work habits may differ. Similarly, errors may result when simplified mental models are applied inappropriately, such as when drivers fail to consider alternate paths through a complicated road network. Because of the potential benefits and risks of mental models, the design of ITS components should consider how the organization of navigation knowledge may influence driver behavior. However, maps and other forms of information must be configured in keeping with drivers' conceptual organization of navigation information. Otherwise drivers may simply reject accurate information from the ITS in favor of their incomplete and inaccurate mental models.

CALIBRATION OF DRIVER ATTITUDES

Although organization of navigation knowledge and mental models of traffic patterns have implications for the way drivers interpret information from ITS components, driver attitudes are likely to have a powerful effect on how that information will be used to guide behavior. Although sophisticated systems may provide drivers with substantial amounts of navigation and congestion information, the driver is still left to make the final decision. For instance, a system may suggest a route, but the driver has the final authority to accept or reject its recommendations. In this situation the decision may depend on perceived system capabilities, trust in the system, and the confidence the driver has in his or her own perceptions, knowledge, or intuitions. Therefore, it seems likely that when driver self-confidence does not match driver capabilities, or when the driver's trust in the system does not match its capabilities, the system will be used when it is inappropriate or be ignored in favor of less optimal driver decisions. Thus, the factors affecting driver calibrations of trust in the system and self-confidence in themselves will help determine how effectively the system will be used.

Trust

Research investigating operator interaction with automation illustrates several cognitive characteristics important in the design of ATIS systems. Several researchers (Halpin, Johnson, & Thornberry, 1973; Lee & Moray, 1992; Muir, 1989; Sheridan & Hennessy, 1984) suggested that operators' trust in automation may play a major role in guiding their reliance or dependence on automation. Just as the relationships between humans are influenced by trust (Barber, 1983; Rempel, Holmes, & Zanna 1985), so trust may mediate the relationship between the supervisory controller and the subordinate machines. Highly trusted automatic controllers may be used frequently, whereas operators may choose to control the system manually rather than engage automatic controllers whom they distrust (Muir, 1989). Accounts of the interaction between automation and operators (Halpin et al., 1973; Muir, 1989; Zuboff, 1988) all suggest that the operator's subjective feeling of trust toward automation plays an important role in its effective use. Although the choice of automatic or manual control probably depends on a number of factors other than trust such as risk, responsibility, and mental workload (Boettcher, North, & Riley, 1989), trust, nevertheless, seems to play a prominent role.

Zuboff (1988) demonstrated the importance of trust in the relationship between operators and the systems they use. She documented operator trust as being a significant factor in the use of automation following its introduction into the workplace. Her case studies revealed two interesting phenomena. First, an operator's lack of trust in the new technology often formed a barrier thwarting the potential that the new technology offered. Second, operators sometimes placed too much trust in the new technology, becoming complacent and failing to intervene when the technology failed.

Halpin et al. (1973) showed similar results, but were able to develop a more specific description of how trust mediates the operator's relationship with automation. They suggested that the relationship depends on both attitudinal and structural factors. With respect to attitudinal factors they cited several examples in which the operator's initial mistrust inhibited the effective use of computer aids. For example, during the introduction of automatic test equipment they found that, "the mistrust may similarly lead to a failure to utilize the available output in favor of some other, richer source of information" (Halpin et al., 1973, p. 167). In contrast, they mention a case in which the initial positive bias led to accepting the computerized system's recommendations, even though it performed far more poorly than the manual system it replaced. In addition to the operators' initial attitudes, structural factors such as the format, timing, and the logic of computer recommendations proved to be important variables in the operators' acceptance of the computer support. The researchers claimed that if the output of the aid fails to conform to the operators' expectations or understanding, then the operators will tend to deemphasize automation's role in the decision process. Halpin et al. (1973) presented an important perspective, emphasizing the effects of the operators' previous attitudes as well as the structure of the interaction on the operators' ability to trust and adapt to automation.

Much research (Barber, 1983; Halpin et al., 1973; Lee & Moray, 1992; Muir, 1989) suggests that trust is a multidimensional construct and cannot be equated with the perception of system reliability. Lee and Moray (1992) reviewed these studies and developed a multidimensional definition of trust that includes four dimensions: foundation of trust, performance, process, and purpose. The first dimension, "foundation of trust," represents the fundamental assumptions of natural and social order that make the other levels of trust possible. "Performance," the second dimension, rests on the expectation of consistent, stable, and desirable performance or behavior.

This dimension corresponds to a user's perception of reliability or efficiency. The third dimension, "process," corresponds to a user's understanding of how the underlying qualities or characteristics that govern the performance of the automation match the current task demands. With humans this might be stable dispositions or character traits. With ATIS/CVO systems, this might represent data reduction methods, rule bases, or control algorithms that govern how the system behaves. The final dimension, "purpose," rests on the user's perception of underlying motives or intentions. With humans this might represent motivations and responsibilities. With machines, purpose reflects the user's perception of the designer's intention in creating the system.

The theoretical description of trust has implications for the appropriate use of ATIS systems. For example, Bonsall (1992) reviewed several studies of user acceptance of route guidance systems. In this review he identified several problems that might inhibit the use of ATIS systems. One problem involved driver acceptance of advice that did not correspond to a driver's immediate goals. Some systems may optimize with respect to all drivers, leading to suboptimal routings for individual drivers. This type of system may undermine driver trust on the dimension of purpose because the intention of the system may not agree with that of the individual. In empirical tests, Bonsall (1992) reported that drivers disregarded navigational advice that did not match their intuitions. Specifically, when a route suggestion deviated significantly from the straight line direction to the destination, it was rejected 44% of the time compared to 9% when it was more closely aligned. This result suggests that the process dimension of driver trust had been undermined by a failure to understand how the advice was generated, leading drivers to reject potentially useful advice. Bonsall (1992) also showed that drivers were sensitive to the objective quality of advice. When advice was close to the theoretical optimum, drivers rejected it only 20% of the time. However, as the quality of advice declined, drivers rejected it more often. When the advice recommended routes that were twice as long as the theoretical minimum, they rejected it 80% of the time. This result suggests that the product dimension of the drivers' trust was undermined by poor performance. Thus, the literature review and empirical evidence compiled by Bonsall (1992) is consistent with the multidimensional definition of trust.

Recent findings of Kantowitz, Hanowski, and Kantowitz (1994) provide an interesting addition to the previous research (Bonsall, 1992; Bonsall & Parry, 1991). This research employed the Battelle Route Guidance Simulator (RGS) to investigate how the quality of information influenced driver trust in the system and reliance on the resulting information. The Battelle RGS incorporated an ATIS device that provided congestion information of the simulated road network. Drivers used the ATIS to estimate congestion, then select the fastest route through the road network. After making a route selection, the drivers viewed a video showing the congestion and their progress along their chosen route through the network. The Battelle RGS was used to investigate several levels of ATIS accuracy, and the results showed that drivers will accept some degree of inaccurate information. When accuracy drops from 100% to 71% the ATIS remains relatively acceptable and useful. However, when accuracy drops to 43% both driver trust and performance suffer. Furthermore, this research showed that driver trust is relatively resilient, recovering after each instance of poor information. However, with a very inaccurate ATIS (43%), trust does not recover as fully compared to a 71% accurate ATIS.

Self-Confidence

Just as operators' trust in automation may influence their reliance on it, so operators' self-confidence may contribute to their dependence on automation. The operators' anticipated performance during manual control (self-confidence) may interact with their trust in the automatic system to guide the allocation policy. Experiments by Lee and Moray (1992) and Lee (1991) showed that the difference between operators' trust in automation and their self-confidence in their own abilities guides the use of automation. This research confirms the need to consider both driver self-confidence and driver trust in ATIS systems to ensure the appropriate reliance on these systems. If operators' self-confidence fails to correspond to their actual abilities, then they may allocate automation inappropriately, just as mistrust may lead to an inappropriate allocation strategy.

Much research has addressed self-confidence, and the results of this research suggest that people often overestimate their capabilities. In other words, people are often overconfident in their abilities, both in forecasting future events (Fischhoff & MacGregor, 1982), and in their use of general knowledge (Fischhoff, 1982). Bankers, executives, civil engineers, clinical psychologists, and psychology graduate students, among others, all displayed overconfidence in their abilities (see Lichtenstein, Fischhoff, & Phillips, 1982, and Yates, 1990 for reviews). Furthermore, in many cases the subjects' overconfidence increased as they received more information during a judgment task. For example, in the context of clinical diagnosis, Oskamp (1982) showed that additional information only slightly improved subjects' judgment accuracy, but it led to a large increase in their confidence in their judgment. In addition, as the evidence accumulated, the number of judgments that changed decreased. Fischhoff (1982) suggested that the general bias toward overconfidence represents a robust effect and is more than an artifact of contrived laboratory conditions. There may be generalized age differences in confidence about abilities. Older adults tend to overpredict their memory performance, whereas young adults either underpredict or are accurate about their performance (Shaw & Craik, 1989).

Because overconfidence may be more than just an artifact of laboratory experiments, miscalibrations in operator self-confidence may prove to be a substantial impediment to driver compliance with automation. If drivers consistently overestimate their capabilities, they will likely reject information produced by the system concerning route suggestions or traffic congestion levels and fail to benefit from the capabilities of the automation. Bonsall (1992) illustrated how self-confidence influenced drivers' use of automatic route guidance systems. His review shows that drivers in unfamiliar areas (those with low self-confidence in their navigation abilities) depend on route guidance systems more heavily than those in familiar areas. Similarly, populations that are likely to be more self-confident (young drivers and high-mileage drivers) were less inclined to accept advice (Bonsall, 1992). These empirical results of driver interaction with automated navigation systems suggest that drivers may not use these systems because of high self-confidence. In some cases this level of self-confidence may be erroneously high.

In contrast to many studies showing unduly high self-confidence, professional weather forecasters often show good calibration with only a small tendency toward either overconfidence or underconfidence (Murphy & Winkler, 1977). The performance of weather forecasters shows that overconfidence can be eliminated under the right conditions. Accurate calibration is also found for expert bridge players when predicting the outcome of a contract that they are about to play (Keren, 1987). One explanation for the lack of overconfidence in weather forecasters' judgments is the clear criterion for their judgment success as well as the immediate feedback they receive. Fischhoff and MacGregor (1986) discussed the problem of overconfidence in terms of a miscalibration in subjects' expectations. They suggested that the subjects' successful interaction with a system depends on system transparency and metatransparency. "Transparency" refers to the ease of interaction with the system and governs the subject's performance. "Metatransparency" refers to how easily performance can be evaluated and how easily subjects can anticipate future performance. Therefore, a transparent system would lead to superior performance by the subject, whereas a metatransparent system would lead to good calibration (no overconfidence or underconfidence). In the design of a system that provides routing and traffic information, transparency refers to characteristics of the interface that facilitate communication of information to drivers, whereas metatransparency refers to the ability of the system to provide feedback concerning the drivers' performance and the value of the information from the system. A system with high metatransparency will ensure a veridical perception of the driver's own capabilities (self-confidence) and the system's capabilities (trust). Calibrating driver trust and self-confidence helps to guarantee that drivers will attend to route suggestions generated to avoid congestion when the system can aid them.

Research investigating driver acceptance of ATIS congestion information has provided some insight into the factors that influence self-confidence and its influence on driver decision making. Kantowitz, Hanowski, and Kantowitz (1994) used the Battelle RGS to show that self-confidence is greater when drivers are navigating a familiar city than when they are navigating an unfamiliar city. This difference in self-confidence is reflected in the number of times drivers diverged from preplanned routes in familiar and unfamiliar cities. In familiar cities, greater self-confidence encouraged drivers to deviate from their initial route when they learned of traffic congestion. In unfamiliar cities, drivers were more likely to stay with their initial route even when faced with congestion.

The Battelle RGS provided only traffic information; the simulated ATIS device did not offer suggestions for alternate routes as has been done in previous research (Allen, Ziedman, Rosenthal, Torres, & Halati, 1991; Bonsall & Parry, 1991). This is an obvious area for additional research. The recent research was aimed at driver acceptance of unreliable information. Future research using the Battelle RGS will also investigate design issues incorporating route guidance. For example, the traffic information presented and route guidance could be varied simultaneously by having suggested routes go through areas in which different levels of traffic congestion are projected. It is questionable whether drivers will accept advice that routes them through heavy traffic areas, even when the system advises that this would be more efficient than taking a less congested minor arterial.

This section identifies drivers' navigation knowledge, use of mental models, and attitudes toward technology as important considerations in the development of ITS. These and other important considerations are summarized in Table 2.3.

CONCLUSION

This chapter does not provide a comprehensive description of driver behavior and information requirements because much of what governs driver behavior is a consequence of the environment, tasks, and goals that influence the driver and not a consequence of any psychological limitations. Function and task analyses will help to identify ways in which these factors influence driver behavior and design requirements. In describing psychological factors affecting driver behavior, this chapter provides a general framework supported with three specific examples. These examples illustrate specific human limitations and characteristics that may influence the success of ITS components. They also illustrate the need to consider the broad range of human behavior that an ITS component must support. Consideration should be given to human limitations and characteristics that fall into each of the

TABLE 2.3
A Summary of Knowledge and Attitude Cognitive Characteristics and Related
Information Requirements

Cognitive Characteristics	Information Requirements
Hierarchical organization of navigation knowledge	Provide information in a form that is compatible with knowledge structure.
Different levels of navigation expertise	Provide people with navigation information tailored to the expertise.
Limited knowledge, even with experts	Provide an accurate representation of street network information.
Tendency toward overly simplistic mental models	Make appropriate mental models apparent in the interface or through the use of analogies.
Potential for inaccurate models	Make appropriate mental models apparent in interface or through the use of analogies.
Reliance on analogies to understand systems	Describe ATIS functions and features using analogies that provide a complete and veridical description of the system.
Trust	Convey an accurate reflection of system performance, mechanisms that guide system, and purpose of system.
Self-confidence	Convey an accurate reflection of driver performance.

categories: perception and action, information assimilation and decision making, and knowledge and attitudes.

ACKNOWLEDGMENT

This research was supported by the Federal Highway Administration Contract DTFH1-92-C-00102.

SUGGESTED READINGS

Michon, J. A. (1993). *Generic intelligent driver support: A comprehensive report on GIDS.* Washington, DC: Taylor & Francis.
Parkes, A. M., & Franzen S. (1993). *Driving future vehicles.* Washington, DC: Taylor & Francis.

REFERENCES

Allen, R. W., Ziedman, D., Rosenthal, T. J., Torres, J., & Halati, A. (1991). *Laboratory assessment of driver route diversion in response to in-vehicle navigation and motorist information systems.* In SAE Technical Paper Series (SAE No. 910701, pp. 1–25). Warrendale, PA: Society of Automotive Engineers.
Ashby, W. R. (1956). *Introduction to cybernetics.* London: Chapman & Hall.

Barber, B. (1983). *Logic and the limits of trust.* New Brunswick, NJ: Rutgers University Press.

Boettcher, K., North, R., & Riley, V. (1989). On developing theory-based functions to moderate human performance models in the context of systems analysis. *Proceedings of the Human Factors Society 33rd Annual Meeting* (pp. 105–109). Santa Monica, CA: Human Factors Society.

Bonsall, P. (1992). The influence of route guidance advice on route choice in urban networks. *Transportation, 19,* 1–23.

Bonsall, P. W., & Parry, T. (1991). Using an interactive route-choice simulator to investigate drivers' compliance with route guidance advice. *Transportation Research Record, 1306,* 59–68.

Caelli, T., & Porter, D. (1980). On difficulties in localizing ambulance sirens. *Human Factors, 22,* 719–724.

Chase, W. G. (1982). Spatial representations of taxi drivers. In D. Rogers & J. A. Sloboda (Eds.), *The acquisition of symbolic skills* (pp. 391–405). New York: Plenum.

Chase, W., & Chi, M. (1979). Cognitive skill: Implications for spatial skill in large-scale environments (Tech. Rep. 1). Pittsburgh, PA: University of Pittsburgh Learning and Development Center.

Fischhoff, B. (1982). Debiasing. In D. Kahneman, P. Slovic, & A. Tversky (Eds.), *Judgement under uncertainty: Heuristics and biases* (pp. 421–444). New York: Cambridge University Press.

Fischhoff, B., & MacGregor, D. (1982). Subjective confidence in forecasts. *Journal of Forecasting, 1,* 155–172.

Fischhoff, B., & MacGregor, D. (1986). Calibrating databases. *Journal of the American Society for Information Science, 37*(4), 222–233.

Gentner, D., & Gentner, D. R. (1988). Flowing waters or teeming crowds: Mental models for electricity. In D. Gentner & A. L. Stevens (Eds.), *Mental models.* Hillsdale, NJ: Lawrence Erlbaum Associates.

Halpin, S., Johnson, E., & Thornberry, J. (1973). Cognitive reliability in manned systems. *IEEE Transactions on Reliability R–22, 3,* 165–169.

Janssen, W. H., Alm, H., Michon, J. A., & Smiley, A. (1993). Driver support. In J. A. Michon (Ed.), *Generic intelligent driver support* (pp. 53–66). London: Taylor & Francis.

Kantowitz, B. H., Kantowitz, S. C., & Hanowski, R. (1994). Driver reliability demands for route guidance system. *Proceedings of the International Ergronomics Association 12th Triennial Congress, 4,* 133–135. Toronto, Canada: Human Factors Association of Canada.

Kantowitz, B. H., Triggs, T. J., Barnes, V. E. (1990). Stimulus–response compatibility and human factors. In R. W. Proctor & T. G. Reeve (Eds.), *Stimulus–response compatibility: An integrated perspective* (pp. 365–388). New York: North Holland.

Keren, G. (1987). Facing uncertainty in the game of bridge: A calibration study. *Organizational Behavior and Human Decision Processes, 39,* 98–114.

Lichtenstein, S., Fischhoff, B., & Phillips, L. D. (1982). Calibration of probabilities: The state of the art to 1980. In D. Kahneman, P. Slovic, & A. Tversky (Eds.), *Judgement under uncertainty: Heuristics and biases* (pp. 306–334). New York: Cambridge University Press.

Lynch, K. (1960). *The image of the city.* Cambridge, MA: MIT Press.

Lee, J., & Moray, N. (1992). Trust and the allocation of function in the control of automatic systems. *Ergonomics, 35*(10), 1243–1270.

Lee, J. D. (1991). The dynamics of trust in a supervisory control simulation. *Proceedings of the Human Factors Society 35th Annual Meeting* (pp. 1228–1232). Santa Monica, CA: Human Factors Society.

Moray, N. (1988). Intelligent aids, mental models, and the theory of machines. In E. Hollnagel, G. Mancini, & D. D. Woods (Eds.), *Cognitive engineering in complex dynamic worlds* (pp. 165–175). London: Academic Press.

Muir, B. M. (1989). *Operators' trust in and use of automatic controllers in a supervisory process control task.* Unpublished doctoral dissertation, University of Toronto, Canada.

Murphy, A. H., & Winkler, R. L. (1977). Can weather forecasters formulate reliable probability forecasts of precipitation and temperature? *National Weather Digest, 2,* 2–9.

Newell, A. (1989). Putting it all together. In D. Klahr & K. Kotovsky (Eds.), *Complex information processing: The impact of Herbert A. Simon* (pp. 399–440). Hillsdale, NJ: Lawrence Erlbaum Associates.

Oskamp, S. (1982). Overconfidence in case-study judgements. In D. Kahneman, P. Slovic, & A. Tversky (Eds.), *Judgement under uncertainty: Heuristics and biases* (pp. 287–293). New York: Cambridge University Press.

Rasmussen, J. (1983). Skills, rules, and knowledge: Signals, signs, and symbols and other distinctions in human performance models. *IEEE Transactions on Systems, Man and Cybernetics, 13,* 257–266.

Rasmussen, J. (1986). *Information processing and human-machine interaction.* New York: North-Holland.

Rempel, J. K., Holmes, J. G., & Zanna, M. P. (1985). Trust in close relationships. *Journal of Personality and Social Psychology, 49,* 95–112.

Sanderson, P. M., & Murtagh, J. M. (1989). Troubleshooting with an inaccurate mental model. *Proceedings of the 1989 IEEE International Conference on Systems, Man, and Cybernetics* (pp. 1238–1243). New York: IEEE.

Shaw, R. J., & Craik, F. I. M. (1989). Age differences in prediction and performance on a cued recall task. *Psychology and Aging, 4,* 131–135.

Sheridan, T. B., & Hennessy, R.T. (1984). *Research and modeling of supervisory control behavior.* Washington DC: National Academy Press.

Siegel, A. W., & White, S. H. (1975). Development of spatial representations. In H. W. Reese (Ed.), *Advances in child development and behavior* (pp. 10–55). New York: Academic Press.

Sorkin, R. D., Kantowitz, B. H., & Kantowitz, S. C. (1988). Likelihood alarm displays. *Human Factors, 30,* 445–459.

Sorkin, R. D., & Woods, D. D. (1985). Systems with human monitors: A signal detection analysis. *Human–Computer Interaction, 1,* 49–75.

Thorndyke, P. W. (1980). Performance models for spatial and locational cognition (Tech. Rep. R-2676-ONR). Washington, DC: Rand Corporation.

Tolman, E. C. (1948). Cognitive maps in rats and men. *The Psychological Review, 55*(4), 189–208.

Triggs, T. J., & Drummond, A. E. (1993). A young driver research program based on simulation. *Proceedings of the Human Factors and Ergonomics Society 37th Annual Meeting* (pp. 617–621). Santa Monica, CA: Human Factors and Ergonomics Society.

Wiener, E. L. (1985). *Human factors of cockpit automation: A field study of flight crew transition* (NASA Contractor Report CR-177333). Moffett Field, CA: NASA-Ames Research Center.

Wilde, G. J. S. (1976). Social interaction patterns in driver behaviour: An introductory review. *Human Factors, 18*(5), 477–492.

Woods, D. D., Potter, S. S., Johannesen, L., & Holloway, M. (1991). *Human interaction with intelligent systems: Trends, problems, new directions.* Cognitive Systems Engineering Laboratory Report. Columbus: Ohio State University.

Yates, J. F. (1990). *Judgement and decision making.* Englewood Cliffs, NJ: Prentice-Hall.

Zuboff, S. (1988). *In the age of the smart machine: The future of work and power.* New York: Basic Books.

3

HUMAN FACTORS DESIGN ISSUES FOR CRASH AVOIDANCE SYSTEMS

Thomas A. Dingus
Virginia Polytechnic Institute and State University

Steven K. Jahns
University of Iowa

Avraham D. Horowitz
General Motors Research and Design Center

Ronald Knipling
National Highway Traffic Safety Administration

INTRODUCTION: PERSPECTIVES ON ITS CRASH-AVOIDANCE OPPORTUNITIES

The Specter of Crash Avoidance Technology

The "electronics revolution" of the past two decades has changed motor vehicles and driving in a myriad of ways. Advanced electronics have been applied to many diverse elements of the automobile, including engine control, transmissions, and instrumentation. Inexpensive and miniaturized application-specific integrated circuits can perform multiple system functions, including sensing, intelligent processing, and communications. Another wave of change to driving is coming as this technology, along with new driver interface designs and vehicle control systems, is applied to collision avoidance. Applications such as forward-path obstacle/headway detection, blind-spot monitoring, driver performance–alertness monitoring, tire–road friction monitoring, and intelligent cruise control are becoming available and affordable. Moreover, the programmable nature of integrated circuits means that system functionality can be refined and adapted to specific circumstances or needs, such as specific vehicle types, environmental conditions, or driver performance capabilities.

This deluge of electronic devices has challenged traditional approaches to crash avoidance research. Traditionally, crash avoidance research has aspired to be *problem*-based rather than *solution*-based. That is, one first identified crash problems, their causes and characteristics, and required functional interventions. Then one identified a technology that could perform the required functions. In short, problems sought solutions, not vice versa.

There is much wisdom in this traditional approach, but there is no denying that "technology push" is jarring researchers from a purely problem-based perspective. The specter of available advanced sensors and processors confronts the automotive industry with an array of solutions that demand consideration in relation to major crash avoidance problems. Crash avoidance research and development is becoming a kind of matchmaking in which crash problems are matched to available technologies (Fancher et al., 1994). Today's approach might be described as both problem-based and solution-based. To be successful, the application of technology to crash avoidance must arise from an integrated understanding of the functional mechanisms of intervention of devices into the sequences of events, human errors in particular, that constitute crash scenarios.

Perspectives on Target Crash Problem Size

All Crashes

In 1990 there were an estimated 16 million U.S. motor vehicle crashes (police-reported plus nonpolice-reported; Knipling, Wang, & Yin, 1993). These crashes resulted in nearly 45,000 fatalities, 5.4 million nonfatal injuries, and 28 million damaged vehicles (Blincoe & Faigin, 1992; and derivations based on 1990 General Estimates System crashes and vehicle involvements). The estimated economic cost of motor vehicle crashes in 1990 was $137.5 billion, greater than 2% of the U.S. Gross Domestic Product (Blincoe & Faigin, 1993). Included in these losses were lost productivity, medical costs, legal costs, emergency service costs, insurance administration costs, travel delay, property damage, and workplace productivity losses. The average cost per crash was about $8,600.

Total 1990 U.S. vehicle registrations were 193 million. The average vehicle operational life is currently estimated to be about 13 years (Knipling et al., 1993). If one makes the simplistic assumption that every crash involves one at fault vehicle (or, more aptly, one vehicle with an at fault driver!), then the average vehicle can be expected to be at fault in about 1.1 crashes during its operational lifetime (i.e., 13 years multiplied by [16 million crashes/193 million vehicles]).

At a discount rate of 4%, the monetary value of each new vehicle's expected at fault crash loss is thus approximately $7,200 based on the 1990 economic cost estimates and these simplistic calculations. From a monetary cost–benefit standpoint, a magical device that would prevent all of a vehicle's at fault crash involvements would thus be worth up to $7,200 added to the cost of the new vehicle. More realistically, a device (or combination of devices) that would reduce each vehicle's at fault crash experience by 25% would be worth up to $1,800 (i.e., 0.25 multiplied by $7,200).

The electronics revolution has resulted in order-of-magnitude decreases in cost and improvements in capabilities of electronic sensors and processors that might be applied to crash avoidance. To many Intelligent Transportation Systems (ITSs) proponents, an electronics budget of $1,800 per vehicle seems an ample amount for achieving a 25% reduction in crashes. The apparent attractiveness of this investment from a purely economic standpoint is helping to drive the current intense interest of industry and researchers in high-technology crash-prevention devices. Moreover, this monetary cost–benefit perspective is conservative because it does not encompass the humanitarian benefits (and the resulting marketing benefits!) that may be achieved through a lessening of the current toll of pain and suffering from motor vehicle crashes.

Crash Types and Subtypes

The typical crash avoidance countermeasure does not act to prevent crashes in a general sense, but rather to prevent a particular type of crash under a particular set of circumstances. Thus, a knowledge of the major crash types is necessary for understanding the prospects for significant crash reduction. Figure 3.1 shows the number of police-reported crashes (1993 General Estimates System) for nine major crash types (Najm, Mironer, Koziol, Wang, & Knipling, 1995). Together, single-vehicle, crossing-path (e.g., 90° crossing path or left-turn across path at intersections), and rear-end crashes represent 75% of the crashes.

Figure 3.1 represents only a top-level classification of crash configurations. One may look at other elements of the crash picture such as fatal crashes or nonpolice-reported crashes. Or one may attempt to delve deeper to classify crashes by cause, countermeasure applicability, or both. In particular, it is revealing to examine crash scenarios from the perspective of countermeasure applicability. Often one finds that only a particular crash subtype is addressable by a countermeasure concept. Understanding the specific nature of relevant crash subtypes is thus essential for a realistic perspective on potential countermeasure benefits.

A clear example of this is provided by backing crashes. Analysis of backing-crash scenarios (Tijerina, Hendricks, Pierowicz, Everson, & Kiger, 1993; Wang & Knipling, 1994) reveals two distinct subtypes: encroachment and crossing-path crashes. *Encroachment* backing crashes involve slow closing speeds and a stationary (or slowly moving) struck pedestrian, object, or vehicle. In contrast, in *crossing-path* backing crashes the backing vehicle collides with a moving vehicle. For example, a vehicle backs out of a drive-

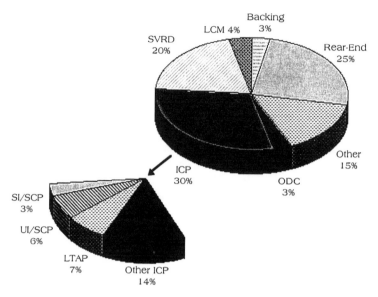

FIG. 3.1. Police-reported crashes by major type (1993 General Estimates System).

way and strikes (or is struck by) another vehicle moving at speed on the roadway. Obviously, crossing-path backing crashes generally involve higher closing speeds. Figure 3.2 illustrates these two backing-crash scenarios. Approximately 43% of all backing crashes are encroachment crashes; the remaining 57% are crossing path crashes.

A promising approach to preventing encroachment backing crashes is proximity detection (e.g., a sensor detects nearby objects in the backing path of the vehicle and warns the driver of its presence). In contrast, crossing-path backing crashes may prove difficult to address with proximity detection devices or other vehicle-based countermeasures in the backing vehicle. Detection of the crossing-path vehicle would require more sophisticated sensors and data processing, and would involve more complex driver human factors issues. Thus, virtually all current research and development on backing crashes is directed, explicitly or implicitly, toward preventing the encroachment subtype only. The 57% of the backing-crash problem represented by crossing-path backing crashes is not addressed by the proximity detection concept, although it may be addressable by other concepts, such as headway detection devices on the nonbacking vehicle.

Consideration of Different Vehicle Types

The crash involvement picture is quite different for different vehicle types. These differences may translate into order-of-magnitude differences in potential benefits, potential cost–benefits, or both from ITS crash avoidance coun-

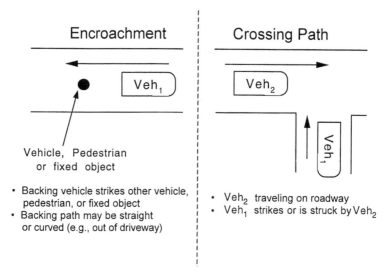

FIG. 3.2. Two major subtypes of backing-crash scenarios.

termeasures. Figures 3.3A and B present numbers of vehicle involvements in crashes and fatal crashes, respectively, for four vehicle type categories: passenger vehicles (cars, light trucks, lights vans), combination-unit trucks (tractor-trailers), medium/heavy single-unit trucks (straight trucks such as dump trucks, delivery trucks, etc.), and motorcycles. In each case, the overwhelming majority of involvements is represented by passenger vehicles.

Note, however, the much larger percentages of combination-unit truck and motorcycle involvements in fatal crashes. For combination-unit trucks, this reflects primarily their crash aggressivity (i.e., the higher damage and injury severities to the other vehicles and their occupants involved in crashes with these trucks). The aggressivity of combination-unit trucks is

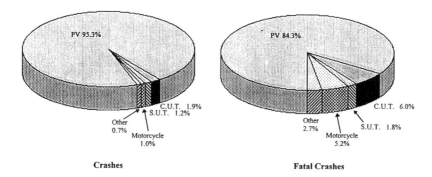

FIGs. 3.3A and 3.3B. Police-reported crash involvements and fatal crash involvements by vehicle type (1991 GES and FARS).

due to their size, weight, and body stiffness. The relatively high number of fatal involvements for motorcycles reflects the high-injury vulnerability of motorcycle riders compared to occupants of closed vehicles.

Figure 3.3 shows that passenger vehicles are the most important platforms for high-technology countermeasures from the perspective of potential total benefits. However, the picture may be different when the potential cost–benefits of countermeasure deployment are considered. In terms of potential percentage cost–benefits, the most promising platform for vehicle-based ITS crash avoidance countermeasures will often be combination-unit trucks. This is illustrated in Fig. 3.4, which shows the expected number of involvements over vehicle life for the four vehicle type categories. Combination-unit trucks have expected numbers of involvements that are more than twice those of the other vehicle types. They are likely to need a crash avoidance countermeasure more times during their operational lives. Thus the payoffs from device installation (whatever its effectiveness) are likely to be greater.

The high likelihood of combination-unit truck involvement in crashes is primarily because of their high mileage exposure; the average truck–tractor compiles about 60,000 miles per year versus about 10,000 miles per year for passenger vehicles. Thus, even though their overall crash rate per vehicle mile traveled (VMT) is less than one half that of passenger vehicles (see Fig. 3.5), their expected number of involvements per vehicle is much greater. For vehicle-based countermeasures that last the life of the vehicle, the latter

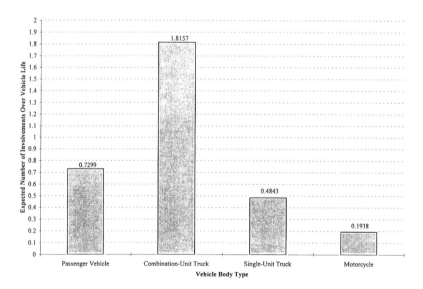

FIG. 3.4. Expected number of crash involvements over vehicle life—four vehicle types.

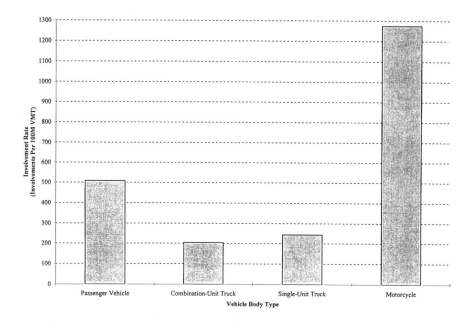

FIG. 3.5. Crash involvement rate per 100M VMT for four vehicle types.

statistic is much more relevant to a determination of potential benefits per unit cost. Ervin (1994) explored this in the context of the "public and private life cycle costs" of operating a truck–tractor.

Motorcycles have relatively high rates of involvement per VMT but relatively low expected numbers of involvements over the vehicle's life. The later is owing to their relatively low VMT per vehicle. As noted earlier, motorcycle crashes are characterized by a relatively high probability of injury/fatality, with the victim(s) most likely to be the motorcycle occupants themselves.

The Causes of Crashes: An Overview

One of the most extensive studies of crash causes was the Indiana Tri-Level Study (Treat et al., 1979). In that study, driver errors were determined to be a definite or probable cause, or a severity-increasing factor, in 93% of crashes. In contrast, environmental factors were cited as certain or probable for 34% of the in-depth cases; vehicle factors were cited in 13%. These percentages total more than 100 because more than one causal or severity-increasing factor could be cited. However, the Tri-Level study found that human factors were the only cause of 57% of crashes.

The human errors that cause crashes include errors of recognition, decision, and performance. These are briefly described and summarized in

Fig. 3.6. Of course, many crashes involve multiple interacting causal factors including driver errors, environmental factors, and vehicle factors.

Recognition Errors

This category includes situations in which a conscious driver does not properly perceive, comprehend, or react to a situation requiring a driver response. It includes inattention, distraction, and improper lookout (i.e., the driver "looked but did not see"). Recognition errors were a definite or probable causal factor in 56% of the in-depth Tri-Level crashes.

Decision Errors

Decision errors are those in which a driver selects an improper course of action (or takes no action) to avoid a crash. This includes misjudgment, false assumption, improper maneuver or driving technique, inadequate defensive driving, excessive speed, tailgating, and inadequate use of lighting or signaling. Decision errors were a definite or probable causal factor in 52% of the Tri-Level crashes.

Performance Errors

A performance error occurs when a driver properly comprehends a crash threat and selects an appropriate action, but simply errs in executing that

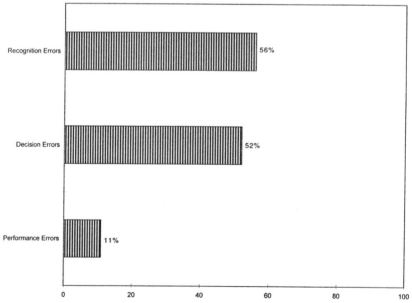

FIG. 3.6. Major human direct causes identified as definite/probable in the Indiana Tri-Level in-depth sample.

action. Performance errors include overcompensation (e.g., oversteering), panic or freezing, and inadequate directional control. Less common than recognition or decision errors, performance errors were apparent in 11% of the Tri-Level crashes.

A general conclusion from the Tri-Level study and most other studies of crash causation is that most crashes are caused by the errors of well-intentioned drivers rather than by overt unsafe driving acts such as tailgating, weaving through traffic, or even alcohol and excessive speed (Najm et al., 1995; Deering, 1994). Unsafe driving acts obviously increase the probability and likely severity of crashes, and thus must be contravened by whatever means are available to society. But unsafe driving acts do not account for a majority of crashes. The majority of motor vehicle crashes are caused by the same kinds of "innocent" or nonvolitional human errors that cause most industrial and other transportation mishaps.

Although the Tri-Level study was a landmark contribution to crash avoidance research, its findings may be misleading if they are not understood from an ergonomic perspective. Attribution of crashes to human error as a category apart from vehicle and environmental causes may lead to a mistaken inference that these human errors are unrelated to vehicle or highway design characteristics. From a prevention standpoint, attribution of crashes to driver error may obscure the fact that safety enhancements to the vehicle (e.g., collision warning systems), the environment (e.g., intelligent signing), or both may make crashes due to driver error less likely to occur.

A recent crash case review study has corroborated the Tri-Level findings and provided a systematic breakdown of principal crash causes by major crash type. Najm et al. (1995) reported the results of a review of 554 Crashworthiness Data System (CDS) and 133 GES crash case files. In this study, experienced crash reconstructionists reviewed accident research case files and made a subjective determination of the probable principal cause based on available information (which, for CDS cases, included driver interviews). The crash sample was large and involved eight specific crash types, but was not wholly representative of these data files or of the national crash picture. Figure 3.7 presents the causal factor taxonomy. Both human error and misbehavior are apparent in these results, with the former (e.g., inattention, looked but did not see, gap judgment errors) accounting generally for the largest share of crashes. Also notable are the many differences among the different crash types in terms of principal causal factors. Every crash type is distinct in its causal profile.

Several caveats are necessary in relation to the Najm et al. results. First, as noted, the study did not address all crash types; the eight types addressed encompassed 71% of the total crash population. Second, the study primarily used cases from the CDS, which includes only passenger vehicle crashes in which at least one vehicle was damaged sufficiently to be towed away. A

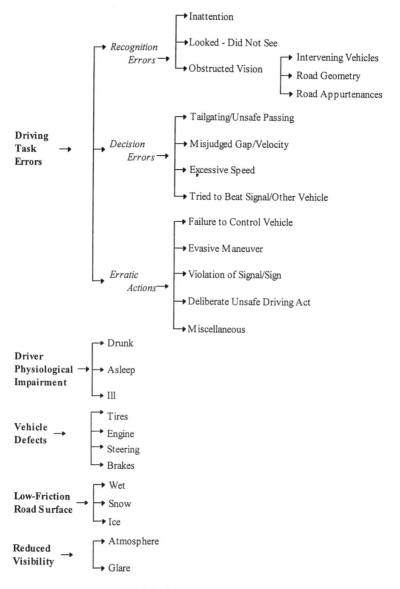

FIG. 3.7. Causal-factor taxonomy.

complex case-weighting scheme was employed to compensate for the un-equal, stratified sampling of cases of various severities and locations in both the GES and CDS.

Finally, both the Tri-Level and Najm et al. studies were studies of crashes, not fatal crashes. Crash severity is a major parameter affecting causal factor profiles. Fatal crashes present a different profile than do crashes in general, showing far greater roles of driver impairment such as alcohol (NHTSA, 1995) and drowsiness (Knipling & Wang, 1995), and misbehaviors such as speeding and other reckless driving (NHTSA, 1991; NHTSA, 1995).

Countermeasure Action: Time-Intensity Function

The preceding discussion shows that, when crashes of all severity levels are considered, most are not caused by deliberate unsafe acts but rather by garden variety mental errors of well-intentioned drivers. Most commonly, an everyday crash threat (e.g., a car stopped ahead, crossing traffic at an intersection) is not perceived in time for the crash to be avoided. In the seconds before a crash, a continuum exists between uneventful normal driving and a crash-imminent situation. The most applicable countermeasure depends on the amount of time available for driver–vehicle response. Specifically, the intensity of action needed to avoid or mitigate the crash increases dramatically as time-to-crash runs out (see Fig. 3.8; NHTSA, 1992). Early in the precrash period a simple advisory may be sufficient to permit drivers to take correct action. Late in the period the crash is unpreventable. At such times, only last-resort crash severity-reduction measures such as preimpact airbag deployment may be possible. Most crash avoidance research addresses the middle region where the crash is potentially preventable through immediate action—by the driver, by vehicle control systems, or by both.

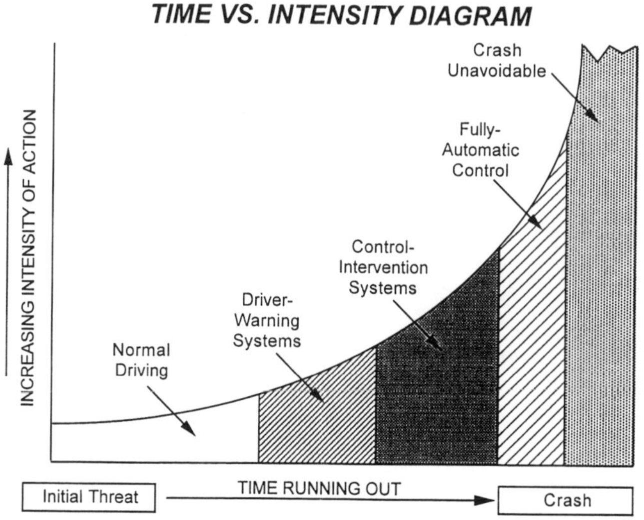

TIME VS. INTENSITY DIAGRAM

FIG. 3.8. Intensity of action needed as time-to-crash runs out.

Crash Countermeasure Payoffs: Frequency and Severity Reduction

The term crash avoidance implies prevention (i.e., reductions in crash frequency). However, crash avoidance countermeasures are likely to yield a second important category of benefits: reductions in crash severity. Figure 3.9 shows this conceptually. Devices that improve the driver–vehicle response to crash threats are likely to affect both the occurrence and the severity (e.g., impact speeds and resulting injuries) of crashes. Earlier driver awareness, faster braking, and other such measures that enable drivers to avoid crashes are likely also to decrease the severity of crashes not prevented. Preliminary analytical modeling of rear-end crash countermeasures indicates that severity reduction would be a significant category of benefits (Knipling et al., 1993).

System-Safety Concerns

An ever-present concern is the safety of crash-avoidance systems themselves. For example, there may be legitimate health concerns relating to the use of emitters such as lasers and radar. Even more insidious is the possibility that some crash avoidance countermeasures, while successfully preventing many target crashes, may also cause some other crashes to occur. For example, collision warning systems may on occasion evoke panic responses by drivers that result in loss of vehicle control, secondary collisions, or both caused by the evasive maneuver to avoid the original crash threat. Such system-safety issues must be addressed for each ITS crash avoidance

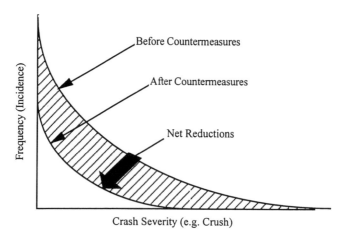

FIG. 3.9. Conceptual model of countermeasure payoffs; decreases in both crash frequency and severity.

countermeasure. Concerns regarding product liability will likely force industry to adopt very conservative deployment strategies for new devices.

Prospects for Crash Avoidance: Perspectives on Crash Types

As noted, most crash avoidance systems are directed toward rather specific target crash types or subtypes. This section highlights some recent research on specific target crash scenarios, causes, applicable countermeasures, and prospects for meaningful crash reduction.

Backing Crashes

The previous discussion on target crash subtype identification showed that encroachment backing crashes are likely to more addressable through the use of high-technology sensors and warning systems than are crossing-path backing crashes. Encroachment backing crashes represent only about 1% of all crashes, but they are an attractive target for crash avoidance because of their short sensor range requirements, slow closing speeds, and the relative simplicity of the needed driver interface (e.g., a warning signal for braking).

Encroachment backing crashes are almost always because the backing driver is not aware of the struck person, vehicle, or object in the path of travel (Tijerina et al., 1993). Rear-mounted sensors (e.g., ultrasound, radar, laser) may be used to detect objects in the backing vehicle's path. A typical range for existing systems is 15 ft, although the effective range is less for relatively small and irregularly shaped targets, such as people (Tijerina et al., 1993). Greater ranges are possible, but at the cost of a higher false or nuisance alarm rate.

The likely effectiveness of rear-obstacle warning systems depends on many different variables (e.g., system range, vehicle rearward speed–acceleration, initial distance between sensor and target, driver reaction time [RT], and braking efficiency). Tijerina et al. (1993) derived effectiveness estimates ranging from 26% to 90% for various encroachment backing crash subtypes and assumptions about backing vehicle motion (e.g., constant speed vs. accelerating).

Rear-End Crashes

Analysis of rear-end crashes has revealed two major subtypes (Knipling et al., 1993). About 70% involve a stationary lead (struck) vehicle at the time of impact, whereas about 30% involve a moving lead vehicle. Typically, the lead vehicle is stopped waiting to turn, perhaps at a traffic signal. The most common causal factor associated with rear-end crashes is driver inattention to the driving task. A second, and overlapping, causal factor is following too closely. One or both of these factors are present in approximately 90% of rear-end crashes (Knipling et al., 1993).

Based on this causal factor assessment, one applicable countermeasure concept appears to be headway detection (HD). HD systems monitor the separation and closing rate between equipped vehicles and other vehicles (or objects) in their forward paths of travel. They must dynamically reduce their functional ranges (i.e., warning distances) at lower speeds to avoid excessive nuisance alarms. There are several large industry research and development (R&D) programs to develop and market this technology.

Analytical modeling of HD system countermeasure action has yielded theoretical effectiveness estimates ranging between 40% and 80% for HD-system-applicable crashes (i.e., those involving driver inattention, following too closely, or both). Figure 3.10 graphically illustrates a small portion of the countermeasure modeling for rear-end lead-vehicle stationary crashes and an HD system with a maximum range of 300 ft. The partial modeling sample shown consists of 100 rear-end crashes arrayed by the precrash speed of the following vehicle (see Knipling et al., 1993 for more details). The line in Figure 3.10 represents one possible design system algorithm for warning distance at different vehicle speeds. Each of the 100 sample points represents a modeling event (i.e., a hypothetical driver/vehicle confronted with the crash situation while aided by the HD system). Each hypothetical driver/vehicle has been randomly assigned a braking RT and deceleration rate per a Monte Carlo simulation designed to approximate the actual popu-

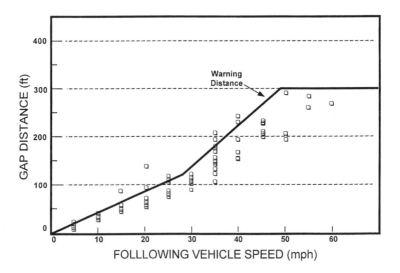

FIG. 3.10. Illustration of 100 sample data points from the HD system modeling for rear-end, lead-vehicle stationary crashes.

lation of drivers and vehicles. Points below the line represent crashes prevented by the countermeasure under these assumptions; those above the line represent crashes not prevented. The full Monte Carlo simulation for this case generated approximately one half million events and yielded an effectiveness estimate of 77%. Farber and Paley (1993) used freeway traffic data on vehicle speed and headway to model the effectiveness of a similar HD countermeasure concept; they estimated potential system effectiveness to be approximately 50%.

Obviously, raising or lowering the warning distance line in Fig. 3.10 would dramatically affect system effectiveness. A task for countermeasure designers is to determine optimal warning distance functions (i.e., those providing meaningful crash reductions but with tolerable nuisance alarm frequencies).

Lane-Change–Merge Crashes

Causal factor assessments (e.g., Najm et al., 1995; Treat et al., 1979) indicate that the majority of lane-change–merge crashes involve a recognition failure by the lane-changing–merging driver. In other words, the driver did not see the other vehicle until the crash was unavoidable. A potential vehicle-based countermeasure to these crashes is a proximity or lateral encroachment warning system (or, possibly, an automatic control system) that would detect adjacent vehicles, especially in the area of the driver's lateral blind zone.

The technology options for lateral proximity detection are substantially the same as for backing crashes. However, the required driver interfaces may be different. The nature of the driver interface for lane-change–merge crash warning systems is more problematic because the driver's steering maneuver to avoid the lane-change crash is likely to be less reliable (and thus more hazardous) than a braking response in a backing situation.

Another important consideration in lane-change–merge crashes relates to the close lateral proximity of vehicles in lane-change situations. Typically, two vehicles in adjacent highway lanes are only 4 to 8 ft apart laterally. To be useful, a system would need to provide the information before the initiation of the lane-change maneuver or early enough in the maneuver to permit successful evasive steering.

Single-Vehicle Roadway Departure (SVRD) Crashes

SVRD crashes (including crashes into parked vehicles) account for 20% of all crashes but nearly 50% of fatal crashes. A review of 100 SVRD crashes reported in Najm et al. (1995) revealed an assortment of causal factors and scenarios, including loss of traction on slippery roads (snow or ice), excessive speed, reckless maneuver, driver inattention/distraction (including evasive maneuver to avoid a rear-end crash), evasive maneuver to external crash threat (e.g., animal, other vehicle encroaching in lane), driver drowsy or asleep at the wheel), gross driver intoxication (often including excessive speed, reckless maneuver, etc.), vehicle failure, and driver illness.

With so many diverse crash causes, no single countermeasure concept emerges for these crashes. One potential countermeasure concept is road edge detection. Such a system would monitor the vehicle's lateral position and trajectory within the travel lane and detect imminent roadway departures. The system could activate a warning to the driver or automatic vehicle control (i.e., corrective steering).

Other countermeasure concepts are applicable to portions of the SVRD problem. For example, the headway detection concept discussed under rear-end crashes would be applicable to SVRD crashes resulting from an evasive maneuver to avoid a rear-end crash. Drowsy driver countermeasures (discussed elsewhere in this chapter) are applicable to a subset of SVRD crashes. Infrastructure-based warning or advisory systems may be applicable to crashes on slippery roads and those involving excessive speeds, especially at hazardous locations such as curves.

The SVRD crash problem is particularly acute for younger drivers. More Americans ages 15 to 24 years die each year from SVRD crashes alone than from homicide, suicide, or disease (National Safety Council, 1991). Drivers ages 15 to 24 years accumulate approximately 15% of all vehicle miles traveled (VMT), yet, they are involved in 38% of SVRD crashes (1991 GES). When one considers such metrics as years of potential life lost or impairment years, the losses suffered due to these crashes are truly staggering.

Head-On Crashes

Most head-on crashes are not because of unsafe passing but rather result from unintended lane departures associated with negotiating curves, loss of control, an evasive maneuver (e.g., to avoid a rear-end crash), or driver impairment or inattention. In addition, a large number are associated with a left-turn-across-path maneuver.

Thus, the countermeasures to head-on crashes may principally be countermeasures to other crash types that have the ancillary benefit of preventing head-on crashes. For example, a headway detection system that prevents rear-end crashes will also prevent head-on crashes occurring as a result of panic evasive maneuvers to avoid rear-end crashes.

Intersection–Crossing-Path Crashes

Of interest here are those crashes that involve crossing vehicle paths, especially crossing straight paths at 90° angles or a left turn of one vehicle across the travel path of another. Crossing-path configurations (illustrated in Fig. 3.11) represent about 30% of all crashes. In addition to crash configuration, a major factor influencing intersection crash causation and prospects

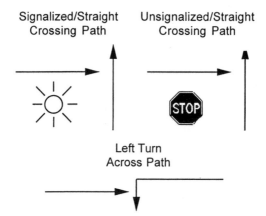

FIG. 3.11.   Four types of intersection–crossing-path crashes.

for mitigation is the presence or absence of traffic control signals. At controlled intersections there is a salient physical indicator of right-of-way privilege, whereas at signed intersections this is determined solely by driver judgment and decision making. Perhaps more important, signalized intersections are already equipped with an infrastructure for advanced electronics systems that might be employed as crash countermeasures.

Crossing-path crashes at signalized intersections almost always involve a signal violation, but these can be of two fundamental types: unintentional, inattention-related violations and intentional signal violations. The latter includes trying to beat the signal change and outright violation of the red light signal. Crashes involving drunk drivers probably include both types. Crashes that result from inattention and intentional signal violations are about equal in prevalence. A red-light warning system might be highly applicable to the inattentive driver, partially applicable to the intoxicated driver and those who try to beat the signal change, but not applicable at all to those who deliberately run the red light.

Similarly, left-turn-across-path crashes may involve a number of possible driver errors such as a recognition failure (e.g., the left-turning driver who looked but did not see the oncoming vehicle), an incorrect gap decision by the left-turning driver, a false assumption about the other vehicle's planned action, and a failure to execute the left turn maneuver in a timely manner. Available data (e.g., Treat et al., 1979) indicate that recognition failure is probably the most common category of driver error for these crashes. Thus, a countermeasure that works by assisting drivers with the gap decision would apply to only a portion of left-turn-across-path crashes.

A particular safety concern relates to involvements of older drivers involvements in intersection crashes. The overinvolvement of very old drivers in crashes is almost entirely related to intersection crashes, and to increased crash involvement as the crossing vehicle; for example, the left-turning vehicle in left-turn-across-path crashes, or the vehicle trying to cross the traffic stream from a stop in an unsignalized intersection crash (Hanowski, Bittner, Knipling, Byrne, & Parasuraman, 1995).

Research Needs

The preceding perspectives on crash avoidance targets of opportunity are largely based on rough first-order analyses of limited data. Such analyses are heuristic rather than definitive. Numerous research data needs must be addressed, including the following:

- Further elucidation of driver error to determine the probable applicability of countermeasure concepts to specific crash types and subtypes.
- Measurement of driver performance parameters such as reaction times (for braking or steering) under various crash threat conditions. This includes the performance of at-risk driver subgroups such as older drivers and commercial vehicle drivers. Major research and development programs are underway to develop advanced research tools such as the National Advanced Driving Simulator (NADS) and standardized measurement protocols to obtain such data (NHTSA, 1992; Clarke, Goodman, Perel, & Knipling, 1993).
- A knowledge base on in situ safety-related driver performance and behavior, including normative, event-related, and component-related driving actions. NHTSA has developed an unobtrusive and inconspicuous vehicle instrumentation suite for obtaining such in situ driver performance data (Knipling & Wang, 1995).
- Data on roadway geometry relevant to specific countermeasure concepts to define performance requirements and assess countermeasure applicability to various roadways. For example, data on the extent of roadway curves and hills are needed to define performance requirements for headway detection systems.

Both Farber, Freedman, and Tijerina (1995) and Najm et al. (1995) have provided more extensive reviews of key ITS human factors research needs. Future research will address the many ergonomic, operational, and technological issues relating to ITSs and crash avoidance in order to formulate countermeasure performance specifications (i.e., recommended functional guidelines for optimal countermeasure performance and effectiveness). Two major research and development challenges will be matching system functionality to the situations and driver errors that actually result in crashes, and designing driver-system interfaces compatible with the sometimes impressive—and sometimes limited—performance capabilities of drivers.

HUMAN FACTORS ISSUES

Human factors issues associated with the design of crash avoidance systems include individual differences due to different population characteristics, cognitive and perceptual considerations, environment of operational use, and a myriad of other factors. Some of these issues are discussed in detail as they relate to warning design principles. The discussion is split into three system classifications: advisory, collision warning, and automated crash-avoidance systems. The categories are not mutually exclusive when it comes to the human factors issues involved in system design. A topic may be relevant for all three system categories, but is discussed specifically for only one of the system categories.

The relation between advisory and collision warnings is conceptually similar to that between preventive medicine and disease treatment. An advisory warning may provide information and draw a driver's attention early in the consequence chain for the prevention of an emergency situation, but a collision warning follows a chain of events close to a crash or to a near-crash that needs immediate treatment. Thus, the potential value of some advisory warnings might be the avoidance of the very need for collision warnings.

Advisory Crash Avoidance Systems

Systems that merely advise the driver of the proximity of vehicles and the likelihood of a crash constitute one form of warning system being considered for future implementation. One particular type is termed a likelihood alarm display (LAD). In an LAD, information about event likelihood is computed by an automated monitoring system and encoded into an alerting signal for the human operator. Sorkin, Kantowitz, and Kantowitz (1988), evaluated operator performance within a dual-task paradigm with two LADs: a color coded visual alarm and a linguistically coded synthetic speech alarm. The results indicated that automated monitoring systems can improve the allocation of attention among tasks, that LADs can improve the allocation of attention among tasks and improve information integrated into operator decisions, and that LADs do not necessarily add to the operator's attentional load.

This type of display and the recommendations brought forth by Horowitz and Dingus (1992), are similar in nature. The idea that warning or potential crash information in general be graded such that the operator or driver does not react to a discrete on/off signal or warning may have a positive effect on the design of warning systems. For example, headway advisory information could be combined with imminent crash warning information in a graded display. The display could even be multimodal with the advisory information presented visually and the warning information presented aurally.

The use of LADs and graded warnings raise an important issue in the design of crash warning systems: It is crucial that the system have sufficient inherent "intelligence" to recognize critical and noncritical situations. If an alarm imposes a high demand on driver attention or annoys the driver to the point where avoidance performance is unchanged or decreases, then the alarm may not even be useful. This can not be emphasized enough.

Several examples of advisory warnings related to driver behavior and the environment are provided in the following list:

Driver

- A headway display suggests to the driver that the distance to the vehicle in front is too short for the speed (tailgating).
- A display tells the driver: "You show signs of fatigue; you may need rest."

Environment

- A display suggests that the driver slow down from the present 60 mph to 40 mph because of a sharp curve 400 ft away.
- When the driver shifts into reverse, a display attracts his attention to a camera view of the rear of the vehicle.
- When the driver turns on the turn signal, a display provides a camera view of that side of the vehicle.
- A display tells the driver: "Turn the lights on; it is too dark for safe driving."
- A road display warns: "Danger of ice formation on road for the next 10 miles."
- A truck ahead displays a unique rear signal following a speed loss without actual braking on an upgrade.

Situation Awareness and Modification of Unsafe Behavior

As previously mentioned, a contributing factor to most vehicle crashes is some form of driver error. In particular, drivers suffer from a variety of recognition errors in which the driver does not properly perceive, comprehend, or react to a situation requiring a driver response. This loss of situation awareness includes inattention, distraction, and situations in which the driver looked but did not see a hazard until it was too late. What causes the complacency that allows drivers to drift from the important visual scanning and vehicle operation tasks at hand?

With respect to rear-end crashes, according to Evans (1991), there are two likely reasons why drivers tend to become comfortable in following at headways that increase the risk of involvement in rear-end crashes. First, a dominant cue when following is the relative speed between the vehicle behind and the one in front. In normal vehicle following, relative speed is very close to zero. There is no risk of a rear-end crash if both vehicles maintain identical speeds, regardless of the speed. Evans believes that the largely static visual impression in vehicle-following tends to lower awareness and concern regarding speed. Second, according to Evans, drivers become comfortable when following too closely because they have learned, from repeated experience, that is safe to do so, in the sense that they have been doing it for years without adverse consequences. Evans also indicated that experience teaches drivers that the vehicle in front does not suddenly slow down very often. The same analogy can be applied to unsafe intersection or lane-change behaviors. Persons who live in rural areas may learn to change lanes with limited mirror and shoulder visual scanning. The decreased traffic density would make this behavior seem very reasonable and not inherently unsafe. However, if this scanning behavior becomes learned to the automatic level, and the driver then operates in a high-traffic-density situation there may be an increased chance of a broadside crash due to ineffective visual checks while changing lanes. For intersections, a driver could become accustomed to running yellow lights or rolling through stop signs, and for years never have an accident caused by this practice. Eventually, these behaviors may lead to a loss in perceived risk, a reduction in situation awareness, and a resulting increase in crash potential.

Advisory displays will be designed with the intention of increasing situation awareness and improving driver response in conditions for which judgment may be difficult. The inexperienced driver and potentially the elderly could certainly benefit from displays of this nature. Systems that compensate for vehicle traction or visibility in snow, darkness, or fog will enhance driver awareness of proximal hazards.

Collision Warning Systems

Warning a driver of an imminent crash compared with giving advisory and proximal location information projects fundamental differences in the design implications of crash avoidance systems. A primary difference is a reduced amount of response time for avoiding the hazard. Advisory and proximity displays serve largely as a continual training tool and possibly as a sensory enhancement tool when conditions of reduced visibility are present. Collision warning displays, however, require a correct and immediate response for crash avoidance. The reaction time that drivers need for making such a response is very diverse, complicating the resulting design of collision warning systems.

Individual Differences in Reaction Time

The correct driver reaction to a potential crash situation is rapid braking or steering. As a major consequence, one of the main factors determining whether the crash will be avoided is the driver's RT. RT and related factors such as maneuver severity or magnitude must be carefully considered in crash avoidance system design.

RT has long been the object of study, but specific mention of reaction time as it relates to in-vehicle braking is somewhat sparse because of the safety implications involved with the study of true emergency response circumstances. More research on reaction times to relevant lead vehicle decelerations is needed. Past research has generally concentrated on reaction to traffic signals and reaction to objects on the road (Olson & Sivak, 1986; Sivak, Olson, & Farmer, 1982). Driver RT estimates vary from 0.9 s for unexpected events with athletes as subjects (Davis et al., 1990) to 1.6 s for 95th percentile drivers reacting to unexpected events using a more representative population (Olson & Sivak, 1986).

In an experiment conducted by Olson and Sivak (1986), it was found that the 50th-percentile RT interval for a population of ordinary drivers confronted with an unexpected roadway obstacle was 1.1 sec, with a range (2nd to 98th percentile) of 0.81 to 1.76 s. Similar results have been found in simulator studies by Carney and Dingus (In preparation). It is to be noted that RT may be shorter for more intimidating test conditions. Taoka (1989) suggested that brake reaction times of unalerted drivers can be represented by a log-normal distribution. He surmised that the log-normal distribution for brake reaction time is more effective than a standard normal distribution because of the skewed log-normal distribution. Most studies have shown that the distribution mean is greater than the median RT because there are more extreme reaction times at the high end of the distribution. Extreme reaction times are more common and realistic because they describe more of the population (age and impairment effect) and take into consideration day-to-day driver episodes of divided attention. Furthermore, detection times may vary depending on the type of signal presented (e.g., auditory or visual).

Wortman and Mathias (1983) conducted an experiment to evaluate the deceleration behavior of drivers at signalized intersections. Time-lapse photography was used at six intersections to determine the approach speeds of the vehicles, the average deceleration rates of the stopping vehicles, the perception–reaction time of the drivers of the stopping vehicles, and the distance that the vehicle was from the intersection at the onset of the yellow interval. Analysis of the data brought forth the following results:

- Mean deceleration rates at various study sites ranged from 7.0 to 13.9 ft/s^2 (2.1 to 4.2 m/s^2). The observed mean deceleration rate for all observations was 11.6 ft/s^2 (3.45 m/s^2).

- Comparison of driver behavior at day and night times did not show significant difference in the mean deceleration rate. But the mean rate decreased from 12.9 to 9.7 ft/s² (3.9 to 3.0 m/s²) at one of the intersections.
- The perception–reaction times of drivers ranged from 1.16 s to 1.55 s with a mean for overall approaches 1.30 s. The 85th percentile value for the time measurement ranged from 1.5 s to 2.1 s.
- Approach speed, distance from the intersection at the beginning of the yellow interval, and the deceleration rate used by the driver had almost no influence on the perception–reaction time.
- There was no significant difference in the perception–reaction times when compared between the yellow only and yellow plus all-red change intervals. Also, analysis of deceleration rates for the two types of change intervals did not yield significant results.

Rice, Dell'Amico, and Rasmussen (1976) applied a clinical approach to the study of driver behavior on the road. Their findings are as follows:

- Males had three times the driving incidents and accidents of females and young drivers over four times that of the over-45 age group.
- From a self-rating questionnaire, 81% of the females rated themselves as average drivers, whereas 54% of the males rated themselves as above average. Of the drivers in the two age groups with highest incident and accident rates, 60% rated themselves as above average! These data show a clear mismatch between driver attitude, risk perception, aggressiveness, and skill that appears to be related to both age and sex.
- In a surprise intrusion maneuver, a barrel was ejected so that a subject with average RT could avoid the obstacle only by steering for speeds up to about 55 mph. The mean reaction time for all drivers was 0.65 s, and it varied from 0.35 s to 1.7 s; 75% of drivers applied brakes initially rather than steering. Significant loss of control after striking the barrel was found in about 27% of the drivers.

Driver RT becomes an important issue when average driver following behavior is analyzed. Wasielewski (1979) found that the average following distance for vehicles is 1.32 s, with a standard deviation of 0.5 s and a median of 1.0 s. This implies that many drivers behave such that they would be unable to successfully react and stop in a large proportion of potential rear-end crash circumstances. However, Davis et al. (1990) found that RT decreases as coupled vehicles (i.e., those traveling relatively close together and nearly the same velocity) draw closer together. This suggests that driver attention increases in relation to how close coupled vehicles are to one another.

Methods for obtaining these RT measures differed for all of the studies in this section. Each study presents a different method to determine a reaction. More research is needed investigating driver reaction to lead vehicles looming quickly with and without brake lights, and driver reaction to various warning displays.

Psychological Refractory Period

Consider a task for which two stimuli (S1 and S2) are presented in close temporal succession, less than 500 ms apart. The earliest study of this task (Telford, 1931) found that RT to S2 was delayed considerably. It was as if the brain was not functioning after receiving the first signal. Thus, the phenomenon was called the psychological refractory period by analogy to the refractory period of a single neuron that will not respond to a second input when it is stimulated by signals that arrive in close succession (see Kantowitz & Sorkin, 1983; Wickens, 1992). As the interstimulus time interval is decreased, human information processing shows more and more signs of overload. The "limited channel model" of attention (Broadbent, 1958) explains the delay in reaction time to S2 as being due to processing of S1. Until S1 clears the "channel," S2 cannot be processed with normal efficiency.

When the interstimulus interval is sufficiently short (less than about 100 ms), a qualitatively different processing sequence occurs: Both responses are emitted together and both are delayed (Kantowitz, 1974). It is as if the two stimuli are occurring so close in time that S2 gets in the channel during the acceptance of S1.

It may be appropriate to apply the psychological refractory period phenomenon to model the effect of an emergency warning. On occasion, the driver does not pay attention to the driving task owing to the presence of an in-vehicle secondary task such as talking to a passenger or controlling an audiotape operation. Consequently, given the presence of a collision warning system, the driver may not always react in time to an emerging dangerous situation, thus triggering the emergency warning (S1). The driver's reaction to S1 in this model is the awareness of the dangerous situation (e.g., the visual stimulus of an approaching vehicle (S2) to which the driver must respond. The time interval between S1 and S2 might be very short, say, shorter than 500 ms. The time reaction to S2 may be delayed compared to that of a simple reaction to S2 without S1.

An emergency warning may prove beneficial by helping to focus the driver's attention on a hazardous situation, or it may prove detrimental by creating attention overload. The main prediction of the multistimulus model is that RT to the collision-warning situation may be larger than that expected from a single stimulus. A dangerous traffic situation coupled with an emergency warning may result in attention overload and delay in reaction, especially when there are alternative possible responses such as braking, steering, or accelerating for the avoidance of a crash. This model requires empirical testing in a driving simulator or other suitable environment where the reaction time to the coupled stimuli—warning and traffic situation—can be successfully measured.

A slightly different information-processing model is required for the case in which the driver, late but surely, prepares to apply the brakes or steer to avoid a danger while an emergency warning is issued at the same time. (In this case S1 and S2 are ordered differently than previously described.) The driver then has to interpret the warning, causing a shift in attention from action to a new stimulus. This situation may also lead to an increase in RT (see Horowitz & Dingus, 1992). Therefore, in assessing the design of a collision-avoidance system, timing of the warning may be critical, not only because a late warning will not allow the driver to respond in time to avoid a crash, but also because an early warning may inhibit the brake-reaction response. Thus, a dangerous traffic situation coupled with an emergency warning may result in attention overload and delayed reaction, especially when there are alternative possible responses such as braking, steering, or accelerating for the avoidance of a crash.

Decision Making Under Stress

Decision errors are those in which a driver selects an improper course of action (or takes no action) to avoid a crash. This includes misjudgment, false assumption, improper maneuver or driving technique, inadequate defensive driving, excessive speed, and inadequate use of lighting or signaling. Treat et al. (1979) found that decision errors were a definite or probable causal factor in 52% of the crashes investigated as part of the Indiana Tri-Level studies. Thus, if a warning can somehow provide additional information that simplifies the decision-making process, it may be beneficial. Such information will be of primary benefit in advisory circumstances (e.g., too fast for curve ahead) because information context is generally required, and there is insufficient time to provide context in imminent crash-warning circumstances.

Attention in Multitask Environments

Drivers must process many sources of visual information while concurrently processing a variety of auditory and kinesthetic signals. There are two distinct methods for allocating attention to the perceptual world: via serial or parallel processing. According to Stokes, Wickens, and Kite (1990) certain environmental conditions force the driver into a serial mode of information processing. Simultaneously, scanning the busy outside world, driving, and switching to information collection from the instrument panel is one example of forced serial processing. Parallel processing also can occur while persons are driving. For example, drivers can listen to and process a radio traffic report while visually scanning the roadway environment.

Serial Processing and Foveal Vision. A driver's visual attention has been said to be analogous to a spotlight in that only those areas that are illuminated are attended to at any one time. In a simplified sense, the driver must initiate a scanning pattern to collect required information because the global environment cannot be captured solely with *foveal vision*, which is the point in the center of the driver's field of view where the eye provides a sharp and clear image on the retina. *Peripheral vision*, which encompasses the areas outside the center of the visual field is also critical to driving for reasons discussed in the next section. Moray (1990) stated that a serial scanning pattern is controlled quite efficiently if the amount of scanned information sources remains small. People can, however, lose track of the last time they sampled an information source. Prioritizing information importance is often observed in driver visual scanning patterns. When the driver is overloaded, less critical information receives reduced attention. Knowledge of this scanning behavior can be utilized in effective display design by grouping important display functions and placing them closer to the primary visual task to reduce switching time. This becomes more important in the design of systems that will inherently be viewed in high-stress, crash avoidance circumstances.

Parallel Processing and Peripheral Vision. Although much of the visual information processing in driving is serial, a high degree of parallel processing occurs as well. A driver constantly relies on motion cues from peripheral vision to maintain lane position. Visual acuity decreases as the image is displaced further from the fovea of the eye. Peripheral vision is primarily used to detect motion and luminosity. This perception of peripheral motion is carried out in parallel, often without conscious effort, along with perception of the car's actual position on the highway using foveal vision (Stokes et al., 1990). As relevant information density increases it is assumed that parallel processing will correspondingly increase. This principle underlies the cautionary progression in interface design of collision warning systems. Display requiring foveal vision may, in fact, misdirect attention away from the very hazard for which the system is trying to alert the driver. Displays that utilize the attention grabbing effect of peripheral motion should be considered for systems warning of the hazards that are, in fact, to the side of the driver. A flashing light or streak of light across the windscreen in the direction of the hazard may orient the driver better than a display in the instrument cluster and possibly better than an auditory display as well. However, such displays must be considered with caution. If attention is diverted because of a false alarm, the overall system effect could be detrimental.

Attention Tunneling

Allocation of attention to perceptual information is also affected by the workload level of the observer. The spotlight of attention analogy discussed previously can effectively be narrowed to a finer point under stressful situations. The driver will selectively scan only critical information, normally in the forward field of view while peripheral cues of reduced importance or salience are filtered out. Thus, it is possible that the cognitive capture of less critical or irrelevant information could lead to a decrease in driver performance. As more and more displays are integrated into the instrument panel area, information will get selectively filtered out as the driver becomes overloaded.

False/Nuisance Alarms and Warning Frequency

False alarms and warning frequency are one of the most important issues that must be dealt in the design of collision warning devices. A false alarm is an alarm activation in which a device does not function as designed (e.g., an electronic sensor interpretation of ambient noise as a signal). Nuisance alarms are similar to false alarms. However, they occur when a system functions as designed but when the situation does not constitute a true crash threat. Problems associated with false and nuisance alarms were demonstrated in the design of the first generation, Traffic-Collision Alerting System (TCAS-I) for commercial aircraft. TCAS-I had such a high nuisance and false alarm rate in congested traffic areas that pilots no longer believed the system was producing a valid alert. This situation is directly transferable to the automotive crash-warning domain.

Historically, warning systems in a variety of domains have used discrete on/off criteria. Such discrete systems are typically based on several parameters (e.g., an assumed RT) that must be estimated in evaluating when and how they should be actuated. These estimates must typically be conservative to address the needs of the majority of users which lead to higher false and nuisance alarm rates. When false alarm rates are too high, the user loses faith in the system and deems it useless. In a paper by Horowitz and Dingus (1992), warning-signal design and the associated human factors issues are discussed. The paper outlines the potential negative effects of collision warning systems in relation to the frequency of warning. Because it is estimated that a driver is involved in a car crash every five years on the average (Evans, 1991), and that rear-end crashes represent about 20% of all crashes, a driver may be involved in a rear-end crash only every 25 years on the average. This indicates that crash warnings theoretically should almost never occur. To overcome the paradox of providing reasonably conservative warnings while minimizing false alarms, four concepts were suggested:

- a graded sequence of warnings
- a parallel change in modality
- individualization of warnings
- a headway-only display

These recommendations are intended to optimize warning displays by reducing the impact of false-alarm rates as much as possible. High false-alarm rates will lead to user annoyance and result in a decrease in reaction performance when a true alarm is displayed.

Perceptual Factors in Rear-End Crashes Specifically

There is considerable evidence suggesting that rear-end crashes may occur because the driver of the striking vehicle does not see the vehicle ahead or because of complex perceptual factors (Mortimer, 1988). Several perceptual factors are present in determining distance and rate of closure information for following vehicles.

When making judgments regarding depth, pictorial cues such as relative size can be one of the strongest depth cues (Levine & Shefner, 1991). As the distance to a lead vehicle decreases, the apparent size of the lead vehicle will increase nonlinearly (Mortimer, 1990). That is to say, when a driver is closing on a vehicle, the relative size increases at a much slower rate initially compared to when the vehicles are close. Thus, it is more difficult for the driver to judge closure rate when a vehicle is some distance away. Considered in light of driver behavior of scanning multiple locations in the environment, it is apparent that crashes with stationary vehicles may be caused partly by a failure of the driver to recognize the high relative velocity.

According to available literature, drivers are accurate and sensitive at judging the direction of relative velocity. That is, drivers are able to judge accurately whether a gap between them and another vehicle is opening or closing (Hoffman, 1974). However, it also appears that drivers base their closure rates heavily on changes in visual angle. Mortimer (1988) found that drivers are able to derive little information on the velocity and relative velocity of the vehicle ahead of their own vehicle. Drivers are also able to make relatively accurate estimates on the distance to the car ahead of them (within approximately 20%) and are reasonably sensitive in determining a change in the headway between their vehicle and the one ahead of them (within an approximately 12% change). Research by Hoffman (1974), found that in many situations, drivers do not have the opportunity to estimate relative velocity because the threshold for human perception of the relative velocity is often not exceeded. It was concluded that unless the relative velocity between two vehicles becomes quite high, drivers will respond to changes in their headway or the change in angular size of the vehicle ahead and use that as a cue to determine the speed they should adopt when following another vehicle. As Hoffman suggested, this implies that rear-end crashes could be reduced if drivers were aided by a display that indicates the velocity of the speed of the car being followed. The display could then relate relative velocity because drivers know the velocity of their own vehicle.

Automated Crash Avoidance Systems

Many human factors issues overlap between system categories. Behavior adaptation and driver acceptance will be issues in advisory and collision warning systems, but they will be prominent concerns with automated systems where driver control is usurped.

Behavior Adaptation and System Reliance

One factor to consider in the design of any collision warning or avoidance system is the long-term effects of system reliance on driving behavior. A basic rule in the design of personnel safety systems in industrial processes is never to allow a guard or protection device to be designed in such a way that it becomes a "stop" used continually by the worker. If the guard fails, the worker will be injured. Automated crash avoidance systems could be relied on in the same way. Drivers could come to rely on the system braking for them, and thereby be at greater risk if the system should fail. Some human factors professionals feel that headway distances are likely to decrease with the use of warning systems because certain users will grow to trust the system to remove the hazard of this behavior, which is called risk compensation. Long-term field evaluations will need to be conducted to empirically determine behavior modification present as the comfort level of the use of a system type increases.

Driver Acceptance

For automated control systems there is a very real concern that drivers will inherently dislike a system that takes over vehicle control. Pilots hold a negative outlook on automation that removes control to this degree. Systems of this nature likely will be costly, and if user acceptance is not addressed carefully, marketing them could prove difficult.

Driver Override

An important general human factors principle states that drivers should not be allowed to disable a warning display. This prevents errors due to incorrect assumptions of system status or failings in memory by the driver. For automated systems, however, the issues are slightly different. In the case of warning displays, an error on the part of the system simply means that a false alarm is given to the driver, but in an automated control system, an error in the system could lead to a crash. Unless system reliability can be proven to be so great that manual overrides are not necessary, some allowance for driver control should be incorporated. Many crash-avoidance circumstances that require automated control will occur so fast that the driver will not have a chance to override the system. If necessary, however, the driver should be able to steer the vehicle, perhaps with a greatly increased resistive force. Overrides will need to be carefully designed, if utilized at all, because there are always situations in which the actions of a driver overriding an automated control input could lead to an accident. For example, if the car is braking automatically, the obvious override would be a depression of the accelerator. However, many accidents occur when the accelerator is mistakenly depressed instead of the brake.

Another related issue is the resumption of control once the automated crash-avoidance maneuver is complete. The driver may need an informational message saying that responsibility for control of the vehicle has been returned to the driver.

WARNING DESIGN PRINCIPLES

There are no established standards for in-vehicle warnings. The following list comprises information that might guide the establishment of human factors standards for in-vehicle information presentation (see also Ross, 1993, for a discussion of the problem of standards in hardware, human–computer interaction, and information display for the current European programs PROMETHEUS and DRIVE; for conventional in-vehicle warning devices see Baber, 1994).

1. *Provide redundancy in system design.* A common approach to improving the reliability of multicomponent mechanisms is to introduce redundancy (Senders & Moray, 1991). If a component's probability of failure is too high, we can use two or more components. The probability of redundant failure is much less than the probability that any single item will fail.

Because inattention is a major cause of crashes, redundancy might be a key design principle in advanced crash avoidance systems. For example, a curve on a freeway exit ramp can be signaled by a regular road traffic sign, and redundantly by a in-vehicle display transmitted by a beacon originated at the road side.

2. *Draw attention to the emergency situation.* To draw attention to an potential hazard, the addition of auditory, or perhaps even tactile warnings may be needed (e.g., a headway detector signaling a tailgating driver that can be supplemented by a steering wheel vibration, analogous to the "stick shaker" signal in aircraft).

3. *Prioritize visual displays by location.* To minimize visual demands of displays, it may be desirable to display information through two main displays: one in the center of the visual scene (foveal vision) for emergency warnings and one off center for advisory warnings. For the driver, the location of the display would become associated with the urgency of information.

4. *Avoid auditory signals for advisory warnings.* The literature on visual and auditory displays in aircraft summarized by Stokes et al., 1990 provides evidence that pilots prefer visual over auditory warnings when there is enough time to react. Drivers also may be startled, annoyed, or both by auditory warnings for such nonemergency situations as "low fuel" or "low windshield fluid."

An alternative approach uses auditory attentional warnings but chooses pleasant sounds such as musical chimes with one or two tones. For a detailed discussion on the relation between warning sound parameters and perceived urgency see Edworthy, 1994, and Edworthy, Loxley, & Dennis, 1991.

5. *Avoid speech displays for attentional warnings.* The British Leyland "Maestro" had a simple voice synthesis system that could announce low oil pressure, high engine temperature or door open (Redpath, 1984). There were, however, a number of complaints that led some drivers to disable the speech system. Baber (1994) suggested that the voice system failed because it did not distinguish between different levels of advisory and emergency messages. Our recommendation is that speech should be used sparingly and mainly for emergency warnings.

6. *Provide unique warnings.* Multiple warnings should be different so that they will be identified easily. For example, different voice types or tones could represent different warnings. Not only do warnings become associated with voices, but switches between messages are emphasized by the change of voice (Leiser, 1993).

7. *Incorporate intelligence in warning presentation dynamics.* Warnings can resemble human dialogue, taking into consideration the driver's reaction to a warning. For example, Leiser (1993) suggested the following dynamic sequence of attentional warnings for tailgating, depending on the driver's reaction to the first warning:

- You are too close to the car in front. *Please* pull back.
- You are still too close. You *must* pull back.
- O.K. Now keep that safe distance.

In addition to message-content intelligence, likelihood-based intelligence is also important in warning-signal design. That is, warning algorithms must consider the feasibility of a set of circumstances when processing signals. For example, the presence of an object at 6 ft from the front bumper may not feasible if the previous several signals did not indicate the presence of any object.

8. *Prioritize driver workload and warning.* In multiple warning or information situations, emergency warnings must get first priority. Also, even when no emergency collision warnings are present, selected advisory warnings should be postponed until the estimated driver workload is not too high. For example, a warning describing the danger of a low-pressure tire should not be given when the driver is braking, managing a curve, or handling other situations of high manual or cognitive load. However, in the same situation, if the driver is planning a secondary task such as asking the phone system to make a call, the system can postpone the task until the driver's workload is reduced. (For a detailed discussion of these issues see Verwey, 1993.)

9. *Individualize warnings.* The novice driver has tutorial needs that differ from those of the experienced driver. In addition, an elderly driver may require different warning timing than a younger driver. Thus, individualization of warnings may be necessary for a successful warning system. An ability to "tune" warning parameters must be carefully considered, however. For example, elderly drivers who mistakenly drive with the warning timing tuned to drivers with a faster reaction time may be at an increase risk of a crash if they rely on the system.

These nine principles are strategic guidelines for advisory and collision warning systems. The development of any specific display also requires detailed human factors testing at the tactical level. Such development requires a substantial, detailed human factors design and evaluation effort.

TECHNOLOGIES AND DISPLAYS BY CRASH AVOIDANCE TYPE

Single Vehicle Road Departure (SVRD)

As shown in the Introduction, SVRD crashes are associated with diverse circumstances related to the driver (excessive speed or reckless driving), to the environment (snow or ice), and to vehicle failure. No single countermeasure concept can alleviate all of these causal factors. Following is a list of potential technologies and displays to address various aspects of SVRD crashes.

Driver Vigilance Monitoring (Impaired-Driver Detection)

Researchers and manufacturers have used a variety of techniques to aid in the measurement of driver alertness. This technology has primarily utilized steering wheel movements (Brookhuis, de Vries, & de Waard, 1991; Dingus, Hardee, & Wierwille, 1987). Other technology that monitors physiological measures such as eye lid closure and heart rate has also been investigated.

Ice Warning

Icing conditions can be predicted by two different technologies: an in-vehicle tire–road friction monitor and an out-of-vehicle system on the road-side that measures air and ground temperature and humidity and calculates dew point to determine the risk of ice formation (Machine Design, 1993).

Curve Approach Advisory

Sensors currently in existence have the capability of measuring the speed of an approaching vehicle. A beacon could then transmit signals from a fixed roadside location to a vehicle to provide warning information.

Lane Departure Warning

One possible SVRD crash-avoidance system would utilize a tactile steer-ing wheel display that vibrates to simulate crossing a center line with raised lane markers. Automation of a steering wheel correction is also a potential candidate. A solution between tactile warning and automated control is semiautomation, in which the steering wheel could offer increasing resist-ance to alert the driver of the error. Brown and Dingus (in preparation) studied semiautomated lateral control systems in a simulator. The systems tested did not show an apparent safety benefit and were not well accepted by subjects.

Avoidance of Obstacles During Low-Visibility Conditions

These systems utilize infrared sensors to detect moving objects (e.g., pedestrians, deer, etc.) and algorithms to assess the potential of a crash. This technology has been implemented in Nissan's AP-X and AQ-X concept cars shown at the 1993 Frankfurt and Tokyo Auto Shows.

Rear-End Crashes

Among multivehicle crashes, the rear-end type is the most common, better understood in terms of causation, and easier to avoid by driver technology assists in comparison to other crashes. As shown in the Introduction, after inattention, the second most common causal factor of rear-end crashes is following too closely, an ubiquitous behavior attributed to habit (see Evans, 1991, pp. 313–316).

Because rear-end crashes result from inattention, following too closely, or both, there is considerable potential for technologies capable of objectively advising the driver of the minimum safe headway, and road conditions. The main output of that technology might be an advisory rather than an emergency warning. The technology would then become an extension of the driver's awareness and senses to correct inattention, bad following habits, and perceptual biases. The human factors challenge is to design a display that will most effectively influence a driver's following distance behavior. Dingus, et al. (in press) determined that the use of graded visual headway information resulted in longer following distances for novice drivers.

Headway-Detection (HD) Systems

By headway detection (HD) we refer to a system with the ability to measure headway and also to relative speed for assessing the immediate danger of a rear-end-crash with the vehicle ahead. Active laser radar and millimeter wave radar are the two main technologies for headway detection systems. Both types of technologies are being tested in Japan, Europe, and the United States. Prototype HD systems are already available commercially. For example, an active laser system for trucks has been produced by the Nissan Diesel Motor Co., an affiliate of Nissan Motor Co. The system simply warns drivers if they follow too closely (an attentional warning). Another example of a commercial application is a microwave radar system produced by VORAD Safety Systems.

Adaptive Cruise Control (an HD Application)

Adaptive Cruise Control (ACC) is an ITS system currently under development, but highly related to headway detection systems. Unlike normal cruise control, ACC systems will control following distance to a lead vehicle, as well as speed, when the system is engaged. Two main types of systems are under consideration and are termed automatic or manual target acquisition systems. An automatic system continually monitors the forward roadway for slower vehicles. When a lead vehicle slows below the designated cruise speed, the system will automatically recognize the difference and switch to a following distance mode of operation. For a manual target acquisition system, the driver must select the "slow target" and switch the system to a following mode. Manual target acquisitions will be much like other HD displays, because of the need to alert the driver of close headways or speed differences.

Advanced Rear Signaling

Avoidance of rear-end crashes is traditionally achieved by drivers watching the signaling—braking, turning signals, and presence lights—of the vehicle immediately ahead and sometimes of more distant vehicles ahead. The literature on rear-signaling suggests that a significant portion of the most common type of rear-end crashes, in which the struck vehicle is stationary, might be avoided by advances in rear signaling. The human factors design challenge is to make the almost stationary and the stationary vehicle more conspicuous to the following driver.

Another major issue in rear signaling is the need to signal speed loss due to an upgrade (especially for trucks) and for all manual transmission vehicles when a driver is downshifting gears. (For a discussion of these issues see Horowitz, 1994.)

Intersections

In the Introduction it was shown that human errors associated with intersection (crossing path) crashes can be divided into three types: inattention, deliberate violation, and intoxication. Advances in advisory warnings have good potential for dealing with the inattention case, which accounts for 36% of all crossing-path crashes. For the other two cases, it appears feasible, although difficult, to develop emergency warnings, automatic intervention, or both.

Because intersection crashes occur in well-defined locations—unlike SVRD and rear-end crashes—beacons transmitting signals from fixed points at the intersection or in its vicinity have potential benefits. After an in-vehicle system receives information on the proximity of the intersection, it will be able to warn the driver about the need for excessive speed.

Drawing the driver's attention to an approaching intersection also might be important for avoidance of rear-end crashes occurring just before the driver reaches an intersection where a vehicle ahead is stopped. That type of crash represents a majority (57%) of all rear-end crashes in which the struck vehicle is stopped.

A more advanced intersection beacon would control an intelligent light signal intersection system. Such a system, called "interactive intersections," has been developed for intersections with low traffic in rural areas by the Prometheus/Pro-Art program in France (de Saint Blancard, 1992). The interactive intersection system detects incoming vehicles and receives a signal when an approaching vehicle signals a turn. Accordingly, the system regulates the traffic by signaling red, green, or flashing yellow. The last signal indicates that the intersection should be negotiated carefully.

Another system tested by the Prometheus/Pro-Art program called "Stopping at Stop Signs" has the goal of encouraging observance of stop signs. When a vehicle approaches the unsignalized intersection, a beacon transmits a message announcing the stop sign and how far ahead it is. The message is relayed to the vehicle's receiver, setting off a buzzer and a visual symbol on the dashboard. If the system calculates that the driver is not responding properly, it activates an automatic stopping procedure: The throttle is closed, the brake is applied until the vehicle comes to a stop, and a message is given to the driver.

No information was provided by de Saint Blancard (1992) on the extent of the human factors investigation of the new intersection concepts developed by Prometheus/Pro-Art.

Lane-Change–Merge

Because of the close lateral proximity of vehicles in lane-changing conditions, it might not be feasible to issue an emergency warning for the avoidance of lane-change–merge crashes. Initial attempts for the development of lateral sensors and emergency warnings have faced the difficulty of false alarms because of lateral objects such as buildings or guardrails. Therefore, the advanced technologies may hold the key promise of providing helpful attentional warnings before the initiation of a lane-change maneuver. However, in a study by Mazzae (1995), it was found that truck drivers performed better and preferred using a convex mirror system relative to two different side-object detection systems.

The envisioned technology for the avoidance of lane-change–merge crashes would render a dynamic display of the surrounding traffic situation including a visual display of vehicles overtaking on either left or right and vehicles at blind locations on either side. Available technology includes a "camera chip" that utilizes simple small lenses and a computer chip for the transformation of the visual image into a visual display.

Head-On Warning System

Because of proximities and approach speeds, a head-on warning system that uses detection of an oncoming vehicle may not be feasible. However, in-vehicle lane-position monitoring and driver impairment monitoring may be practical in the future, as discussed for SVRD technologies. These technologies may help to alleviate head-on crashes in addition to SVRD crashes.

Backing

The technology used in backing crashes is similar to that used for forward-crash hazard warning: primarily radar and laser. For stationary objects such as pedestrians, animals, or tricycles, the sensors can work well at alerting the driver. Detection of moving objects has the associated design hindrances of slow driver response time, driver attention orientation, and sensor capability limitations. Another solution being considered is the use of cameras to provide a better view out the rear of the vehicle, showing objects in the blind-spot areas or those approaching from the rear and to the side.

SUMMARY

The goal of advisory crash avoidance and collision warning system designers is to increase the situation awareness of the typical driver to a level at which common performance errors will be reduced, thereby reducing the overall crash rate of vehicles. To accomplish this, great care will be required in all phases of the design, and a number of issues will have to be addressed. Long-term use of the systems and their effect on driving behavior will have to be closely monitored. It is possible that behaviors such as driver over-reliance could result in a crash rate increase for particular designs. Technology has given system designers an opportunity to make great strides in crash reduction and improvements in transportation safety. However, it must never be forgotten that technology in this application is a double-edged sword that must be wielded with care.

REFERENCES

Baber, C. (1994). Psychological aspects of conventional in-car warning devices. In N. Stanton (Ed.), *Human factors in alarm design* London: Taylor & Francis.

Blincoe, L. J., & Faigin, B. M. (1992). *The economic cost of motor vehicle crashes, 1990* (NHTSA Tech. Rep. DOT HS 807 876). September.

Blincoe, L. J., & Faigin, B. M. (1983). Economic impact of motor vehicle crashes—United States, 1990. *Morbidity and Mortality Weekly Report (MMWR)*, Centers for Disease Control, Vol. 42, No. 23, June 18.

Broadbent, D. E. (1958). *Perception and communication.* London: Pergamon Press.

Brookhuis, K. A., de Vries, G., & Waard, D. (1991). The effects of mobile telephoning on driving performance. *Accident Analysis and Prevention, 23,* 309–316.

Brown, T., & Dingus, T. A. (in preparation). *An examination of active control interventions in prevention of run-off-road accidents.* Manuscript in preparation.

Carney, C., & Dingus, T. A. (in preparation). *The effects of collision warning sensor range and timing on the driver's ability to respond to lead vehicle braking and following situations.* Manuscript in preparation.

Clarke, R. M., Goodman, M. J., Perel, M., & Knipling, R. R. (1993). Driver performance and IVHS collision avoidance systems: A search for design-relevant measurement protocols. *Proceedings of the 1993 Annual Meeting of IVHS America,* IVHS America, pp. 241–248.

Davis, D., Schweizer, N., Parosh, A., Lieberman, D., & Aptor, Y. (1990). *Measurement of the minimum reaction time for braking vehicles.* Israel: Wingate Institute for Physical Education and Sport.

de Saint Blancard, M. (1992). PROMETHEUS/Pro-Art: A synthesis on studies related to image processing and intelligent vehicle applications. Presented at the General Motors Research and Development Center, Warren, MI, July 2.

Deering, R. K. (1994). General Motors Safety Center, Crash avoidance technologies to assist the driver, Presentation at the American Society of Civil Engineers Conference, *Innovations in Highway Safety—A Broad Perspective,* May 17.

Dingus, T. A., Hardee, H. L., & Wierwille, W. W. (1987). Development of models for on-board detection of driver impairment. *Accident Analysis and Prevention, 19*(4), 271–283.

Dingus, T. A., McGehee, D. V., Manakkal, R., Jahns, S. K., Carney, C., & Hankey, J. (in press). Human factors field evaluation of automotive headway maintenance/collision warning devices. *Human Factors.*

Edworthy, J. (1994). *The design and evaluation of warnings and signals.* London: Taylor & Francis.

Edworthy, J., Loxley, S., & Dennis, I. (1991). Improving auditory warning design: Relationships between warning sound parameters and perceived urgency. *Human Factors, 33,* 205–231.

Ervin, R. D. (1994, June). Linking truck design to public and private life-cycle costs. *Conference Proceedings of the International Symposium on Motor Carrier Transportation,* Williamsburg, VA. National Academy Press, Washington, DC.

Evans, L. (1991). *Traffic safety and the driver.* New York: Van Nostrand Reinhold.

Fancher, P., Kostyniuk, L., Massie, D., Ervin, R., Gilbert, K., Reiley, M., Mink, C., Bogard, S., & Zoratti, P. (1994). *Potential Safety Applications of Advanced Technology* (Tech. Rep. FHWA-RD-93-080).

Farber, E., Freedman, M., & Tijerina, L. (1995). Reducing motor vehicle crashes through technology. ITS America, *ITS Quarterly, III*(1), 12–21.

Farber, E., & Paley, M. (1993). Using freeway traffic data to estimate the effectiveness of rear-end collision countermeasures. *Proceedings of the 1993 Annual Meeting of IVHS America.* IVHS America, pp. 260–268.

Hanowski, R. J., Bittner, A. C., Jr., Knipling, R. R., Byrne, E. A., & Parasuraman, R. (1995, March). *Analysis of older driver safety interventions: A human factors taxonomic approach.* Paper presented at the Fifth Annual Meeting of the Intelligent Transportation System Society of America (ITS America), Washington, DC.

Hoffman, E. R. (1974). Perception of relative velocity. In *Studies of Automobile and Truck Rear Lighting and Signaling Systems.* Ann Arbor: University of Michigan, Highway Research Institute (Rep. No. UM-HSRI-HF-74-25).

Horowitz, A. D. (1994). Human factors issues in advanced rear signaling systems. *Proceedings of the 14th Enhanced Safety of Vehicles Conference,* Munich, Germany: NHTSA.

Horowitz, A. D., & Dingus, T. A. (1992). Warning signal design: A key human factors issue in an in-vehicle front-to-rear-end collision warning system. *Proceedings of the Human Factors Society 36th Annual Meeting* (pp. 1011–1013). Santa Monica, CA: Human Factors Society.

Kantowitz, B. H. (1974). Double stimulation. In B. H. Kantowitz (Ed.), *Human information processing.* Hillsdale, NJ: Lawrence Erlbaum Associates.

Kantowitz, B. H. & Sorkin, R. D. (1983). *Human factors.* New York: Wiley.

Knipling, R. R. (1993). Could advanced technology have prevented this crash? IVHS America. *IVHS Review,* pp. 23–44. Fall.

Knipling, R. R., Mironer, M., Hendricks, D. L., Tijerina, L., Everson, J., Allen, J. C., & Wilson, C. (1993). *Assessment of IVHS countermeasures for collision avoidance: Rear-end crashes* (NHTSA Tech. Rep. DOT HS 807 995).

Knipling, R. R. & Wang, J. S. (1995, October). Revised estimates of the U.S. drowsy driver crash problem size based on General Estimates System case reviews. *39th Annual Proceedings, Association for the Advancement of Automotive Medicine,* Chicago.

Knipling, R. R., Wang, J. S., & Yin, H. M. (1993). *Rear-end crashes: Problem size assessment and statistical description* (NHTSA Tech. Rep. DOT HS 807 994).

Leiser, R. (1993). Driver vehicle interface: Dialogue design for voice input. In A. M. Parkes & S. Franzen (Eds.), *Driving future vehicles* (pp. 275–294). London: Taylor & Francis.

Levine, M. W. & Shefner, J. M. (1991). *Fundamentals of sensation and perception, second edition.* Monterey, CA: Brooks/Cole.

Machine Design (1993). Black box sends ice warnings. *Machine Design.* October 22, p. 17.

Mazzae, E. (1995). Field test of side obstacle detection systems. Unpublished master's thesis. Wright State University, Dayton, OH.

Moray, N. (1990). Designing for transportation safety in the light of perception, attention, and mental models. *Ergonomics, 33,* 1201–1213.

Mortimer, R. G. (1988). Rear-end collisions. Chapter 9 in *Automotive Engineering and Litigation* (Vol. 2, pp. 275–303). New York: Garland Law Publishers.

Mortimer, R. G. (1990). Perceptual factors in rear-end collisions. *Proceedings of the Human Factors Society 34th Annual Meeting*, Vol. 1, 591–594.

Najm, W. G., Mironer, M., Koziol, J. S., Jr., Wang, J. S., & Knipling, R. R. (1995). *Examination of Target Vehicular Crashes and Potential ITS Countermeasures*. Report for Volpe National Transportation Systems Center, May (DOT HS 808 263, DOT-VNTSC-NHTSA-95-4).

National Highway Traffic Safety Administration National Center for Statistics and Analysis. (1991). *National Accident Sampling System General Estimates System* (DOT HS 807 796).

National Highway Traffic Safety Administration. (1991). *Commercial Motor Vehicle Speed Control Devices*. Report to Congress (Publication No. DOT HS 807 725). NHTSA Office of Crash Avoidance Research.

National Highway Traffic Safety Administration. (1992). *NHTSA IVHS Plan* (Publication No. DOT HS 807 850). NHTSA Office of Crash Avoidance Research.

National Highway Traffic Safety Administration. (1995). *Traffic Safety Facts 1994* (Publication No. DOT HS 808 292). NHTSA National Center for Statistics and Analysis.

National Safety Council. (1991). *Accident Facts, 1991 Edition*. ISBN 0-87912-159-9.

Olson, P., & Sivak, M. (1986). Perception-response time to unexpected roadway hazards. *Human Factors, 28*, 91–96.

Redpath, D. (1984). Specific applications of speech synthesis. *Proceedings of the First International Conference on Speech Technology*. Bedford, England: IFS.

Rice, R. S., Dell'Amico, F., & Rasmussen, R. E. (1976). *Automobile Driver Characteristics and Capabilities: The-Man-Off-the-Street*. Presented at the SAE Automotive Engineering Meeting (pp. 7–15). Deerborn, MI, October.

Ross, T. (1993). Creating new standards: The issues. In A. M. Parkes & S. Franzen, (Eds.), *Driving Future Vehicles* (pp. 347–358). London: Taylor & Francis.

Senders, J., & Moray, H. (1991). *Human error: Cause, prediction, and reduction*. Hillsdale, NJ: Lawrence Erlbaum Associates.

Sivak, M., Olson, P. L., & Farmer, K. M. (1982). Radar measured reaction time of unalerted drivers to brake signals. *Perceptual and Motor Skills, 55*, 594.

Sorkin, R. D., Kantowitz, B. H., & Kantowitz, S. C. (1988). Likelihood alarm displays. *Human Factors, 30*(4), 445–459.

Stokes, A., Wickers, C., & Kite, K. (1990). *Display technology: Human factors concepts*. Warrendale, PA: Society of Automotive Engineers.

Taoka, G. T. (1989). Brake reaction times of unalerted drivers. *ITE Journal, 59*, March 19–21.

Telford, C. W. (1931). The refractory phase of voluntary and associative responses. *Journal of Experimental Psychology, 14*, 1–36.

Tijerina, L., Hendricks, D. L., Pierowicz, J., Everson, J., & Kiger, S. (1993). *Examination of Backing Crashes and Potential IVHS Countermeasures* (DOT HS 808 016).

Treat, J. R., Tumbas, N. S., McDonald, S. T., Shinar, D., Hume, R. D., Mayer, R. E., Stansifer, R. L., & Catellan, N. J. (1979). *Tri-Level Study of the Causes of Traffic Accidents: Final Report Volume I: Causal Factor Tabulations and Assessments*. Institute for Research in Public Safety, Indiana University (DOT Publication No. DOT HS-805 085).

Verwey, W. B. (1993). How can we prevent overload of the driver? In A. M. Parkes, & S. Franzen, (Eds.), *Driving future vehicles* (pp. 235–244). London: Taylor & Francis.

Wang, J. S., & Knipling, R. R. (1994). *Backing Crashes: Problem Size Assessment and Statistical Description* (NHTSA Tech. Rep. No. DOT HS 808 074).

Wasielewski, P. (1979). Car following headways on freeways interpreted by the semi-Poisson headway distribution model. *Transportation Science, 13*, 36–55.

Wickens, C. D. (1992). *Engineering psychology and human performance, second edition*. New York: HarperCollins Publishers.

Wortman, R. H., & Matthias, J. S. (1983). Evaluation of driver behavior at signalized intersections. *Transportation Research Record, 904*, 117–139.

4

COMMERCIAL VEHICLE-SPECIFIC ASPECTS OF INTELLIGENT TRANSPORTATION SYSTEMS

William A. Wheeler
John L. Campbell
Rhonda A. Kinghorn
Battelle Human Factors Transportation Center

Commercial vehicles represent a unique portion of Intelligent Transportation System (ITS) consumers. This chapter describes some of the necessary human factors considerations in the design of ITSs that will support this portion of the driving public. These considerations are the result of both physical and operational differences that exist between commercial vehicles and the general driving population. The discussion concentrates on commercial vehicle operations that have the greatest potential impact on ITS design.

CHARACTERISTICS OF COMMERCIAL VEHICLES

Most of us recognize the eighteen-wheeler that seems ever present along the freeway as a commercial vehicle. A closer examination of the term, "commercial vehicle," however, reveals that the distinguishing characteristic of a commercial vehicle is as much a matter of fiscal regulation[1] as of vehicle construction. Under an economic, rather than a regulatory definition of

---

[1]The U.S. Department of Transportation defines a Commercial Motor Vehicle as a "motor vehicle used in commerce to transport passengers or property if the vehicle (a) has a gross combination weight rating of 26,001 or more pounds inclusive of a towed unit with a gross vehicle rating of more than 10,000 pounds, or (b) is of any size and used in the transportation of materials found to be hazardous for the purposes of the Hazardous Materials Transportation Act, which requires that the motor vehicle be placarded under the Hazardous Materials Regulations (Code of Federal Regulations, Part 383.5).

commercial vehicles, everything from the bicycle used by message delivery services to the multiaxle trailers used to move a house would qualify as a commercial vehicle, although they would obviously have very different needs so far as the ITS is concerned. More practically, however, commercial vehicle applications of ITS can be defined as applying to those vehicles whose physical characteristics or operational uses suggest different needs for functions and features within the ITS than are normally used by a private driver.

Under this definition the general characteristics of commercial vehicles include the use of the vehicle on public roads or highways, where the use of ITS technology is likely to address one of the major goals of ITS (i.e., improving the environment, increasing economic productivity, improving the quality of life, or increasing safety). One more goal should probably be included in connection with ITS use in commercial vehicles (i.e., improving the performance of the system in which the vehicle is used). Although an ITS could be developed with little or no attention paid to commercial vehicle operations, the goals of the program suggest that this would not be done.

TYPES OF COMMERCIAL VEHICLE OPERATIONS

Commerce

Perhaps the most characteristic and most easily identified type of commercial vehicle operations is that associated with commerce. Such operations include all of the operations that involve the movement of goods and materials from one location to another. The forerunner of vehicle use in commerce is the ox and wagon that goes back perhaps 3,000 or 4,000 thousand years. The major difference between these earlier operations and those of today is probably the role played by time in the success or failure of the transportation enterprise. In earlier days, including the early days of the automobile and the internal combustion engine, commercial traffic could claim a significant portion of the road by virtue of the size and bulk of the equipment involved. The increased popularity of the private automobile soon resulted in competition between commercial vehicles and automobiles for highway space. As highways become choked with an ever greater number of vehicles, commercial vehicles find themselves increasingly at the mercy of delays beyond their control. To a delivery person or cross-country trucker, delays mean only one thing—a loss of money. Because the standard method of payment for truck drivers is based on the number of miles driven, delays can also cause considerable stress to the driver as such delays are effectively losses to income. It is, therefore, not only a matter of convenience for commercial vehicles operators to find more efficient ways to get from one place to another. It is a matter of economic necessity, one that ITS technologies can help facilitate.

Personal Transport

Livery services such as taxis, limousines, and buses have long provided people with quick, convenient, and economical ways of getting from one place to another. Modern personal transport includes the variety of vehicles and systems that allow a person to travel with minimum personal investment in equipment and maximum assurance that the trip will be as short and economical as possible. Personal transport includes short trips provided by the neighborhood cab or city bus and longer trips such as those taken on a cross-country or chartered bus. Although the vehicles are different, the same principal concern exists for the carriage of people as it does for the transport of goods: How can the trip be made as quickly and cost-effectively as possible. Some types of vehicles in the class of commercial operations also are required to make scheduled arrivals at predesignated locations along a fixed route. Although this is discussed in another context later, the constraints of a time schedule and a specific route impose added burdens, (e.g., conflicting with driver duty and rest periods, forcing layover days due to dock closures and limited of operations for intermodal transfer points) on the Commercial Vehicle Operation (CVO) system for which ITS technology may provide at least a partial answer.

Emergency Response Operations

Emergency response operations include all activities that require rapid movement of one or more emergency vehicles (i.e., ambulance, fire truck, police cruiser, emergency utility repair truck) to a designated point. The movement of these vehicles is often a coordinated activity involving vehicles from different locations converging on a specific point or area.

Police. Police activities require the ability to move very rapidly from one part of a city to another. Traditionally, this is done is by allowing police vehicles exemption from many of the normal rules of the road and by the imposition of rules for other drivers to pull over and clear a path for the police vehicle when either a siren or flashing red lights are used. Increasingly, however, traffic conditions on some highways are so bad that clearing a path for the police vehicle is simply not practical. ITS technologies will assist both in providing police dispatchers and emergency equipment drivers with the most efficient route to a particular location and in supplying route guidance to the scene based on actual traffic conditions. ITSs will also provide integrated traffic management equipment to clear the route as much as possible by the use of priority traffic signals and related management.

Fire. Fire operations may involve either paramedic vehicles or fire apparatus. Paramedic vehicle operations usually are confined to the response of one relatively small vehicle to an accident or other life-threatening event. This involves locating the appropriate address or scene of the accident and may involve the need to move through congested highway systems. After the injured person has been treated and picked up, the paramedic unit often must transport the individual to a hospital. In paramedic operations, speed of movement to the victim and then to the hospital is essential.

In fire operations the movement of apparatus to the scene of the fire involves both large vehicles and coordinated activities between different vehicles and sometimes vehicles coming from different stations. The traditional "lights and siren" approach to clearing traffic for emergency vehicles depends on public compliance to ensure that the vehicles reach the fire or accident scene quickly. This approach takes care of one aspect of getting quickly to the scene; the other relies on drivers familiar with the location, routes, and other characteristics of the highways (e.g., turn radius limitations for long-ladder trucks, construction, and other temporary obstructions) that affect getting to the scene. As cities become more complicated, local fire departments more consolidated, and traffic more congested, these approaches will begin to break down. ITS technologies will assist in providing both fire dispatchers and drivers with the most efficient route to an accident or fire location by providing both situation-based routing that avoids traffic congestion and flexible traffic management that clears the route that fire vehicles would take to the scene.

Ambulance Operations. Ambulance operations are much like those of paramedics. They have the same requirements for rapid movement through the traffic structure, accurate identification of the destination, and necessary navigation to get there. ITS technologies that relate to navigation and avoidance of traffic backups will be key to these types of activities.

Utility Emergency Operations. Emergency operations involving utilities such as power, water, or traffic signal control are usually not the same as life-threatening situations encountered by police, fire, or ambulance services. As a consequence, they usually do not have the same ability to use lights and sirens to clear traffic so they can get to the place they are needed. These types of services, when operating in an emergency mode, are likely to be faced with traffic backups and closures of normal routes. The priority for this type of emergency vehicle is therefore accurate navigation to a required location and routing around traffic backups and other obstructions that may have been caused by a natural disaster or other condition.

Tow Trucks and Road Clearance. A majority of the delays encountered on today's urban highways are caused by disabled vehicles or those involved in an accident. Rapid removal of such vehicles is a major problem of most traffic management systems. Both commercial and government tow trucks often find themselves stuck in the same traffic backup that they are trying to correct. ITS management could provide both routing information and route clearance services to assist in this vital work (e.g., identifying the nearest on ramp to the traffic obstruction and, if necessary, clearing the route to allow the tow truck to run against the normal flow of traffic).

TASKS ASSOCIATED WITH COMMERCIAL VEHICLE OPERATIONS

Once they are fully implemented, ITS will enhance the safety and efficiency of the current transportation system. In general, however, they are designed to enhance or improve the system, not significantly alter the way things are done. Before considering the impact that ITS may have on the tasks that drivers and dispatchers perform, it is necessary to understand the general scope and characteristics that such tasks have now.

Route Planning

Route planning is the first in a sequence of tasks performed in Commercial Vehicle Operations (CVOs). Such planning is primarily concerned with determining how best to get from the present location to single or multiple destinations. The particular route plan and the considerations used to make up that plan are largely dependent on the type of operations being conducted.

Over-the-Road or Long-Distance Route Planning. Route planning for over-the-road or long-distance operations requires balancing a number of different considerations and constraints, but it is inevitably aimed at developing a route to optimize the efficiency of the truck and driver. Typically, this means planning a route that minimizes travel time and distance. The route-planning process includes three major considerations. The first is where the pickup and delivery points are in relation to one another. Their location determines what potential routes exist. The second consideration is for any routes that must be discounted because either the load or equipment is incompatible for the route, or there are other restrictions on the route (e.g., restricted hours of operation for commercial vehicles such as are found in the Los Angeles area). These two considerations restrict the specific routes that may be taken and require a detailed review of the various federal, state, and local regulations that may apply to each possible route. The third consideration is for conditions along the route that may cause a delay when compared to those of other possible routes. The potential causes for delays include weather, terrain, road construction, and the likelihood of significant traffic backups on the route. The sources for this type of information are varied and include the drivers' previous experience, long- and short-range weather forecasts, and notification of construction activity from state highway departments. Availability of ITS technology, particularly that which would integrate the traffic monitoring and prediction capabilities of an Intelligent Traffic Management System (ITMS) with a working In-Vehicle Routing and Navigation System (IRANS), would provide the driver with real time and projected information concerning such delays.

Special Cargo Operations Such as Wide Loads, Explosives, and Toxic Chemicals. Planning the movement of trucks hauling oversize loads or hazardous materials requires more careful planning than is done with most types of cargo. There are two specific considerations for such trips. The first is that a route may not be physically suited to accommodate the vehicle and its load. Lasting evidence of failures to adequately plan for these limitations is declared by the battle scars we see along bridge abutments and the tops of underpasses. The second is that the route must comply with various regulations imposed by local or state authorities concerning use of the road. The basis for these regulations is usually, though not always, safety or environmental concerns. For example, a common regulatory practice excludes explosives and flammable materials from tunnels such as the Holland Tunnel in New York City. Such a restriction is risk based on the concern that an accident involving such a cargo would have unacceptable consequences. Environmentally based restrictions are becoming more and more common, a fact indicated by trucking restrictions recently introduced in Los Angeles, California, that are primarily aimed at reducing air pollution during commuting times by reducing the number of commercial vehicles on the road during those times.

Local Delivery to Optimize Delivery Schedules. Local delivery of everything from packages to groceries represents a significant portion of CVOs. The efficient use of a commercial vehicle in such work requires accurate planning of the most optimum route that the vehicle will travel. Local delivery consists of several distinctly different types of operations. The most obvious, and perhaps the simplest, is the scheduled route. This type of delivery services resembles a bus route in which the driver has a fixed delivery route, destinations known at the beginning, scheduled pickup and delivery times, and a fixed sequence for making pickups and deliveries. Such schedules are common in a variety of businesses and allow the development of a route and schedule based almost entirely on experience. The second type of operation is an on-call operation characterized by flexibility in pickup and delivery points, times, and sequence. Optimizing a flexible schedule is a major challenge for operations involved in providing such a service.

Dispatch

Dispatch tasks involve the coordination of resources (i.e., vehicles, drivers, and time) to achieve the desired goal of the operation.

Commercial Long-Haul Operations. Some of the basic tasks performed by dispatch in commercial long-haul operations include these:

- Assigning drivers for specific loads
- Arranging pickup times with the shipper
- Determining routing of the load
- Assigning trucks, trailers, and other carrier equipment for specific loads
- Leasing trucks, trailers, and other needed equipment
- Preparing shipping documents
- Supervising driver performance and fitness for duty
- Arranging for optimum use of the equipment and drivers
- Relaying emergency information to the driver on the road
- Arranging repair or replacement of equipment that breaks down on the road
- Verifying drivers' log books
- Arranging trans-shipment of loads from one driver to another

These tasks usually are performed by a specially trained person who has specific knowledge of carrier operations and knows how best to coordinate those operations. To perform these tasks, the dispatcher is often in contact with the shipper, drivers, carrier maintenance organization, carrier sales representative, dispatchers in other locations, and receiver. He or she may also need to be in contact with state regulators and vendors of specific services such as wrecking or road repair.

Local Delivery Service. The primary tasks performed in the dispatch of local delivery operations are those associated with optimizing the route to ensure that deliveries are made to each delivery point on time and that the route minimizes losses in efficiency resulting from either traffic congestion or delays at the loading dock. A local delivery dispatcher may thus use information gathered from fleet drivers and other systems (e.g., traffic management system terminals) to direct drivers around congestion. An operation that has an on-call operation, of course, will have a much more complex and dynamic dispatch task than one that operates fixed schedule routes.

Emergency Vehicle Operations. The dispatch tasks involved in emergency vehicle operations include the following:

- Receiving an emergency call
- Assigning a vehicle to respond to the call
- Monitoring the development of the emergency situation

The tasks involved in receiving a call include taking the call for assistance and recording specific information (e.g., location of the emergency, type of assistance required, and other information that the emergency responders will need to know about the situation). After a call has been received, the dispatcher decides what resources (e.g., vehicles, personnel, special equipment) are needed, which are available, and which allow the fastest and most appropriate response. The dispatcher then contacts the resources and assigns them to the emergency. During the assignment, the dispatcher briefs the driver or other emergency response personnel on the situation and may also provide support services to the emergency response by contacting the traffic management system to clear a path for the vehicle or other agencies with whom the response needs to be coordinated (e.g., police units may be notified of a fire response). Following assignment of response vehicles to an emergency, the dispatcher will monitor the progress of the response and provide additional assistance as needed.

The tasks carried out to support the vehicle monitoring functions include monitoring the location and status of emergency vehicles within a particular area. Those carried out in support of the call assignment functions are largely decision-making tasks involving a determination of what resources are needed, what resources are available, and where the resources are located. This function also involves communicating instructions to the vehicles involved and describing the emergency involved, the location, and expected support from other units.

Driver Tasks Performed Within the Vehicle

Long-Haul Operations. The tasks performed by the driver involved in long-haul trucking operations include these:

- Inspecting the truck and trailers (if any)

- Navigating to the pickup point for the shipment
- Supervising or actually performing the loading and unloading of the truck or trailer
- Tying down or securing the load
- Selecting the optimum route to the destination
- Driving the vehicle
- Navigating the vehicle to the destination
- Supervising the unloading of the truck or trailer
- Preparing the necessary paperwork
- Maintaining the vehicle on the road
- Preparing the driver's log
- Dealing with state officials
- Dealing with customs and border-crossing officials
- Coordinating his or her activities with the carrier dispatcher

To perform these tasks, the driver must interact with a number of different people including the shipper and receiver representatives, loading dock hands, carrier dispatchers, state regulatory and enforcement personnel, and representatives of the various service industries that support the trucking community.

Local Delivery Service. Drivers involved in local delivery service may be called upon to perform an number of different tasks:

- Inspecting the truck and trailers (if any)
- Selecting a route that will meet the required pickup and delivery schedule
- Supervising or actually performing the loading and unloading of the truck or trailer
- Organizing, staging, and securing the load
- Driving the vehicle
- Navigating the route
- Preparing the driver's log
- Collecting and distributing shipment paperwork
- Coordinating his or her activities with the carrier dispatcher

Although the basic tasks associated with local delivery do not vary much from what a long-haul driver does, the distribution of tasks and their relative difficulty depends to a greater extent on the immediate environment. The relatively compressed pickup and delivery schedule, increased traffic and use of mixed roads including surface streets all contribute to make local delivery driving a significantly different experience than long-haul driving.

Emergency Vehicle Operations. The tasks required of emergency vehicle drivers are significantly different from those of any other group of commercial drivers. The tasks involved may include these:

- Controlling the vehicle at high speeds
- Selecting a route that will minimize travel time
- Driving against the flow of traffic and out of the normal traffic lane
- Operating radios and other vehicle equipment (e.g., sirens, warning horns, emergency flashers, public address systems)
- Driving the vehicle
- Gathering information concerning the situation and event
- Planning actions to be taken at the emergency location
- Coordinating information and plans with others over the radio
- Gathering and recording information in official reports

In emergency vehicles with more than one person some of these tasks (e.g. communications, planning and record keeping) may be done by the second person.

The primary difference between the tasks performed by drivers of emergency vehicles and those operating other types of commercial vehicles is that such operations often involve critical time limits. As a consequence, drivers may be required to operate their vehicles in a manner significantly different from that of a normal driver. To reduce the risk of such operations, both technological and regulatory means are usually used to give priority and a degree of protection to emergency operations (e.g., priority traffic signals and requiring nonemergency vehicles to pull off the road and stop when emergency vehicles approach with emergency signals operating).

Dealing With Regulations

CVOs, particularly those involving the trucking industry, necessarily involve a hierarchy of federal, state, and sometimes local regulatory bodies. The regulations that each jurisdiction enforces are primarily related to two activities: generating revenue for the state or municipality and enforcing safety-related activities. These regulations concern several different areas:

- Generating logbooks and records to substantiate that the vehicle is being operated as the regulations require

- Maintaining the vehicle in a condition that ensures safe operation of vital safety systems such as brakes, tires, and lights
- Loading of the vehicle so that its driving characteristics allow its safe control and minimum damage to the roadway
- Licensing of the vehicle to regulate specific loads and routes
- Assessing and paying taxes related to cargo carriage

Each of these areas is accompanied by enforcement and monitoring actions such as those performed at inspection and weigh stations. The driving time lost at these stations, as well as the hazards associated with heavy vehicles leaving and entering the highway, have resulted in a number of ITS-related projects designed to demonstrate the practicality of automating these functions, most notably the HELP/Crescent,[2] Advantage I-75,[3] and PASS[4] projects.

Necessary Adaptations of Advanced Traveler Information Systems for CVOs

Advanced Traveler Information Systems (ATIS) are likely to be among the first major ITSs developed. Their obvious attractiveness to the motoring public as a potential tool for dealing with some of the frustrations associated with driving in both unfamiliar places and heavy congestion provides a potentially large group of customers for such equipment. This, combined with the relative accessibility of many of the technologies necessary to perform at least some ATIS functions, makes early commercial development of operational ATIS a real possibility.

The strategic plan for ITS in the United States (IVHS America, 1992) identified four major subsystems to ATISs. Although each of these subsystems may have common functions and characteristics for both CVOs and private drivers, in many respects the system will need to contain CVO-specific information.

---

[2] Heavy Vehicle Electronic License Plate Program, a multinational effort to design and test an integrated heavy vehicle monitoring system incorporating Automatic Vehicle Identification (AVI), Automatic Vehicle Classification (AVC), and Weigh in Motion (WIM) technology (U.S. Department of Transportation, Intelligent Vehicle Highway System Projects, February 1993).

[3] An operational test that will allow transponder-equipped and properly documented vehicles to proceed along the I-75 corridor without stopping at weight and inspection stations (U.S. Department of Transportation, Intelligent Vehicle Highway System Projects, February 1993).

[4] An integration of AVI, WIM, AVC, and on-board computers (OBC) to identify, classify, and direct selected heavy vehicles in advance of weight stations and ports of entry at Oregon's Ashland Port-of-Entry. This is an operation test of the ability to sort and direct such traffic at highway speeds on an interstate highway (U.S. Department of Transportation, Intelligent Vehicle Highway System Projects, February 1993).

In-Vehicle Routing and Navigation Systems (IRANS). Appropriate route selection and navigation are important activities for commercial vehicle drivers, no matter what type of operation they serve. The IRANS goal of providing the driver with guidance information that provides the most efficient way to get from the present location to a selected destination or destinations is common for both the private and the CVO driver. The considerations that must be made by an IRANS device supporting CVOs are very much determined by the characteristics of the specific CVO vehicle it supports (i.e., large and heavy trucks will require a great deal more support than smaller vehicles such as emergency vehicles).

Besides the obvious need for IRANS to use a map database specifically designed for specific types of CVO vehicles, the system must be able to accommodate changes in load, weather, or other dynamic situations. Table 4.1 illustrates some of the normal and situation-specific considerations that need to be built into an IRANS device for it to be useful in heavy vehicle CVOs.

Although most of the static considerations of an IRANS device can be built into the system database, at present the required data and an easy means for dealing with the data demands of a three-dimensional structure have lagged behind the development of the most commonly used geographic databases. Quality control is a significant problem in the development of map databases commonly used with vehicle-based mapping systems. Rapid growth and changes in much of the highway infrastructure make it necessary for database providers to verify the detailed accuracy of their databases constantly. Although significant efforts have been made to improve the quality of these databases, the means available for updating and verifying the maps still must rely, to a large extent, on the labor-intensive process of actually driving the route. Inclusion of CVO-specific information will significantly increase the difficulty and expense of this task.

Until a CVO-specific map database is developed, IRANS-equipped CVO drivers will probably need to limit the full use of IRANS routing to long-haul operations and designated truck routes. Drivers could, of course, use the routes provided by IRANS to assist them in finding local destinations, but such use would still rely heavily on the driver's knowledge of possible restrictions

TABLE 4.1
CVO Specific Considerations for IRANS

Static Considerations	Dynamic Considerations
Regulatory route restrictions	Hazardous material restrictions
Underpass/bridge height	Unusual weight or size
Underpass/bridge width	Highway surface condition (snow, ice)
Road geometry (bumps, hollows)	Driver hours of service remaining

on the proposed route and early identification of conditions that might cause him or her problems. In this sense IRANS becomes little more than an electronic map for the driver. As such, the system would need to provide the driver with an easy-to-use, complete preview of the proposed route. In such use, the method of presentation would be critical to the design and would possibly favor the use of map-based presentations because these usually are more familiar to commercial drivers than turn-by-turn presentations.

In-Vehicle Motorist Services Information Systems (IMSIS) for Commercial Vehicle Operations. IMSIS will provide information to a CVO driver similar to what they might supply for a private driver. The type of information required by a CVO driver is considerably different and possibly less extensive than what a private driver might want. A recent series of focus groups (Barfield et. al., 1993) studying commercial truck drivers presented results indicating that IMSIS would probably be the least useful of the ATIS functions. The practical reason for this involves the limited number of services that cater to CVOs and the ease in getting such information from other drivers or company dispatchers. To locate services such as food and motels that cater to CVOs, the drivers often use their CB radios to get recommendations from other drivers. For services such as fuel and repair, they usually use the company dispatcher, particularly because many companies have service contracts with specific providers.

To the extent that IMSIS are to contain useful information for CVOs it will need to be purged of services that do not support the extra length and maneuverability considerations of commercial trucks. Given the cost of developing such a system, the relatively limited number of services that support CVOs, the relative ease with which a CVO driver can obtain services information from other sources, and the special considerations of providing truck routing off from established truck routes, it appears unlikely that this area will receive much attention among system designers and developers. Should such designs be undertaken, however, one characteristic of CVO operations will need to be incorporated into the design. When seeking services, a CVO driver will usually make plans to use the services well in advance of actually doing so. In practice IMSIS, which for private cars will usually focus on identifying services within a limited radius, will be required to identify CVO services 50 or 100 miles further down the road.

In-Vehicle Sign Information Systems (ISIS) for CVOs. ISIS are intended to provide drivers with relevant information similar to the regulatory, caution, and information signs commonly posted along the sides of streets and highways. Information such as posted speed limits, railway crossings, and street names are candidates for inclusion in such systems. Although ISIS has not received much attention to date in the development of ATIS systems, it has several potentially important attributes for CVO use. Among these would be the possibility of developing an ISIS that would be sensitive to the specific needs of CVO vehicles. Examples of what such a system might provide CVO include these:

- Recommended maximum speed for trucks descending mountain slopes based on vehicle weight, equipment, and the configuration of the highway
- Warnings when exceeding maximum allowable speed limits for trucks for specified conditions (e.g., night driving)
- Warnings concerning restrictions on the use of specific equipment when such is activated (e.g., use of unmuffled jake brakes)
- Warnings about restrictions of road use for loads of specific cargos (e.g., explosive or flammable cargos about to enter a restricted tunnel)

If developed and supported in this way, ISIS would, in effect, be a job performance aid to the CVO driver, alerting the driver of incompatible conditions between his actions and either roadway or regulated conditions. Three primary characteristics make ISIS attractive. First, it may be based on a relatively simple broadcast beacon or transponder technology and thus need not have the expensive vehicle location and map database technology of IRANS and IMSIS. Second, decoding and presentation of information to the driver can be done with relatively simple, self-contained, on-board equipment. Finally, the ISIS information provided to a driver can be limited to that which actually applies to the situation. This can be done according to vehicle proximity to the event rather than the availability of convenient places to put the road sign. For example, an operational ISIS would provide emergency vehicles operating at high speeds with more timely notification of warning information.

In-Vehicle Safety and Warning Systems (IVSAWS) for Commercial Vehicle Operations. IVSAWS are intended to provide proximity warning of impending, and usually temporary, hazards such as construction zones, ice on the highway, and an accident ahead. Like ISIS, such systems could be developed around beacons or transponders. The major characteristic of IVSAWS other than the type of information presented is that at least part of the infrastructure is likely to be portable and temporary. Besides the possibility of presenting CVO-specific hazards and the possible desirability of presenting the warning at a greater distance from the hazard so that commercial vehicles are given a greater distance to react, IVSAWS are not likely to be significantly different for CVO drivers than for private drivers.

CVO-Specific Systems

Besides the basic ATIS subsystems, at least two other information-related areas have been identified as of major interest to CVOs. These systems are so important to the CVO community that it is likely that many of them were being developed well before there were any serious discussions of ITS.

Advanced Communications Systems and CVOs. Having timely and reliable communications between CVO drivers and their dispatchers has long been a major concern for CVOs. The mobile nature of truck operations with flexible pickup and delivery points makes such communications a key to productivity and profit in the industry. A truck on the road without a paying load is an unacceptable overhead expense whether its driver cannot be contacted or it is due to some other reason. It should come as no surprise, then, that mobile communications systems have long been a high priority for the CVO community. The development of ITS technology has merely spurred that interest, particularly in terms of the compatibility of information systems, including driver-to-driver and driver-to-dispatcher systems, with the basic requirements of safe driving.

Initial mobile communications for CVO vehicles were provided by various radio systems whose major characteristic was that they demanded immediate attention by the driver. Use of voice radio is still very much in use by CVOs, particularly in local operations in which dedicated company radios are used between delivery fleets and dispatchers, and in over-the-road operations in which citizen band radios are commonly found in trucks and are used for both recreational and informational exchanges between drivers. As with the private driver, cellular telephones also are used by many CVO drivers.

By their nature CBs, business radios, and cellular telephones have several characteristics that limit their desirability for CVO use. The major disadvantage is that they are all demand systems requiring, or at least encouraging, immediate response from the driver when a call is received. Such a system represents added workload for the driver. As a result, drivers commonly turn off their radios in heavy workload situations (Wheeler et. al., 1995). The practice of turning off or ignoring the radio to reduce workload highlights the other major disadvantage of such communications: lost communications. The realtime nature of radio communications frequently may result in failure of the driver to receive the information. Unless the driver actually completes the communications, a company dispatcher has no idea if important information has been received and will result in action.

Advanced communications systems tend to reduce the disadvantages of real-time radio systems by using message storage systems in the vehicle. In most cases the messages are transmitted and stored as text messages, although graphic or voice information can be dealt with in the same way. The major characteristics of advanced communications are that they are stored by the vehicle system and can be retrieved by the driver as desired, and that they have a built in capability to (a) assure the dispatcher that the message has been received and stored by the on-board equipment, and (b) give the dispatcher feedback when the driver has read and acknowledged the message. Such systems are very similar to electronic message systems used in many office local area networks, providing many of the same functions, such as priority delivery and receipt information.

The major disadvantage and human factors design concern for most of the advanced communications systems is the method used by the driver to enter or respond to a message and the way that a message is displayed. In some designs a standard, computer-type keyboard is used by the driver to compose messages; in others a set of quick keys is used (Burger, Smith, & Ziedman, 1989). To provide the maximum degree of flexibility and efficiency in the formulation of messages, the standard keyboard is the favored method. Burger et al. estimate that use of these systems resulted in high visual, motor, and processing loads similar to those experienced by a fixed-base dispatcher, and that their use thus would be incompatible with operation of the vehicle. This often presents a significant intrusion into the driver's workstation because there is no easy place to put the keyboard, and there is no real work surface on which to place it to compose a message. Keyboard use while driving is obviously a difficult thing to do, although drivers may attempt it. The message display itself appears to present a lesser problem because it frequently can be confined to a light-emitting diode display with a few lines of scrolling text and can be mounted more conveniently somewhere on the dash of the vehicle. Human factors considerations for the display type, content, and location are obviously similar to those of message displays associated with other ATIS equipment in both commercial and private vehicles.

One type of advanced communications system now in common CVO use is the Automatic Vehicle Location (AVL) system. The major function of AVL systems is to provide dispatchers with information on the vehicle's actual location. Such equipment may use a variety of techniques to determine location, including satellite navigation systems such as the Global Positioning System (GPS), triangulation systems that use ground-based transmitters, on-board inertial or distance and direction of travel measurements, and ground-based sensors. AVL system operations are largely transparent to the CVO driver although their existence may influence his or her actions. The sense that a dispatcher or supervisor is constantly "looking over my shoulder" is a fairly common complaint by drivers. The fact that AVL systems are sometimes used with performance monitoring systems that provide reports of vehicle speed, fuel economy, driving hours, and other performance measures to the dispatcher amplifies this feeling. However, not all of the capabilities of AVL systems result in negative consequences for the driver, because they also are capable of providing a very rapid and accurate way to get assistance in the event of an emergency and can assist in rapid recovery of hijacked vehicles, a fairly common concern in CVOs.

Advanced Systems of Regulation and CVOs. One of the most significant potential advantages of ITS technology to the CVO community is the possibility of reducing the amount of time that drivers themselves spend dealing with state and country regulatory agencies. Besides the travel time lost in stopping for a weigh or inspection station one of the more stressful parts of the driver's job is the uncertainty that he or she may feel when dealing with regulators (Wheeler et al., 1993). The amount of paperwork associated with modern commercial trucking combined with the complexity of modern commercial vehicles means that drivers expect to encounter at least some difficulty with almost any inspection. Current industrial and regulatory practices result in the driver having to take much of the responsibility for solving any problem discovered with either the paperwork or vehicle. Because inspection facilities are often in relatively remote areas, a failed inspection can result in significant delays, inconvenience, and expenses to the driver.

The most commonly advanced regulatory ITS system is probably Weigh in Motion (WIM). This system allows a driver to proceed at nearly full highway speed while his or her truck is weighed. If the truck is within its allowable weight and all other prerecorded information checks out well, the driver is allowed to continue without stopping at the weigh station. Such a system is virtually transparent to the driver and only requires signing to indicate in some way that he or she either has to pull into the station or is free to continue.

A second type of ITS technology coming into more common use is the use of the smart card, a usually passive transponder that provides a roadside-based reader with positive identification of the vehicle and a selected set of characteristics. The most common application is for the automatic collection of highway or bridge tolls, and in such services smart cards can be used by both commercial and private vehicles.[5] The technology also has been demonstrated for the automatic clearance of commercial vehicles at state ports of entry and other places where permits, tax records, and related administrative paperwork are checked. Again, as with WIM, smart card use is nearly transparent to the CVO driver except for the display or signal

---

[5]Smart card toll collection lanes have been installed on the Autopistas in Barcelona, Spain; the 730-kilometer Cofiroute in southwestern France; the 250-kilometer Esterel-Cote D'Azur in southern France; the Golden Gate Bridge in San Francisco; the Crecent City Connection Bridge in New Orleans; Georgia State Route 400 outside of Atlanta; The Aberdeen Tunnel, Hong Kong; the Lake Pontchartrain causeway, New Orleans; and all 10 turnpikes in the state of Oklahoma.

required to indicate that he or she needs to either stop or may continue. The advantages of such a system, particularly for records clearance, are obvious both in terms of the travel time saved by not having to stop to produce records and in terms of the assurance that a driver can have that if he successfully passed one such inspection, he probably will not have to be concerned with subsequent inspections. Such a system may even allow the company to provide prescreening of the smart card before the driver leaves the terminal to ensure that all necessary information on the shipment is up to date and in the state Department of Transportation (DOT) or Public Utility Commission (PUC) database.

One of the most difficult regulation-based systems that probably will be developed in connection with ITS is automatic vehicle condition monitoring. Such a system would use sensors on the vehicle to monitor the condition of safety-related vehicle systems and components (e.g., brake adjustment, lights, tire pressure). Such sensors would provide the driver with information warning of safety-related problems similar to that now provided by air pressure gauges, coolant temperature gauges, and fuel-level gauges. However, the required sensor technology is considerably more complex. From a regulatory standpoint, ITS presages the time when such information would be provided to the inspection station and possibly even a highway patrol vehicle as the truck drives at highway speed. As with WIM and the smart cards in collecting tolls and verifying records, the vehicle would be allowed to continue on its way unless the system indicated that there was a problem. In general, CVOs have been resistant to automatic vehicle condition monitoring in connection with regulatory monitoring inspections because of concerns about legal self-incrimination. Regulators also have been resisting the technology because of what they perceive to be the limited ability for defining and monitoring all appropriate safety-related systems automatically.

Beyond doubt, the most difficult of potential regulation-based systems will be those associated with automatic driver fitness-for-duty monitoring. The most simple system would use the information available from automatic trip recorders or logbooks to provide regulators with information for determining whether drivers have complied with hours of service regulations. Significantly more complex systems may one day be developed to provide fitness-for-duty indications based directly on driver performance or physiological state. Once such information is available, it could be transmitted easily to inspection stations and highway patrol. Obviously, there are serious technical, legal, and privacy issues associated with providing fitness-for-duty information directly to state patrol troopers or inspection stations. As a consequence, at this time it appears unlikely that such systems will be incorporated into ITS in the near future except as they might be applied to automated logbooks.

Weigh-in-motion, automatic toll-taking, and automatic vehicle-clearance systems all have been demonstrated and are in commercial use in selected areas. Automated vehicle-condition monitoring and inspection have yet to be developed.

COLLISION WARNING SYSTEMS
FOR COMMERCIAL VEHICLES

The increasing frequency of collisions and near-collisions continues to be a key concern within the CVO industry. Motor vehicle crash data compiled in 1993 from the Fatal Accident Reporting System (FARS) and the General Estimates System (GES) indicate that large trucks (gross vehicle weight rating >10, 000 lbs) were involved in an estimated 395,000 crashes (National Highway Traffic Safety Administration, 1994). Large trucks and buses accounted for over 4% of the fatal crashes in 1993; in all, 136,000 persons were injured and 4,849 were killed. In particular, combination-unit trucks average more than four times the per-vehicle annual mileage of single-unit trucks, yet are disproportionately represented in truck crashes. For example, combination-unit trucks accounted for only 27% of medium/heavy truck registrations, but represented 59% of all truck crash involvements and 75% of fatal truck crash involvements in 1991 (NHTSA, 1993).

In response to these concerns over commercial vehicle safety, the automotive electronics industry has initiated a number of efforts to develop in-vehicle countermeasures capable of detecting objects in close proximity to a vehicle and providing a warning to the driver. For example, Eaton Vehicle On-board Radar (VORAD) Technologies manufactures a forward-looking collision avoidance and warning systems device for truck and bus applications. The radar-based system has a detection range of 350 ft and tracks the movement and relative distance of objects ahead. Drivers are alerted with a yellow light when a target is detected, a red light when either closing speed or headway is unsafe, and an auditory signal when there is danger of a collision. Mitsubishi Motors Corporation also manufactures a distance-warning system incorporating a 3-beam laser for its line of large trucks (Bulkeley, 1993). The system computes relative distance between the truck and an object ahead and, then considering the speed of the truck, issues a warning to the driver if relative distance is too short.

Similar devices are being developed for CVO applications by a host of electronics and automobile manufacturers. These efforts to develop collision avoidance systems encompass a wide range of applications and sensor technologies. With respect to applications, systems are being developed to sense and to warn the CVO driver of forward, rear, and side collisions. Forward or headway systems are intended primarily to address driver inattention, distraction, or misperception concerns such as when the driver approaches a congested portion of the freeway at high speeds or fails to notice that a stoplight ahead has turned from green to red. Rearward systems mainly help drivers avoid collisions while backing their vehicles at very low speeds (Garrott, Flick, & Mazzae, 1995) Side-sensing systems (or blind spot sensors) are principally used as supplements to outside side/rear mirror systems, and to aid drivers during lane-change, merging, and turning maneuvers (NHTSA, 1993). This range of collision avoidance capabilities provides a number of benefits to the CVO community. For example, forward systems may aid drivers of emergency vehicles in verifying that an intersection is clear prior to crossing it. Rearward systems can help the CVO driver identify objects immediately behind the commercial vehicle that are not reflected in side-view mirrors, whereas drivers can use side-object systems to verify that sufficient clearance is available prior to making a lane change.

Both active and passive sensor technologies are being used as the core hardware technology of collision avoidance systems. Active technologies include radar (millimeter wave or microwave), ultrasonic, and laser devices. Although the physics for each of these devices is very different, the basic principles of operation are similar. Active collision avoidance sensors in production or under development consist of one transmitter or more and receiver units mounted on the front, side, rear, or top of the vehicle. These units send data to a central processor that determines if a collision is about to take place or if an object is detected and, if so, displays a visual and/or auditory warning to the driver. Passive sensors, which include infrared and video devices, send the infrared or video signal directly to a display mounted inside the vehicle, where the external scene to the front, rear, or side of the vehicle can be viewed in realtime by the driver. Although collision avoidance devices that would automatically take control of a vehicle (e.g., steering, braking, or acceleration) in the event of an impending collision are technologically feasible and being considered (Mironer & Hendricks, 1994; Verway et. al., 1993), informing and warning the driver of a potential collision is still the method of choice.

Despite the availability of first-generation collision avoidance systems (Bulkeley, 1993), relatively little is known about driver vehicle interface (DVI) requirements for these complex human–machine systems (Mazzae, Garrott, & Cacioppo, 1994). The DVI is important because it affects the ability of drivers to detect, understand, and correctly respond to the warning information. The DVI includes such characteristics as alerting mechanisms and designs, alert logic and timing, system accuracy and false alarm rates, and required control functions. Across the range of driving situations and tasks, the implications of driver perception, performance, and preferences on the design of these DVI characteristics are largely unknown. In the CVO environment in particular, a number of critical design and implementation issues remain to be addressed. Key DVI issues associated with collision warning systems for use in the CVO environment are briefly identified and discussed in the following section.

Driver Response Time to Alerts

A key issue in the development and design of collision avoidance systems is the perception–reaction times of drivers to collision warning alerts. Although this issue has been extensively studied within the human factors community, a range of findings have been reported. Wortman and Matthias (1983) measured the nighttime braking response times of 839 drivers to the onset of an amber signal at an intersection and reported mean values ranging from 1.09 to 1.55 s. Chang, Messer, and Santiago (1985) conducted a similar study, but in daytime as well and on both dry and wet roadways, and reported mean response times of 1.3 s and a 95th percentile value of 2.5 s. Lerner (1993), measured the perception–response times of both younger and older drivers to a simulated on-the-road emergency and reported a mean reaction time of 1.5 s and an 85th percentile value of 1.9 s. AASHTO (1990) uses a design reaction time of 2.5 s to determine stopping distance when designing roadway elements such as signs, road curvatures, and traffic signal visibility and timing. Fundamentally, an alert must be presented early enough in the total time frame of the collision event for the driver to perceive and understand the alert, and take appropriate action. Thus, assumptions made about driver capabilities to perceive and to respond to collision alerts affect virtually every design parameter of the collision avoidance system, from requirements for sensor range and scanning rate to limits on system processing time and the optimum modality for the collision alert.

Uncertainties About Alert Design

Collision alerts may be auditory, visual, tactile, or some combination of these three modalities. Within the CVO environment, each of these three modalities is associated with some advantages and disadvantages. For example, auditory alerts reduce the visual load on the driver (Wolf, 1987) and are well-suited to a collision-warning situation in which immediate action is required. However, their attention-getting abilities can become annoying to the CVO driver if the alerts occur frequently or are associated with a high false alarm rate. Furthermore, the ambient noise levels within the cab of a commercial vehicle can approach 100 decibels, leading to concerns about required intensity levels of the alerts and possible auditory masking (McCormick & Sanders, 1982). Visual alerts are less intrusive than auditory alerts, and the location of the display in the commercial vehicle can be used as a cue to the direction of the impending collision (e.g., colocated with a side-view mirror). However, driving a commercial vehicle requires a great deal of visual scanning just to maintain proper lane position and situational awareness of surrounding traffic conditions, and a visual alarm alone may not draw the driver's attention to the hazard. Also, using a visual display to present collision avoidance alerts introduces yet another visual task at precisely the same time that the driver's attention should be external to the vehicle. Tactile displays typically provide stimuli in the form of mechanical vibration or electrical impulses (McCormick & Sanders, 1982). In the CVO environment, tactile alerts might be transmitted through the seat back, the steering wheel, the accelerator pedal, or even by an automatic braking function. For example, Janssen and Nilsson (1990) conducted a simulator study using an "intelligent gas pedal" that applied a counterforce to the driver's foot as a collision alert. They found that the alert was associated with a reduction in headway on the part of their subjects. Importantly, headway reduction was not accompanied or offset by any inappropriate steering, acceleration, or braking behaviors.

Control Requirements by Drivers

It is not clear which collision avoidance system parameters should be adjustable by the CVO driver. Options for driver control include switching between alert modalities (e.g., auditory vs. visual presentation of alerts), modifying the intensity of the alert (e.g., loudness or brightness), and adjusting the sensitivity of the system and timing of alerts. For example, to better reflect their own driving styles and prevailing driving conditions, drivers may want to adjust the headway setting or time-to-collision parameter of a collision avoidance system. Such an adjustment would have the practical consequence of allowing drivers to select either a more or less conservative timing logic for the system, thus changing the timing of alert presentation in response to a potentially unsafe driving condition. Such a control function might increase user acceptance of the system and reduce false alarms. However, it also increases the likelihood that the alerts will be presented too late for the driver to make an appropriate response within a given collision scenario.

False Alarms

A false alarm occurs when a signal or target is said to be present when in fact no such signal or target is present. In the context of collision avoidance systems, two types of false alarms are relevant. First, a "real" false alarm occurs when a collision alert is presented to the CVO driver in the absence of any crash-relevant obstacle or event. Second, a "perceived" false alarm occurs under circumstances in which the driver is already aware of the hazard or feels that the alert itself or the urgency associated with the alert is incorrect or inappropriate. In either case, false alarms will reduce the trust and confidence that the driver places in the system, thus reducing system effectiveness. In general, users are most reluctant to rely on equipment they do not trust (Lee & Moray, 1992). When trust in the device is too low and an alarm is presented, drivers may spend additional time verifying the problem, thus slowing appropriate collision avoidance actions. Alternatively, they may choose to ignore the alarm, thus completely defeating the purpose of the system.

Inappropriate Levels of Driver Trust in the System

Just as having too little trust in a collision avoidance system can reduce system effectiveness, so having too much trust in the system can lead to a host of other performance concerns. For example, putting too much trust in an imperfect system may lead to a false sense of security on the part of the CVO driver. Similar problems have occurred in comparable domains. In the aircraft environment, for example, inappropriately high levels of trust in automated flight systems can cause pilots to ignore other sources of flight data or to forego established and prudent flight procedures (Danaher, 1980). In general, when trust in the device is too high, CVO drivers may assume that the system will detect any impending collision and therefore reduce their own vigilance levels or be willing to accept higher levels of driving risk and, for example, neglect typical collision avoidance behaviors.

Increased Training Requirements

CVO drivers will need time and training to become familiar with the various capabilities and modes of operation of collision avoidance systems. Despite their potential for increasing CVO safety, these devices represent a new on-board system that drivers must learn to use while driving. In this context, Mazzae and Garrott (1995) compared the performance and subjective opinions of CVO drivers using on-board side-object detection systems (SODS) and standard mirror systems. The collision avoidance system did not improve object detection performance over that obtained from a fender-mounted convex mirror, but drivers reported their thinking that the system was beneficial. Thus, Mazzae and Garrott (1995) speculated that drivers needed more time to become comfortable with the device and familiar with using it. Collision avoidance warning system devices share many of the human factors concerns associated with the implementation and use of automated systems. In particular, the introduction of such devices changes the human operator's role from one of continuous manual control to one with decreased manual control but increased monitoring requirements. Such changes suggest the need for changes in the amount and nature of CVO driver training for taking full advantage of the benefits that collision avoidance systems can provide.

Summary of Issues

In summary, collision avoidance systems have the potential to increase significantly both the safety and efficiency of the CVO industry. However, as seen previously, the introduction of collision avoidance technologies into the CVO environment is not issue-free. As with any new automotive technology involving the driver, many issues associated with CVO driver perception, performance, and preferences must be addressed. Key research questions associated with the introduction of collision avoidance systems into the CVO environment include the following:

- What expectations about CVO driver response time to a collision avoidance alert are reasonable? What are the implications of such expectations on system development, design, and subsequent performance?
- What alert modalities and configurations will be associated with the highest levels of system efficiency and driver safety? What methodological approaches are most likely to produce valid and reliable empirical data that can support collision avoidance system alert design?
- What collision avoidance system functions should be adjustable or under the control of the CVO driver?
- What levels of system reliability are demanded by CVO drivers? What are the consequences of false alarms?
- What are the implications of overconfidence in a collision avoidance system? How can system design address this issue?
- What new training requirements are introduced by collision avoidance systems? Must drivers simply learn how to use the system, or are fundamental changes in the driving task required?

AUTOMATED VEHICLE CONTROL SYSTEMS FOR COMMERCIAL VEHICLES

Automatic vehicle control systems have been a dream, sometimes a nightmare, of vehicle designers and traffic engineers for a number of years. Visions of a time when the entire family could enjoy a game of cards while their car quickly and safely carried them along America's highways by following a guide wire buried in the pavement were popular by at least the late 1950s. Development of safe and economical systems have not followed such dreams, however, and are well beyond the immediate horizon of ITS. For CVOs, such systems are likely to be very late in development simply because they must be able to accommodate a wide range of characteristics that result from day-to-day changes in the load and configuration of each particular truck. There are several potential automated vehicle control systems that may have early applications to CVOs.

Driver Fitness-for-Duty

One of the most difficult safety issues in CVOs is how to determine when a driver is no longer fit to drive because of fatigue, illness or some other cause. Projects aimed at developing reliable ways of detecting when a driver is experiencing dangerously lowered states of alertness have been going on for a number of years. Proposed approaches range from physiological measurement (e.g., eyeblink, heart rate, and electroencephalogram) to performance monitoring (e.g., secondary task performance, lane tracking, and speed variability). Although there is no currently accepted standard or best approach to determining driver fitness-for-duty, particularly as regards fatigue and alertness, sufficient effort is being made to solve this problem that a solution will undoubtedly be found in the next 5 to 10 years.

Once the technology is available to determine accurately when a CVO driver should no longer drive, the problem will become one of how to implement the technology. Information gathered from both accident investigations and anecdotal evidence indicate that a large proportion of drivers, both private and commercial, will continue to drive well after they are so tired that they should not do so. There is good reason to believe that the mere availability of a driver-fitness-for-duty gauge, will not keep all drivers from pushing well beyond the limits of safety.

Automated vehicle control systems linked to a system that will determine when a driver is becoming, or has become, unable to drive will probably follow the same approach used by collision avoidance systems (i.e., first warning, then assisted control of the vehicle followed by actual control of the vehicle). Systems that will bring commercial vehicles to a completely automatic stop and do so safely will require significant technological sophistication. In the meantime, fitness-for-duty systems that alert and possibly assist the driver in bringing the vehicle to a safe stop appear to be a more likely possibility.

Rollover Warning and Control

Unlike private automobiles, commercial vehicles often carry loads that significantly change the driving characteristics of the vehicle. The way a load is placed on the trailer, the type of load, and the characteristics of the trailer mean that each load presents different problems to the driver. The use of articulated tractor-trailer units, particularly those involving more than one trailer, further complicates this problem by removing much of the feel or feedback that a driver could use to judge the reaction of the vehicle to different road conditions and speeds.

Of particular concern is the tendency of a commercial vehicle to roll over as a result of the combined effects of a higher than normal center of gravity, the configuration of the road, and the speed of the vehicle. The dynamics of a rollover accident are complex, but measurable, and systems can be designed to alert drivers that they are approaching a rollover condition. At some point in the rollover process, however, the control actions required to balance the forces involved and prevent an accident are likely to exceed the driver's capabilities, particularly because he or she receives little information on the forces acting on the trailer.

A rollover warning and control system would assist the driver in much the same way that antiskid brakes work to increase braking efficiency by helping optimize and balance the forces necessary to bring the vehicle safely through the incident. Unlike antiskid braking, however, the vehicle controls involved are likely to include brakes, steering, and engine controls, and the required actions may be complex. For example, a potential rollover in a right-hand-turn situation might involve all of these maneuvers: the partial application of trailer brakes to provide drag at the rear of the combination, a reduction in the steering wheel angle to reduce the rate of turn, and an increase in throttle to increase the pull on the trailer and right the combination.

Brake Enhancement Systems

The most common brake enhancement systems currently in use are antiskid braking systems. These computer-controlled systems detect when a braked wheel has stopped moving and release the brake in a way that ensures that maximum friction is maintained between the tires and the road. In a car or similar four-wheel vehicle, such systems also ensure that the vehicle stops within the minimum distance.

The braking response of articulated vehicles, particularly those with more than one trailer, can be complex, and the braking efficiency of different trailers in such a combination may not be equal. To help ensure that the combination will stop in a straight line, brakes or trailer weights may be adjusted to provide the greatest braking in the rearmost trailer and the least in the foremost one. Such adjustments do not ensure that the truck will stop in a straight line under all conditions, however, and generally result in greater stopping distances than would be possible if all brakes were allowed to operate as a combined unit. Sensors, not only of wheel rotation, but of trailer forces and road geometry, can be combined to achieve the most efficient braking possible under all conditions.

Platooning and CVO Heavy Vehicles

Platooning is an ITS concept that involves two or more vehicles traveling very closely in the same direction. In such a combination, the lead vehicle provides speed and lateral control to the vehicles following behind. The vehicles are linked electronically to function as though they were trailers behind the lead vehicle. Platooning could significantly increase the traffic-carrying capacity of existing highways. Heavy commercial vehicles such as tractor-trailer trucks are seldom included when the effect of platooning is considered, but their use as lead vehicles would have several significant advantages:

- A vehicle driven by a professional driver who by the nature of his or her employment is subject to stricter licensing, fitness-for-duty, and physical standards than are private drivers
- A vehicle with greater mass and thus longer stopping distances than most automobiles and light trucks, thus providing an increased margin of safety from rear-end collisions during emergency or panic stops
- A vehicle with greater mass, thus greater ability to absorb the impact of a collision, which would reduce the damage to following vehicles
- A vehicle with greater size that would thus "break wind" for the following vehicles to increase both fuel efficiency and reduce engine wear on those vehicles

ADVANCED TRAFFIC MANAGEMENT CENTERS AND CVO OPERATIONS

Operators within Advanced Traffic Management Centers (ATMC) will be responsible for a broad range of operations and functions. By the use of ITS technologies, realtime data will be sent to ATMCs from a variety of sources, and ATMC operators will be required to integrate and respond to these data in an accurate and timely manner. The ATMC will also be responsible for issues involving public safety and highway maintenance that relate to the use of surveillance systems, phone systems, and radio communications. ATMC operators will serve as the prime contact point for law enforcement, regional highway maintenance, emergency dispatch centers, and a host of other transportation-related agencies and operations. Key ATMC responsibilities will include monitoring, analyzing, and disseminating information on

- traffic patterns and conditions,

- the condition of the roadway systems,
- the location and nature of roadway incidents,
- maintenance activities,
- weather conditions, and
- emergency situations.

CVO vehicles will represent a special set of challenges and requirements for future ATMC operation. For example, CVO-related roadway incidents can have a significant effect on travel times and traveler safety, especially during peak hours. In general, issues such as traffic flow separation, managing special purpose vehicles, and rerouting CVO vehicles during accidents are relevant.

The specific role that CVOs will play in ATMCs is still being developed. Sensors that can differentiate between larger commercial vehicles and automobiles have been developed already, and their data will probably be used as input to the complex traffic models that will be the basis of ATMCs. In addition, commercial vehicles, particularly vehicles such as buses or route delivery vehicles that follow fixed routes, may be used as probe vehicles to determine highway transit times through various sectors.

One clearly attractive concept of ATMCs is the separation of CVO traffic from automobile traffic (McCallum, Lee, Sanquist, & Wheeler, 1993). Such separation will require designated CVO routes or traffic lanes and a means of informing the driver when these routes must be used.

INTERFACE DESIGN ISSUES

Physical and Cognitive Characteristics of CVO Drivers

In general, the physical and cognitive characteristics of CVO drivers tend to be more homogeneous than they are for private drivers. This stems from the fact that a CVO driver must pass more stringent health, knowledge, and experience requirements to obtain a commercial driver's license (CDL) than required for a private driver's license. In addition, CVO drivers usually fall within the normal range of working adults and thus do not include either young adults or the elderly. Moreover, as a result of much more serious consequences (i.e., loss of the CDL and livelihood), they are less likely than a private driver to drive under the influence of alcohol or drugs. Although men constitute the majority of CVO drivers, women are entering the field in increasing numbers (Waller, 1992). Because CVO drivers represent a restricted subset of the population for the eventual use of ITS rather than a different population, there is little incentive for designing ITS specifically to accommodate the characteristics of the group.

Vehicle Characteristics That Influence Interface Design

Workstation Design for CVO Drivers

The console and cab environment of commercial vehicles represents a distinctly different type of workstation than does that of the automobile or light truck. Because most vehicle-based ITS equipment, particularly ATIS controls and displays, will need to be compatible with this environment, it is important to understand at least some major characteristics of the most common types of CVO vehicles.

Buses. Buses represent a mixed workstation because the drivers in most cases must be able to communicate with passengers and perform service-related functions such as announcing approaching stops. Extensive use of diesel engines, air brakes, and other systems may result in a significant number of controls and displays, particularly in buses intended for long-distance highway use.

Bus drivers may face extensive instrument and control layouts, but they are spared many of the environmental conditions commonly encountered by commercial trucks. Because most buses have engines placed in the rear of the vehicle and the passenger area is usually well-insulated, engine noise usually is not a factor. Ambient noise from passenger conversations may be high, however, and could present particular difficulty as a masking noise for audio displays. Other conditions (e.g., vibration, heat or cold) normally found in truck driving are not a likely design concern for ITS equipment in buses. The effect of the load on driving characteristics also merits little consideration, although bus drivers must pay particular attention to the smoothness of their driving and avoid rapid control inputs as much as possible.

Trucks. For installing ITS equipment, trucks represent the most difficult of the normal environments. The driver's cab has a very high noise level, often exceeding 90 dB(A) for extended periods of time. Truck drivers must deal with an extensive number of controls and may need to monitor more than 15 or 20 separate displays and condition lights. Visibility is severely restricted in all but the forward direction. Furthermore, the location of the engine either under the cab or directly in front of it, in addition to the stiff suspension needed for heavy loads, results in severe engine and road vibrations when the vehicle is operating.

Besides the cab requirements of commercial trucks that severely limit the design options available for ATIS controls and displays, the operating characteristics of the vehicle will need to be considered as well. The fact that each load and trailer configuration may significantly change the way the vehicle drives and reacts to both the driver and the road has major implications for the design of collision avoidance and vehicle-control systems. In particular, such designs must be sufficiently flexible to allow their use over a wide range of loads and configurations.

Emergency Vehicles. Outfitting emergency vehicles with ITS technology, particularly ATIS, may present some of the most difficult ITS design challenges in the entire array of possible vehicles. (Motorcycles may be a similarly difficult task, although for different reasons.) There are two major reasons why emergency vehicles present problems. First, they are already saturated with special-use information equipment such as computers, and radios. Second, there are virtually no standards for the way auxiliary equipment is to be added to the vehicle. With the exception of ambulances and fire trucks, emergency vehicles are usually conventional cars, vans, or light trucks to which the additional equipment is simply added. Although large fleets such as those found in major cities may reflect some standardization in the way that the auxiliary radios and other features are installed, the equipment is nearly always added to an existing console rather than integrated into the various systems. Each system stands alone and in competition with the others. Adding another information system to this arrangement will undoubtedly result in competing demands for the driver's time. In the best of all possible worlds, the demands are shared by a another person in the vehicle, but in many cases this is not done.

The CVO Driver's "Awareness Envelope"

Commercial vehicles, particularly truck-trailer combinations, provide a different view of the world than that experienced in an automobile. The view from the cab of a commercial truck is usually much better than from a car because of its increased height. Thus a truck driver can often see and understand what is happening on the road ahead sooner than a private driver can.

Driver awareness to the rear and right side of the cab is restricted by the size of the cab, its height, and its layout. If the tractor is pulling a van-type trailer, there is a significant blind spot immediately beyond the vehicle that can extend for several hundred feet.

Information on what is happening behind the vehicle and on the right side is almost exclusively obtained by the use of mirrors. To provide coverage of the area, a tractor may have as many as four or five different types of mirrors on either side, and many of these will be convex, thus giving the driver a distorted view of what is happening.

Unlike automobiles, which can easily keep up with the flow of traffic and therefore have limited concern for overtaking traffic, commercial vehicles must often be concerned with passing traffic. The lack of physical feedback from the trailers, particularly when there are more than one, also means that the driver must rely on his mirrors to determine what is happening to his load and trailers. Recent video tapes of CVO drivers indicate that they use the mirrors in several different ways:

- Quick glances at the main mirror to get a general idea of what is happening to the rear
- Detailed and time-consuming study of one or more mirrors to obtain specific information
- Pass-monitoring to track the progress of a passing car or truck or the progress of the driver's own truck in passing another vehicle

A commercial driver's situational awareness is driven by the size of his vehicle, the extent of blind spots to his vision, the lack of maneuverability and turning characteristics of the vehicle, and the distance required to stop or take evasive action. It is by increasing this situational awareness that ITS technologies are likely to make the greatest impact on the way a driver does his job. This is obvious in the case of crash avoidance technologies such as side-looking radar and warnings that might be provided by IVSAWS, but it is equally true of route-guidance systems and similar capabilities if they can provide the driver with timely and detailed warning of actions that must be taken in the near future.

Workload: A Major CVO Constraint

Driving commercial vehicles is a labor-intensive job that requires both physical and mental effort. Unlike the modern automobile that requires very little of the driver other than lateral speed control, getting maximum speed and efficiency from a commercial vehicle requires that the driver constantly monitor the road ahead, the way that his engine is operating, and his speed. If the vehicle is a truck it may have between 9 and 18 forward gears, and each of these must be manually shifted to get the most out of diesel engines with very narrow power bands. When not shifting to climb a hill, drivers are concerned about controlling the speed downhill to save their brakes. Such control may require either downshifting or the manual activation of engine brakes by throwing a switch or pressing a special pedal.

In addition to monitoring speed and engine performance, CVO drivers need to be focused constantly on maintaining lateral control of their vehicles to keep them in the appropriate lane. This is not always easy with a vehicle that may be as long as 105 feet and have as many as three independent trailers moving around, particularly if the road is rutted or damaged. Controlling lateral position of both tractor and trailers when the only indication of the trailers' positions is the image in a mirror is no easy matter.

Drivers needing to take care of these activities, while at the same time monitoring the tractor and trailer for obvious mechanical or load securement problems, have little time or inclination to deal with ITS unless the use of such a system aids them in performing the basic driving task.

Drivers of emergency vehicles (e.g., ambulances, and police and fire vehicles) must also deal with much greater workload than that commonly experienced by a private driver. They must frequently operate their vehicles at higher speeds than traffic or road conditions would normally allow. In addition, they must often perform secondary tasks while driving (e.g., controlling emergency signals, communicating on the radio, planning for actions that they will need to take at the scene of an emergency). Although many emergency vehicles employ multiperson teams, this does not eliminate entirely the additional burdens associated with emergency response. The workload effects under such conditions may well be amplified by the psychological stress of an emergency situation (Wheeler & Toquam, 1988).

Even the relatively benign commercial operations of bus and taxi drivers may involve increased workload considerations because both require frequent, and often uncontrolled, interaction with passengers.

System Characteristics That Influence Interface Design

CVOs are commercial ventures for which important considerations must be made if the trucking industry is going to be a major player in ITS. These considerations are primarily organizational, but still need careful consideration in the design of ITS that will include CVOs.

Proprietary Information Requirements

One major concern of trucking companies and other commercial vehicle operators is that information provided to or gathered by ITS systems remain confidential. If, for instance, permit data used by regulators to expedite commercial loads through ports of entry were generally available, competitors would be able to use the information to develop marketing or rate strategies against the company. In a highly competitive business like trucking such information, particularly in the form of the compiled databases that necessarily will be a part of ITS, can be of particular importance.

A number of the proposed ITS systems, particularly in ATIS, could involve the electronic transfer and storage of proprietary or business sensitive information from the company to ITS-related databanks. This and the means used to safeguard the information are particular concerns to the industry.

Keeping the Driver Honest

Pay for commercial truck drivers is often based, at least in part, on the number of miles driven. In a minority of cases, pay may be based on an hourly rate. In the past, truck driving has been a largely unsupervised activity. After the trucker left the company yard, no one paid too much attention to how he got to his destination so long has he did so quickly. The use of mileage payments, equivalent to piecework in private industry, helped to ensure that drivers did not go sightseeing or fishing with the company truck. Pure and simple, mileage was money, and the more miles you could drive in a week, the more you made. The only companies that considered using an hourly wage rate were those with fixed routes and a well-established historical record of how long it took to drive from one point to another.

The financial incentive provided by milage payments encourages drivers to drive extended and often illegal hours. This, coupled with the generally accepted impression among both drivers and their employers that the hours of service regulations established in the 1930s are obsolete, has led to the commonly accepted practice of drivers' falsifying their logbooks to reflect fewer driving and duty hours than were actually done, often with at least the tacit approval of the company.

ITS-related technology such as Automated Vehicle Location systems, automated logbooks, trip recorders, and the computerized records of WIM and automated records checking are going to provide independent and objective data of the number of hours that a driver has been on duty or on the road. At the same time, the availability of IRANS and ATMC data on traffic flow and delays may give drivers potent new arguments when they need to explain to their dispatchers why it took them so long to make the run.

CONCLUSIONS

The systems and subsystems that will be part of ITS may prove to be of significant benefit to the CVO community as well as to the general driving public. It would be a mistake to conclude, however, that CVOs will be able to use the same systems in the same ways as does the private driver. CVOs are significantly different from those of a homemaker doing the weekly shopping, a business woman driving a rented car in a strange city, a youth making a pizza delivery, or families taking cross-country trips in their RVs. The driving location and general environment may be the same, but the organizational, technical, and administrative situations are certainly different.

The demands placed on CVO drivers by the size, design, and operating characteristics of their vehicles place a significant workload on the driver requiring that designers make careful decisions in designing ITS systems that reduce rather than increase that workload. The operating environment of the vehicle cab, with its high ambient noise, vibration, and competing visual displays will require special design considerations if ATIS systems are to be at all useful. Furthermore, the restricted range of services required by CVOs, and the limitations on the routes that they may legally or practically use will require the development of CVO-specific map and services databases if the ATIS is to be of much use. The size, weight, and slow control response of CVO vehicles require that special designs be developed for crash avoidance and automated vehicle-control systems if they are going be installed in the larger commercial vehicles. The business-sensitive nature of commercial vehicle load and traffic information will require special consideration in the design of CVO regulatory systems such as WIM and automatic vehicle-clearance systems.

REFERENCES

AASHTO (1990). *A policy on geometric design of highways and streets.* Washington, DC: American Association of State Highway and Transportation Officials.

Barfield, W., Bittner, A. C., Jr., Charness, N., Hanley, M., Kinghorn, R. A., Landau, R., Lee, J. D., Mannering, F., Ng, L., & Wheeler, W. A. (1993). *Development of human factors guidelines for Advanced Traveler Information Systems (ATIS) and Commercial Vehicle Operations (CVO) components of the Intelligent Vehicle Highway System (IVHS); Task F working paper: Identify ATIS/CVO users and their information requirements* (DTFH61-92-C-00102). Seattle, WA: Battelle Human Factors Transportation Center.

Bulkeley, D. (1993). The quest for collision-free travel. *Design News, 10-4-93,* 108–112.

Burger, W. J., Smith, R. L., & Ziedman K. (1989). *Supplemental electronic in-cab displays: An inventory of devices and approaches to their evalutation. PB89-215081.* U.S. Department of Transportation, Feburary 1989.

Chang, M. S., Messer, C. J. , & Santiago, A. J. (1985). Timing traffic signals change intervals based on driver behavior. *Transportation Research Record, 1027,* 20–30.

Danaher, J. W. (1980). Human error in ATC systems operations. *Human Factors, 22,* 535–545.

Garrott, W. R., Flick, M. A., & Mazzae, E. N. (1995). *Hardware evaluation of heavy truck side and rear object detection systems* (SAE Technical Paper Series No. 951010).

Heavy Duty Trucking (1993). Hi-tech crash protection. *Heavy Duty Trucking.* Irvine, CA: November 1994.

IVHS America (1992, April). *Strategic plan for intelligent vehicle-highways systems in the United States* (final draft). Washington, DC: Department of Transportation.

Janssen, W. H., & Nilsson, L. (1990). *An experimental evaluation of in-vehicle collision avoidance systems* (Drive Project V1041). The Netherlands: Traffic Research Center.

Lee, J., & Moray, N. (1992). Trust and the allocation of function in the control of automatic systems. *Ergonomics, 35,* 1243–1270.

Lerner, N. D. (1993). Brake perception–reaction times of older and younger drivers. *Proceedings of the Human Factors and Ergonomics Society 37th Annual Meeting,* San Diego: Human Factors and Ergonomics Society, pp. 206–210.

McCallum, M. C., Lee, J. D., Sanquist, T. F., & Wheeler, W. A. (1993). *Development of human factors guidelines for Advanced Traveler Information Systems (ATIS) and Commercial Vehicle Operations (CVO) components of the Intelligent Vehicle Highway System (IVHS); Task B working paper: ATIS and CVO development objectives and performance requirements* (DTFH61-92-C-00102). Seattle, WA: Battelle Human Factors Transportation Center.

Mazzae, E. N., Garrott, W. R., & Cacioppo, A. J. (1994). Utility assessment of side object detection systems for heavy trucks. *Proceedings of the Human Factors and Ergonomics Society 38th Annual Meeting,* San Diego: Human Factors and Ergonomics Society, pp. 466–470.

Mazzae, E. N., & Garrott, W. R. (1995). *Human performance evaluation of heavy truck side object detection systems* (SAE Technical Paper Series No. 951011).

McCormick, E. J., & Sanders, M. S. (1982). *Human factors in engineering and design.* New York: McGraw-Hill.

Mironer, M., & Hendricks, D. (1994). *Examination of single vehicle roadway departure crashes and potential IVHS countermeasures.* Springfield VS: NTIS.

National Highway Traffic Safety Administration (NHTSA, 1994). *A compilation of motor vehicle crash data from the fatal accident reporting system and the general estimates system.* Washington, DC: Author.

Verway, W. B., Alm, H., Groeger, J. A., Janssen, W. H., Kuiken, M. J., Schraagan, J. M., Schuman, J., van Winsum, W., & Wontorra, H. (1993). GIDS functions. In J. A. Michon (Ed.), *Generic intelligent driver support* (pp. 113–146). London: Taylor & Francis.

Waller, P. F. (1992). *Truck transportation productivity: Responding to the changing workforce.* Paper presented at the Motor Vehicle Manufacturers Association's National Truck Transportation Productivity Seminar, April 8, 1992, and published in *UMTRI Research Review, 22*(6).

Wheeler, W. A., Kinghorn, R. A., Stone, S. R., Raby, M., Kantowitz, B. H., & Bittner, A. C., Jr., (1995). *An investigation of driver stress and fatigue for drivers of long combination vehicles (LCV) (draft final report).* Seattle, WA: Battelle Human Factors Transportation Center.

Wheeler, W. A., Lee, J. D., Raby, M., Kinghorn, R. A., Bittner, A. C., Jr., & McCallum, M. C. (1993). *Development of human factors guidelines for Advanced Traveler Information Systems (ATIS) and Commercial Vehicle Operations (CVO) components of the Intelligent Vehicle Highway System (IVHS); Task E draft working paper: Task analysis of ATIS/CVO functions.* Seattle,WA: Battelle Seattle Research Center.

Wheeler, W. A., & Toquam, J. L. (1988). *Control room evaluation system* (Battelle HARC Tech. Rep. BHARC-700/88/026). Seattle, WA: Battelle Human Affairs Research Centers.

Wolf, L. D. (1987). The investigation of auditory displays for automotive applications. *Society for Information Display Symposium of Technical Papers, XVIII,* 49–51.

Wortman, R. H., & Matthias, J. S. (1983). Evaluation of driver behavior at signalized intersections. *Transportation Research Record, 904,* 10–20.

5

HUMAN FACTORS DESIGN OF
AUTOMATED HIGHWAY SYSTEMS

Lee Levitan
Honeywell Technology Center

John R. Bloomfield
The University of Iowa

Using the most recent notions, an automated highway system (AHS) is defined in general as a system that combines vehicle and roadway instrumentation to provide some level of automated ("hands off /feet off") driving. The vehicles, all dual mode in nature, are capable of functioning on both normal and automated roadways, which generally are roadways equipped to control the speed and steering of an appropriately instrumented vehicle. As envisioned, automated roadways will use existing rights-of-way, and will evolve from current highway structures. Benefits of an automated highway system include the following:

- Improved safety for vehicles in automated lanes,
- Increased efficiency,
- Predictable trip times,
- Reduced environmental pollution due to a decrease in fossil fuel consumption and emissions, and
- Reduced stress for those traveling in an automated lane (as compared with manual driving) (National Automated Highway System Consortium, 1995, pp. 13–14).

There have been numerous visions or scenarios of a proposed AHS that vary along several dimensions, including the following (see Tsao, Hall, Shladover, Plocher, & Levitan, 1993, for detailed examples of scenarios that vary along these dimensions):

- *Degree of automation.* Development of an AHS is often seen as evolutionary. The final implementation currently is envisioned to be capable of fully controlling speed and steering for mixed vehicles—cars, vans, buses, trucks of all sizes, and so on—on the same roadway, with complete coordination of all vehicle movement (e.g., merging into the automated lane, lane keeping and lane changing, collision avoidance). The system will be told at what exit from the automated roadway the driver wishes to get off—which the driver retains the right to change later—and the system will drive the vehicle to the designated exit. While under automated control, the driver, with no driving-related responsibilities, will be free to engage in any reasonable activity. In the event of a medical emergency, the driver (or passenger) will be able to communicate this to the system with a single, simple control action, and the system will immediately drive the vehicle to a safe place and (concurrently) inform emergency medical services personnel of the problem, provide vehicle description information (e.g., make, model, and license number), and tell where the vehicle is located.

At the other end of the evolutionary path, though perhaps not typically thought of in this way, is simple cruise control that provides automated control of vehicle speed. Along the path toward full automation will be "smarter" versions of cruise control that will maintain a set distance from the vehicle ahead, collision warning and avoidance systems, and those that provide additional but less-than-full automation.

- *Type of infrastructure and vehicle equipment needed.* For full automation, a highly sophisticated and intelligent roadside system will communicate with all vehicles on the automated roadway to control and coordinate their activities. Each vehicle will have equipment to communicate with the roadside controller; sensors to detect other vehicles ahead, behind, and to its side; actuators to implement steering, braking, and acceleration commands; and an interface that allows the system and driver to transfer information. The degree of "intelligence" in the roadside system will depend on the specific implementation: In one example, the roadside system will tell vehicles what their maximum speed should be and what gap (distance between the front bumper of one vehicle and the rear bumper of the vehicle immediately ahead) they should maintain, but the vehicle itself will be responsible for maintaining the speed and gap; in a different version, the roadside system will directly control speed and gap for the vehicles.

In contrast to full automation, earlier implementations along the evolutionary path will have much more intelligence on the vehicle itself. Indeed, in what Honeywell calls a free agency/self-contained scenario (Tsao et al., 1993), there is no roadside system at all. The vehicle on-board systems will do automated lane keeping and maintenance of a set distance from the vehicle ahead. It also will provide for warnings about potential collisions with vehicles (or perhaps other objects) ahead or to the side. However, it will not do automated lane changes—the drivers will have to change lanes themselves. Moreover, there will be no coordination among vehicles; each will be a "free agent." In this scenario, the driver will be expected to remain fully alert to the driving task throughout.

• *The degree of separation between automated and unautomated traffic.* In some full-automation scenarios, the automated lanes will be physically segregated from the other lanes. There will be separate on and off ramps for the automated lanes, and traffic will not be able to move between automated and unautomated lanes. In other scenarios, traffic will move between the two types of lanes, and there may be barriers separating them. Where there are barriers, gaps between barriers will allow vehicles to move into and out of the automated lanes. In any case, the roadside system will control movement between types of lanes, thus minimizing any disruption to the flow of traffic and preventing accidents (to the extent possible).

• *Vehicle control rules.* With full automation, the roadside system may control either groups of vehicles—some call them platoons, some call them strings—or individual vehicles. Strings will consist of several vehicles (the number varies by author) closely spaced—gaps as short as 1 m (3.3 ft) are envisioned (e.g., Tsao et al., 1993)—and traveling at relatively high speeds. The National Automated Highway System Consortium (1995) specifies a design speed limit for the AHS of 200 km/h (124 mph).

In this chapter we report on work done for the Federal Highway Administration (FHWA) under a contract to investigate human factors aspects of automated highway systems design. The objective of the contract was to provide human factors input during the conceptual stages of AHS development to affect its design and implementation and to lay the foundation for future advancement of the AHS. There were two parallel and interweaving tracks in the contract, one analytical and one experimental. The primary purposes of the analytical track were to develop a picture of what the AHS might look like under various combinations of the dimensions just discussed above, and to determine the range of human factors issues that must be resolved to ensure a successful implementation of any AHS. The purpose of the experimental track was to investigate human-factors issues in a high-fidelity simulation environment in which the appropriate variables could be controlled and the variables of interest could be systematically manipulated.

SOME HUMAN FACTORS ISSUES ADDRESSED DURING THE CONTRACT

Regardless of the particular vision, there are many human factors issues that need to be addressed to ensure a successful AHS implementation. Important issues determined during our FHWA contract, and for which we have at least partial answers based on that work, include the following:

- Virtually all scenarios include close-following vehicles at relatively high speeds when under fully automated control. What psychological impact will travel under such conditions have on vehicle occupants? Will they accept the small gaps?—the high speeds? What effect will there be when another vehicle enters the lane ahead, and the gap between the two vehicles shortens until the appropriate intervehicle gap has been achieved?
- How should control be transferred from the driver to the AHS, and from the AHS back to the driver?
- What effects will automated travel have on manual driving behavior?—on such things as speed, lane-keeping performance, and size of the gap into which a lane change is made?
- What role can the driver be expected to play when a failure occurs?

Of course, a large number of other issues must be resolved to ensure that any implementation of the AHS is successful. Some additional human-factors issues are listed in the last section of this chapter: Future Research Issues. Other issues from different domains include such things as how to handle the increased volume of traffic as the vehicles leave the automated lanes and enter the city streets; what level of reliability in subsystems (e.g., brake actuators) will be required to achieve the overall safety goal, and if it can be achieved within reasonable per vehicle cost and weight constraints; whether a special AHS license will be required and, if so, what will the driver have to do to get one; and how equity will be assured in terms of how the system is paid for versus who is able to use it. Although these latter concerns are clearly beyond the scope of this chapter, the reader must remember that presently there are far more questions about automated highway systems than there are answers to particular issues. The discussion in the remainder of this chapter provides some answers to the human factors issues noted earlier.

Seven experiments were conducted during the first stage of the contract. We review them first, before discussing some of the analytical work done on the contract.

Stage I Experiments

All seven experiments were run in the Iowa Driving Simulator located at The University of Iowa. A picture of the simulator is shown in Fig. 5.1. It has a moving-base hexapod platform covered with a projection dome. The inner walls of the simulator act as screens onto which imagery is projected by an Evans and Sutherland image generator. For these experiments, correlated imagery was projected onto a 3.35-rad (192°) section of the inner wall in front of the simulator vehicle (a midsize Ford sedan) and onto a 1.13-rad

FIG. 5.1. The Iowa Driving Simulator.

(60°) section to its rear. The driver of the simulator vehicle viewed the front section through the windshield and side windows of the car, and the rear section either by looking over his or her shoulder, by using an inside rearview mirror, or by using a left side mirror mounted on the outside of the vehicle. There was no right side mirror on the vehicle and no right-side rear view. Details of the simulator and its operating system can be found in Kuhl, Evans, Papelis, Romano, and Watson (1995) and Kuhl and Papelis (1993), respectively.

In all seven experiments, a three-lane expressway was used, with the left lane primarily though not exclusively for automated vehicles, the center lane primarily though not exclusively for unautomated vehicles, and the right lane exclusively for unautomated vehicles. There were no barriers between any pairs of lanes, and there was no dedicated transition lane. Driving was done at midday with clear weather and a dry, standard road surface. Conventional 3.7-m (12-ft) lane widths were used. While the vehicle was under automated control, the steering wheel did not move, the accelerator pedal reflected vehicle acceleration, and the brake pedal was disconnected. Simulated vehicles in the unautomated lanes traveled at an average speed of 88.6 km/h (55 mph). When the driver's car was under automated control, the system controlled the speed and steering and maintained a predetermined gap.

Results are available so far for only five of the experiments.

Experiments 1 and 2: Human Factors Aspects of the Transfer of Control From the Automated Highway System to the Driver

The objective of the first two experiments was to determine the effect of several variables on the transfer of control from the automated highway system (AHS) to the driver (Bloomfield et al., 1994).

Independent Variables. The independent variables were these:

• *Design speed in the automated lane.* Speeds were 104.7 km/h (65 mph), 128.8 km/h (80 mph), and 153.0 km/h (95 mph). Speed was a between-subjects variable. (Note: With a *between-subjects variable,* each subject receives only one value or level of the variable. Thus, in this case, each subject drove at only one of the three speeds.) The speed of vehicles in the automated lane is important in determining trip lengths, among other things, and the driver's psychological reaction, especially to the higher speeds used in this experiment, was of particular interest in helping to decide whether such speeds could be used on an actual AHS. Of equal importance was whether the speed in the automated lane had any effect on manual driving performance following automated travel. Determining the existence (if any) of such carryover effects and quantifying them is of critical importance: To the extent that driving after automated travel is adversely affected, steps must be taken in the design of the AHS either to counteract the carryover effects or to minimize their magnitude.

• *The gap between vehicles within a string in the automated lane.* (In this experiment, a string consisted of three vehicles.) Gap times used for these experiments were .0625 s, .25 s, and 1.0 s. The combinations of design speeds and gap times produced the gap distances shown in Table 5.1. Gap was a within-subjects variable. (Note: With a *within-subjects variable,* each subject receives every value or level of the variable. Thus, in this case, each subject drove with each of the three gap times on different trials.) Intrastring gap will partly determine what density of traffic the automated lane(s) will support; at issue in this experiment was whether there were differential carryover effects as a function of gap size, and what gaps drivers actually will accept.

• *Density of traffic in the unautomated lanes.* This was a within-subjects variable. Densities were 6.2 and 12.4 vehicles/km/lane (10 and 20 vehicles/mi/lane). At the lower density, traffic flows freely; at the higher density, the

TABLE 5.1
Gap Distances for Experiments 1 and 2

	Gap Time		
Design Speed, km/h (mph)	.0625 s	.25 s	1.0 s
104.7 (65)	1.8 m (6.0 ft)	7.3 m (23.8 ft)	29.1 m (95.3 ft)
128.8 (80)	2.2 m (7.3 ft)	8.9 m (29.3 ft)	35.8 m (117.3 ft)
153.0 (95)	2.7 m (8.7 ft)	10.6 m (34.8 ft)	42.5 m (139.3 ft)

presence of other vehicles is noticeable and there is a slight decline in the freedom to maneuver. This variable is important to the design of the AHS to the extent that carryover effects are different as a function of traffic density.

• *Age.* The effect of this variable was determined by comparing the results across the two experiments. In Experiment 1, there were 36 subjects ages 25 to 34 years, half male and half female. In Experiment 2, there were 24 subjects ages ≥ 65 years. In each age subgroup, 65 to 69 years and ≥ 70 years, there were six males and six females. Age is important because of its potential influence on drivers' reactions to such things as design speed and intrastring gap, and because of the possibility that carryover effects are different for drivers of different ages. If the AHS is to be successful, it must accommodate drivers of all ages, so this experiment was designed to provide data on possible age-related effects.

The Driving Task and Dependent Variables. Prior to the first experimental trial, each driver had two 2- to 3-min practice trials: The first was on a rural roadway with no other traffic present, and the second was on a freeway in low-density traffic. At the start of an experimental trial, the driver was in the middle vehicle of a three-vehicle string, under automated control in the left lane. After 2 to 4 min under automated control, and 60 s before the driver would reach a selected exit if his or her vehicle continued at the design speed, the AHS issued an Exit advisory—a tone followed by an auditory, verbal message stating that the driver's exit was approaching. The driver's task was to take control of the vehicle, drive into the center lane, then move into the right lane and leave the freeway at the specified exit.

Dependent variables were as follows (see Fig. 5.2):

• *Response time:* the time between the issuing of the Exit advisory by the AHS and the time when the driver took control of the vehicle

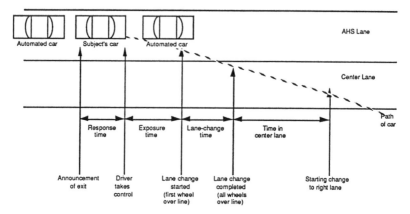

FIG. 5.2. Key moments and times from Experiments 1 and 2.

- *Exposure time:* the length of time the driver stayed in the automated lane after taking control of the vehicle, measured as the time between when the driver took control of the vehicle and when the first wheel touched the line between the left and center lanes on a completed lane change

- *Lane-change time:* the time it took the driver to drive from the left lane to the center lane, measured as the time between when the first wheel touched the line between the left and center lanes and when the fourth wheel crossed the same line

- *Exit speed:* the speed at which the driver's vehicle was traveling when it left the automated lane

- *Center-lane time:* the length of time the driver stayed in the center lane after completely leaving the automated lane, measured as the time between when the fourth wheel crossed the line between the left and center lanes and when the first wheel touched the line between the center and right lanes on a completed lane change

- *Delay time:* the time the vehicle immediately behind the driver's vehicle in the automated lane was delayed owing to the behavior of the driver's vehicle. For example, if the driver reduced speed, the vehicle behind the driver's vehicle was slowed down, as was the string behind the one in which the driver's vehicle was located, and so on. The delay time was measured as the time between when the driver took control of the vehicle and when the driver's vehicle had moved completely out of the left lane and the system had accelerated the vehicle that was immediately behind the driver's back to the design speed

- *Collisions and incursions* (incomplete changes) from the left lane to the center lane

TABLE 5.2
Experimental Plan for Experiments 1 and 2

Design Speed, km/h (mph)	Intrastring Gap (Seconds)	Unautomated Traffic Density	
		6.2 vehicles/km/lane (10 vehicles/mi/lane)	12.4 vehicles/km/lane (20 vehicles/mi/lane)
104.7 (65)	.0625	Group 1	Group 1
	.25	Group 1	Group 1
	1.0	Group 1	Group 1
128.8 (80)	.0625	Group 2	Group 2
	.25	Group 2	Group 2
	1.0	Group 2	Group 2
153.0 (95)	.0625	Group 3	Group 3
	.25	Group 3	Group 3
	1.0	Group 3	Group 3

The Experimental Conditions. A factorial experimental design was used. Each group of subjects drove in all six combinations of gap by density; one group was assigned to each design speed. The plan is shown in Table 5.2.

In both experiments, drivers were given questionnaires following the experimental sessions with questions about the realism of the simulator and the driver's views about some aspects of the AHS, as well as some of the details of the experiments.

Results. Analyses of variance (ANOVAs) were first run on each dependent variable (as appropriate) using the Experiment 1 data, the Experiment 2 data, and the combined data from the two experiments. If a statistically significant main effect or interaction was found, post hoc testing was done to investigate further. Key findings are presented as follows:

• The speed at which drivers left the automated lane—the exit speed—was slower at 104.7 km/h (65 mph) than at 128.8 or 153.0 km/h (80 or 95 mph):

—At 104.7 km/h (65 mph):   91.4 km/h (56.8 mph)
—At 128.8 km/h (80 mph):   104.4 km/h (64.9 mph)
—At 153.0 km/h (95 mph):   110.9 km/h (68.9 mph)

Given that the speed limit in the unautomated lane was 88.6 km/h (55 mph), the data show an apparent carryover effect of the automated travel on speed: At the two higher design speeds, the driver entered the unautomated lane at a speed about 16 to 22 km/h (10 to 14 mph) faster than the average speed in that lane. The fact that their exit speed at 104.7 km/h (65 mph) virtually matched the speed limit in the unautomated lane indicates that these drivers do not simply drive $\geq$ 16 km/h (10 mph) above the speed limit.

• The younger drivers had six incursions (incomplete lane changes) and one collision in 216 trials; the older drivers had eight incursions and two collisions in 144 trials. All three of the collisions occurred in the 120 trials at a design speed of 153.0 km/h (95 mph). Both the incursion rate (14 in 360 trials, or a 3.9% rate) and the collision rate (3 in 120 trials at the highest speed, or a 2.5% rate) are unacceptably high when projected onto a real-world freeway system. Thus, merely announcing to the driver after a period of automated travel that his or her exit is approaching and then giving the driver back control of the vehicle going at least 153.0 km/h (95 mph) is not workable, and some other method of transferring control from the AHS to the driver will have to be found.

Findings from the questionnaire included the following:

• Drivers would have preferred larger gaps between vehicles.
• Drivers preferred the automated lane at all speeds, with the strongest preference at 153.0 km/h (95 mph) and with the younger drivers showing a stronger preference than the older drivers.
• At 104.7 km/h (65 mph), drivers would have preferred going faster; at the two higher speeds, drivers showed no preference for higher or lower speeds.
• Drivers believed that an automated highway system will reduce the stress of driving.

Thus, it appears that the small gaps used in these experiments, at least when combined with relatively high speeds, were not acceptable to the drivers. However, the high speeds did not lead, on the average, to any discomfort.

Experiment 3: Human Factors Aspects of the Transfer of Control From the Driver to the Automated Highway System

The objective of this experiment was to determine the effect of several variables on the transfer of control from the driver to the AHS (Bloomfield, Buck, Christensen, & Yenamandra, 1994).

Independent Variables. The independent variables were these:

• *Design speed in the automated lane.* This was a within-subjects variable. Speeds were 104.7 km/h (65 mph), 128.8 km/h (80 mph), and 153.0 km/h (95 mph).

- *The gap between strings of vehicles in the automated lane.* This gap was based on the minimum gap time that would allow a vehicle entering that lane to accelerate to the design speed without affecting traffic flow in the automated lane:

  —At 104.7 km/h (65 mph): 2.0 s and 2.4 s. The actual minimum gap time was .4 s. Because that would not allow even sufficient time for the driver to make the lane change, 2.0 s was used instead. The higher value is 2.0 s longer than the minimum.

  —At 128.8 km/h (80 mph): 2.0 s and 4.0 s.

  —At 153.0 km/h (95 mph): 5.5 s and 7.5 s.

The combinations of design speeds and gap times produced the gap distances shown in Table 5.3. Gap was a within-subjects variable.

- *Method of transferring control from the driver to the AHS.* This was a between-subjects variable, and two methods were used: manual: When the driver's vehicle was fully in the automated lane, the driver pressed a button to tell the system to take control; and partially automated: When the driver's vehicle was fully in the automated lane, the system automatically took control. How control is transferred from the driver to the AHS is important because it may have an impact on the efficiency of traffic flow in both the unautomated and automated lanes. In addition, drivers must be comfortable with whatever method is finally adopted if the AHS is to be successful.

The Driving Task and Dependent Variables. At the start of a trial, the driver's vehicle was on a freeway entry ramp. The driver's task was to drive into the right lane of the freeway and then into the center lane. After about 15 s in the center lane, the driver was given a countdown followed by an Enter command, with the command coming just as the rear bumper of a string of vehicles in the automated lane cleared the front bumper of the driver's vehicle. Upon hearing the command, the driver was to drive immediately into the left (automated) lane, where control was transferred by one

TABLE 5.3
Interstring Gap Distances for Experiment 3

Design Speed, km/h (mph)	Interstring Gap Time	
	Shorter	*Longer*
104.7 (65)	2.0 s; 58.1 m (190.7 ft)	2.4 s; 69.7 m (228.8 ft)
128.8 (80)	2.0 s; 71.5 m (234.7 ft)	4.0 s; 143.1 m (469.3 ft)
153.0 (95)	5.5 s; 233.6 m (766.3 ft)	7.5 s; 318.5 m (1045.0 ft)

of the two methods. In either case, after the driver's vehicle was under automated control, the system moved it into the lead position of the string of vehicles approaching it from behind. The gap between vehicles within a string was .0625 s.

Dependent variables were as follows (see Figs. 5.3 to 5.6):

- *Entering response time:* the time between when the system issued an Enter command and when the driver started to move into the automated lane

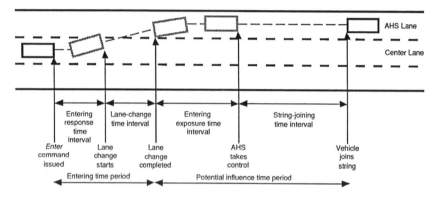

FIG. 5.3. Track of the entering vehicle during critical moments and time periods.

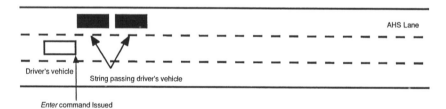

FIG. 5.4. Position of the oncoming string of automated vehicles when the *Enter* command was issued in Experiment 3.

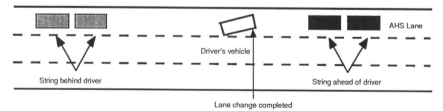

FIG. 5.5. Position of the oncoming string of automated vehicles when the lane change was completed in Experiment 3.

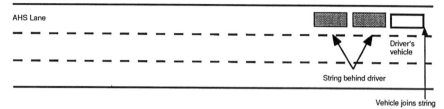

FIG. 5.6. Position of the oncoming string of automated vehicles when the driver's vehicle joined the string in Experiment 3.

- *Lane-change time:* the time it took the driver to drive from the center lane to the left lane, measured as the time between when the first wheel touched the line between the center and left lanes and when the fourth wheel crossed the same line
- *Entering exposure time:* the time between when the lane change was completed (i.e., when the fourth wheel crossed the line between the center and left lanes) and when the AHS took control of the vehicle
- *String-joining time:* the time from when the system took control of the driver's vehicle until it became the lead vehicle of the string just behind it
- *Delay time:* the amount of time it took a vehicle traveling at the design velocity to travel the delay distance, which was the difference between the distance actually traveled by the lead vehicle of the string immediately behind the driver's during the entering exposure time plus the string-joining time, and the distance that the lead vehicle would have traveled if the driver's vehicle had not entered the automated lane
- *Collisions*

The Experimental Conditions. A factorial experimental design was used. There were 24 subjects, all 25 to 34 years old; half were male, half were female. Each group of subjects drove in all six combinations of gap by design speed; one group was assigned to each method of transferring control. The plan is shown in Table 5.4.

All subjects were given a questionnaire following the experimental session with questions about the realism of the simulator, about the driver's views concerning some aspects of the AHS, and about some details of the experiments.

Results. Analyses of variance (ANOVAs) were first run on each dependent variable (as appropriate). If a statistically significant main effect or interaction was found, post hoc testing was done to investigate further. There were two key findings:

TABLE 5.4
Experimental Plan for Experiment #3

	104.7 km/h (65 mph)		128.8 km/h (80 mph)		153.0 km/h (95 mph)	
Method of control transfer	2.0-s gap	2.4-s gap	2.0-s gap	4.0-s gap	5.5-s gap	7.5-s gap
Manual	Group 1	Group 1	Group 1	Group 1	Group 1	Group 1
Partially automated	Group 2	Group 2	Group 2	Group 2	Group 2	Group 2

- The delay time was less for the partially automated method of transferring control to the AHS than for the manual method.
- There were no collisions.

Thus, a preliminary answer to the question regarding how control should be transferred from the driver to the system is that the AHS should take control automatically when the vehicle is fully in the automated lane.

The pattern of results in Experiment 3 for the questions asked in Experiments 1 and 2 was very similar to that from those earlier experiments:

- Drivers would have preferred larger gaps between strings of vehicles (not between vehicles within a string, as in Experiments 1 and 2).
- Drivers preferred the automated lane.
- Drivers believed that an AHS will reduce the stress of driving.
- Drivers would have preferred higher speeds. Note that there was a general preference for higher speeds, whereas in Experiments 1 and 2, there was a preference for higher speed only when the design speed was 104.7 km/h (65 mph).

Experiments 4 Through 7: Introduction

These experiments were run together as a multiple experiment, and four separate issues were investigated:

- A comparison of three methods of transferring control from the driver to the AHS
- The acceptability of decreasing separations between the driver's vehicle and a vehicle ahead during the transition of the vehicle ahead into the automated lane
- The effect on driving behavior of having to take control when the system is unable to perform all of its functions

- The effect of automated travel on normal driving behavior

Each of the 60 subjects participated in six trials, each of which had three sections of travel. For all the trials, unautomated traffic density on the freeway was 6.2 vehicles/km/lane (10 vehicles/mi/lane). An overall description of the trials follows.

- *Trial 1, section 1—preautomated driving for Experiment 7.* The driver started on a two-lane roadway with no other traffic present, then drove onto the freeway and continued in the center or right lane or both throughout the trial. The vehicle was always under manual control.
- *Trials 2 to 6, section 1—Experiment 4.* At the start of each trial, the vehicle was on the hard shoulder to the right of the freeway. The driver drove into the right lane and then to the center lane. From the center lane, the vehicle moved into the automated (left) lane using one of three methods of transferring control from the driver to the AHS. When in the automated lane, the vehicle was moved into the lead position of a string of vehicles approaching it from behind. When it became the lead vehicle, Experiment 4 ended.
- *Trials 2 to 5, section 2—Experiment 5.* This experiment began with the driver's vehicle under automated control and in the lead position of a string. A second vehicle then entered the automated lane ahead of the driver's vehicle, moving into the gap in the automated lane either early or late. As the entering vehicle accelerated to the design velocity in the automated lane, the driver's vehicle approached it from behind until the entering vehicle became the new lead vehicle of the string, with the driver's vehicle then second. Throughout these trials, the driver moved a lever to indicate comfort or discomfort with the situation. When the entering vehicle became the lead vehicle in the string, Experiment 5 ended.
- *Trials 2 to 5, section 3—Experiment 6.* This experiment began with the driver's vehicle second in a string. After some time, the driver received an advisory that the vehicle was approaching a freeway segment on which the AHS could not support either steering or speed control or both. There were two conditions for transferring control to the driver: when the driver was ready or when it simply was given up by the system. The driver performed the function not supported by the AHS for .8 km (.5 mi) and then reached a freeway segment in which full automation was again available. Transfer of control back to the system was done in one of two ways: The system simply took control, or the driver gave back control when he or she was ready. When the AHS was again fully in control of the vehicle, Experiment 6 ended.
- *Trial 6, section 2—postautomated driving for Experiment 7.* At the beginning of this trial, which marked the end of Experiment 4, the driver's vehicle,

under AHS control, was the lead vehicle in a string. After some period of time, the driver received an advisory that the vehicle was approaching a freeway segment on which the AHS could not support either steering or speed control or both. There were two conditions for transferring control to the driver: It was simply given up by the system, or the driver took control when he or she was ready. The driver was later informed that the system would not resume control of the vehicle, and was instructed to drive out of the automated lane. The driver then drove into the center lane and continued to drive for some period of time. Experiment 7 ended with the driver not in the automated lane and in control of the vehicle.

Table 5.5 shows where data were collected for each experiment. Each subject was given a questionnaire following the experimental session with questions about the realism of the simulator, about the driver's views concerning some aspects of the AHS, and about some details of the experiments.

Experiment 4. Human Factors Aspects of Transferring Control From the Driver to the Automated Highway System With Varying Degrees of Automation. The objective of this experiment was to investigate different methods of transferring control from the driver to the AHS (Bloomfield, Christensen, Peterson, Kjaer, & Gault, 1995).

Independent Variables. All of the following were analyzed as between-subjects variables:

• *Age.* There were 30 subjects ages 25 to 34 years, and 30 subjects ages ≥ 65 years.

TABLE 5.5
Where Data Were Collected for Each of the Experiments 4-7

Trial	Section 1	Section 2	Section 3
1	Experiment 7--preautomated travel baseline	Not applicable	Not applicable
2	Experiment 4	Experiment 5	Experiment 6
3	Experiment 4	Experiment 5	Experiment 6
4	Experiment 4	Experiment 5	Experiment 6
5	Experiment 4	Experiment 5	Experiment 6
6	Experiment 4	Experiment 7--postautomated travel	Not applicable

- *Gender.* Half of the drivers in each age group were female and half were male. In the older age group, to provide a balance of ages by gender, there were eight males and seven females ages 65 to 69 years, and eight females and seven males ages ≥ 70 years. Data from the two age subgroups in the older age group were analyzed together. As with age (noted earlier), gender may influence a driver's reaction to such things as speed and between-vehicles gap, and may lead to differences in carryover effects. Understanding gender differences, if there are any, will allow an AHS design that maximizes acceptance of the system by both women and men.

- *Design speed in the automated lane.* Speeds were 104.7 km/h (65 mph), 128.8 km/h (80 mph), and 153.0 km/h (95 mph).

- *The gap between strings of vehicles in the automated lane.* This gap was based on the minimum gap time that would allow a vehicle entering that lane to accelerate to the design speed without affecting traffic flow in the automated lane:

—At 104.7 km/h (65 mph): 2.0 s and 2.4 s. The actual minimum gap time was .4 s. Because that gap time would not even allow sufficient time for the driver to make the lane change, 2.0 s was used instead. The higher value is 2.0 s longer than the minimum.

—At 128.8 km/h (80 mph): 2.0 s and 4.0 s.

—At 153.0 km/h (95 mph): 5.5 s and 7.5 s.

The combinations of design speeds and gap times produced the gap distances shown in Table 5.6.

- *Method of transferring control to the AHS.* Three methods were used:

—Manual. The driver, situated in the center lane (adjacent to the automated lane on the left), requested entry into the automated lane, and the system issued a countdown followed by an Enter command. At the

TABLE 5.6
Interstring Gap Distances for Experiment 4

Design Speed, km/h (mph)	Interstring Gap Time	
	Shorter	*Longer*
104.7 (65)	2.0 s; 58.1 m (190.7 ft)	2.4 s; 69.7 m (228.8 ft)
128.8 (80)	2.0 s; 71.5 m (234.7 ft)	4.0 s; 143.1 m (469.3 ft)
153.0 (95)	5.5 s; 233.6 m (766.3 ft)	7.5 s; 318.5 m (1045.0 ft)

Enter command, the driver steered into the left (automated) lane, then pushed a button to relinquish control of the vehicle.

—Two-stage partially automated. The driver, situated in the center lane, requested entry into the automated lane. The system then took control of the vehicle's speed but not its steering, then issued a countdown followed by an Enter command. At the Enter command, the driver steered into the left lane. When the vehicle was completely in that lane, the system automatically took control of the steering also.

—Fully automated. The driver, situated in the center lane, requested entry into the automated lane. The system then took control of both the vehicle's speed and steering while in the center lane and drove it into the left lane.

Figures 5.7 to 5.9 show the driving situation for the different methods of transferring control to the system.

The Driving Task and Dependent Variables. This experiment began with the driver's vehicle on the right shoulder of the freeway. The driver drove into the right lane and then into the center lane. While in the center lane, the driver requested entry into the automated lane by activating a control in the vehicle. After the request, the driver stayed in the center lane and maintained a speed of about 88.6 km/h (55 mph); (if the driver violated either condition, a message was sent indicating same). Control was then transferred to the AHS by one of the three methods. When the driver's vehicle was in the left lane and fully under the control of the AHS, the system accelerated the vehicle to the design speed and made it the lead vehicle of the string approaching it from behind.

The dependent variables were the following:

- *Entering response time:* the length of time between when the Enter command was issued and when the first wheel of the driver's vehicle touched the line between the center and left lanes
- *Lane-change time:* the length of time between when the first wheel of the driver's vehicle touched the line between the center and left lanes and when the fourth wheel crossed the same line
- *Entering exposure time:* the length of time between when the fourth wheel of the driver's vehicle crossed the line between the center and left lanes and when control was completely transferred to the system
- *String-joining time:* the length of time between when control was completely transferred to the system and when the vehicle became the lead vehicle of the string immediately behind it
- *Possible delay time:* the time that the string immediately behind the driver was delayed due to the activities of the driver's vehicle
- *Collisions*

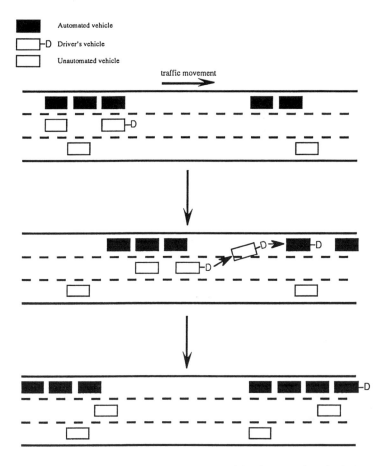

FIG. 5.7. The sequence of events in Experiment 4 for the manual method of control transfer.

The Experimental Conditions. The method of transferring control, gender, and age were between-subjects variables. Design speed and interstring gap were only partially within-subjects variables because there were six combinations of the two variables, but data were collected from each subject on only five trials. For analysis purposes, speed and interstring gap were treated as between-subjects variables.

Results. Analyses of variance (ANOVAs) were run on the dependent measures. If a statistically significant main effect or interaction was found, post hoc testing was done to investigate further. Key findings are presented as follows:

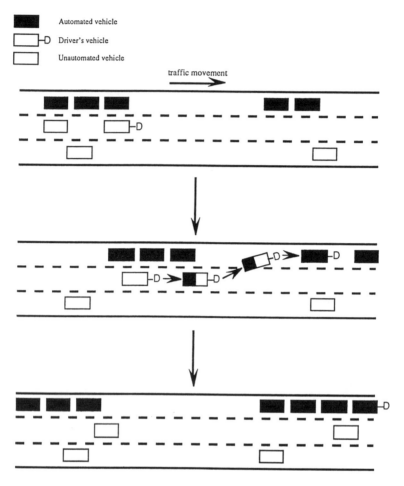

FIG. 5.8. The sequence of events in Experiment 4 for the two-stage partially automated method of control transfer.

- The time between the instruction to enter the automated lane and when the fourth wheel of the car crossed the line between the center and left (automated) lanes was much less for the fully automated method of transferring control to the AHS than for either the two-stage partially automated or manual methods.
- The time the vehicle just behind the driver's vehicle in the automated lane was delayed owing to the driver's vehicle entering that lane was longer for the manual method than for either of the other two methods.
- There were no collisions.

Thus, when vehicles must enter the automated lane from an adjacent unautomated lane, it appears that having the AHS take complete control of

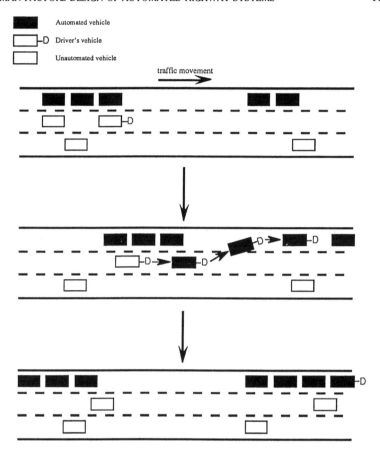

FIG. 5.9. The sequence of events in Experiment 4 for the fully automated method of control transfer.

the vehicle when it is still in the unautomated lane is the best method for transferring control from the driver to the system, because this will cause the least disruption of traffic in the automated lane. (It is difficult to specu-late on the likelihood of collisions on a real AHS with the various transfer methods, because that will depend on such things as communications and actuator reliability as well as the goodness and reliability of the control algorithms used, none of which were tested in these experiments.)

The questionnaire results again confirmed the findings of earlier experi-ments:

- Drivers would have preferred longer gaps between strings of vehicles.
- Drivers preferred the automated lane.
- Drivers believed the AHS will reduce the stress of driving.

- Drivers were relatively satisfied with the speeds at which they traveled when under automated control.

Experiment 5. The Driver's Response to Decreasing Vehicle Separations During Transitions Into the Automated Lane. The objective of this experiment was to determine the driver's comfort level when the driver's vehicle was the lead vehicle in a string and no vehicles were entering the automated lane ahead, and during the time when another vehicle was joining the driver's string from ahead and becoming the new lead vehicle (Bloomfield, Christensen, Carroll, & Watson, 1995).

Independent Variables. All of the following were analyzed as between-subjects variables:

- *Age.* There were 30 subjects ages 25 to 34 years, and 30 subjects ages ≥ 65 years.
- *Gender.* Half the drivers in each age group were female and half were male. In the younger age group, there were 15 males and 15 females. In the older age group, there were eight males and seven females between the ages of 65 and 69 years, and seven males and eight females ages ≥70 years. Data from the two age subgroups in the older age group were analyzed together.
- *Design speed in the automated lane.* Speeds were 104.7 km/h (65 mph), 128.8 km/h (80 mph), and 153.0 km/h (95 mph).
- *The gap between strings of vehicles in the automated lane.* This gap was based on the minimum gap time that would allow a vehicle entering that lane to accelerate to the design speed without affecting traffic flow in the automated lane:

  —At 104.7 km/h (65 mph): 2.0 s and 2.4 s. The actual minimum gap time was .4 s. Because that would not allow even sufficient time for the driver to make the lane change, 2.0 s was used instead. The higher value is 2.0 s longer than the minimum.
  —At 128.8 km/h (80 mph): 2.0 s and 4.0 s.
  —At 153.0 km/h (95 mph): 5.5 s and 7.5 s.

The combinations of design speeds and gap times produced the gap distances shown in Table 5.7.
- *Whether the vehicle entering the automated lane entered the interstring gap early or late.* When the vehicle entered early, it entered close to the last vehicle of the string ahead of it and far from the string in which the driver's vehicle was traveling, thus reducing the need for the system to lower the

TABLE 5.7
Interstring Gap Distances for Experiment 5

Design Speed, km/h (mph)	Interstring Gap Time	
	Shorter	*Longer*
104.7 (65)	2.0 s; 58.1 m (190.7 ft)	2.4 s; 69.7 m (228.8 ft)
128.8 (80)	2.0 s; 71.5 m (234.7 ft)	4.0 s; 143.1 m (469.3 ft)
153.0 (95)	5.5 s; 233.6 m (766.3 ft)	7.5 s; 318.5 m (1045.0 ft)

speed of the driver's vehicle. When it entered late, it entered closer to the driver's vehicle, thus causing a considerable slowdown in the driver's vehicle. The place in a gap where a vehicle enters the automated lane will have an impact on throughput in that lane; it also may have an impact on the driver's psychological comfort. This experiment was designed to determine the latter.

The Driving Task and Dependent Variables. This experiment began with the driver's vehicle under automated control and in the lead position of a string, at which position it continued for .5 to 4 min. During this time, the driver moved a lever mounted behind the shift stick to indicate comfort with the current situation: Pushing the lever forward indicated relative comfort; pulling it backward indicated relative discomfort. Absolute comfort and discomfort were defined as forward and backward, respectively, of the center (neutral) point of the lever.

After the .5 to 4 min, a second vehicle entered the automated lane ahead of the driver's vehicle, moving into the gap in the automated lane either early or late. As the entering vehicle accelerated from 88.6 km/h (55 mph) to the design speed in the automated lane, the driver's vehicle approached it from behind until the entering vehicle became the new lead vehicle of the string (i.e., until the distance between it and the previous lead vehicle of the string was equal to the intrastring gap), with the driver's vehicle then second. Throughout this time, the driver again moved the lever to indicate comfort or discomfort with the situation. When the entering vehicle became the lead vehicle in the string, Experiment 5 ended. Figure 5.10 shows the driving situation for this experiment.

The dependent variable was the driver's comfort level.

The Experimental Conditions. Two experimental designs, both factorial, were used in this experiment. For the analyses, all variables were treated as between-subjects variables.

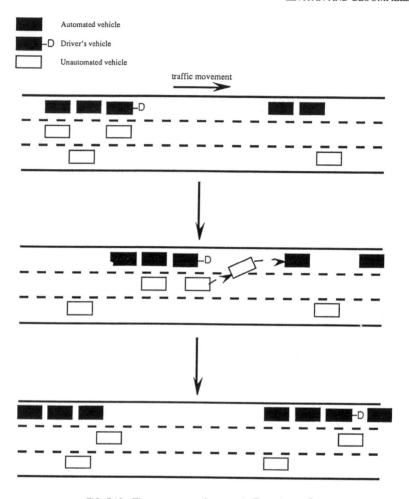

FIG. 5.10. The sequence of events in Experiment 5.

Results.   The comfort-level scale range was between +1.0, indicating maximum comfort, and −1.0, indicating maximum discomfort. Examination of comfort levels plotted against time revealed that before the second vehicle entered the automated lane (ahead of the driver's vehicle), the average was +.54. On 89.9% of the trials the comfort level was positive; on 3.2% it was neutral; and on 6.9% the comfort level was negative. Thus, across all design speeds by interstring gap combinations, drivers were relatively comfortable being under automated control.

Next, analyses of variance (ANOVAs) were run on the comfort-level data. If a statistically significant main effect or interaction was found, post hoc testing was done to investigate further. Key findings were these:

• Mean comfort level before the second vehicle entered the automated lane ahead of the driver's vehicle was higher for male drivers (+.63) than for female drivers (+.46) (see Fig. 5.11).

• As the second vehicle entered the automated lane ahead of the driver's vehicle, average comfort level dropped from +.54 to −.52 (see Fig. 5.11). Decreases were seen on 86.2% of the trials, whereas on the other 13.8%, the comfort level stayed approximately constant throughout the experiment. It should be noted that after the second vehicle entered the automated lane, the driver's comfort level became negative on 71.6% of the trials; on the other 28.4%, the comfort level decreased but remained positive.

• The data for trials on which comfort level decreased were examined to determine the low point of the initial decrease, and those low points were analyzed statistically. It was found that younger drivers had less discomfort

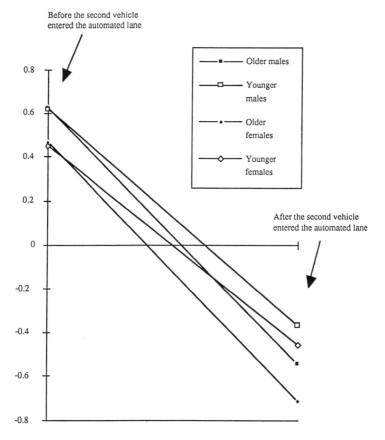

FIG. 5.11. Comfort level of drivers before and after a second vehicle entered the automated lane.

than older drivers, with average low points of −.41 and −.63, respectively; and male drivers had less discomfort than female drivers, with average low points of −.46 and −.58, respectively.

Thus, drivers were not comfortable when a second vehicle entered the automated lane ahead of them. Analyses indicated that a decrease in comfort level may have been triggered by a time to collision estimate, but although that mediator explains most of the relevant results, it does not explain them fully. Further investigation is needed to clarify this issue.

On the questionnaire, drivers indicated that they felt their uses of the lever accurately reflected their comfort levels.

Analytical Results

Regarding a possible role for the driver under failure conditions, computer simulations were used to observe dynamic vehicle behavior over time following control actuator failures. The simulations, which were run on a Sun workstation, modeled the steering mechanization of an AHS vehicle and simulated its motion. (For more detail about the simulation, the interested reader is referred to DeMers et al., 1994.) The dependent variable was the time from the onset of the failure until the vehicle diverged from its lane, and each failure was run under various speeds and roadway configurations, as follows:

- Speeds of 88.6 km/h (55 mph), 104.7 km/h (65 mph), 120.8 km/h (75 mph), and 153.0 km/h (95 mph)
- Straight or curved roadway, with standard highway curvature and superelevation
- Lane widths of 2.6 m (8.5 ft) and 3.7 m (12 ft)

Only single failures on single vehicles were considered. Three failures were simulated, and the results are as follows:

1. The vehicle is lane following (not lane changing) on a straight roadway when a "hard-over" steering failure occurs. Initial modeling showed that a hard-over steering failure at even the lowest speed tested (88.6 km/h [55 mph]) caused the vehicle to skid out of control. Thus, to constrain the situation to one in which there would be at least some possibility of recovery, steering failure was limited to one in which skidding was prevented. The results are shown in Table 5.8.

By comparison, the 95th percentile steering choice-reaction time to a visual stimulus for the alert driver fully in control of the vehicle is 1.2 s

TABLE 5.8
Time to Diverge From the Lane for Simulated Failure #1

| Speed, km/h (mph) | Lane Width | |
	2.6 m (8.5 ft)	3.7 m (12 ft)
88.6 (55)	.40 s	.60 s
104.7 (65)	.43	.64
120.8 (75)	.45	.67
153.0 (95)	.48	.70

(Lings, 1991). Clearly, it would be impossible for this "design" driver to play any reasonable role in dealing with this type of failure.

2. The vehicle is lane changing on a straight roadway when a hard-over steering failure occurs. When the failure occurs, the vehicle is midway between two adjacent lanes. Again, the situation was constrained to one in which skidding was prevented. The results are shown in Table 5.9. As with the first failure, the relatively long choice reaction time of the driver (1.2 s) would preclude any meaningful response on his or her part.

3. The vehicle is lane following on a curved roadway when a neutral steering failure occurs. The results are shown in Table 5.10. Comparison of the lane divergence times with the 95th percentile choice reaction time of 1.2 s offers the possibility of the driver being of some assistance in this type of failure situation. However, the driver would have to be alert and attentive, and it is best to assume that this will not be the case. Also, the authors of the original report point out that if the capability is provided for the driver to assume control in an emergency—based on data such as those for failure

TABLE 5.9
Time to Diverge From the Lane for Simulated Vehicle Failure #2[a]

| Speed, km/h (mph) | Lane Width | |
	2.6 m (8.5 ft)	3.7 m (12 ft)
88.6 (55)	.60 s	.75 s
104.7 (65)	.63	.78
120.8 (75)	.67	.82
153.0 (95)	.70	.85

[a]Shown are the times to diverge from the lane into which the vehicle is moving.

TABLE 5.10
Time to Diverge From the Lane for Simulated Vehicle Failure #3[a]

	Lane Width	
Speed, km/h (mph)	2.6 m (8.5 ft)	3.7 m (12 ft)
104.7 (65)	1.8 s	3.0 s
120.8 (75)	1.1	1.8
153.0 (95)	.6	1.0

[a]At the highway design speed of 88.6 km/h (55 mph), the vehicle would follow the roadway through the curve. Because there would be no lane deviation, data for that speed are not shown in the table.

#3—then less alert and less attentive drivers will have the same capability. Moreover, a driver in such a state could worsen the situation when there is a relatively minor nonneutral steering failure by making an exaggerated control input, thus unintentionally steering the vehicle into the adjacent lane or causing a skid and reducing braking effectiveness (Plocher & De-Mers, 1994).

There certainly may be situations in which the driver will be able to provide a meaningful response in an emergency (e.g., if braking is lost due to a communication failure rather than an actuator failure, it might be reasonable to transfer braking-control-only to the driver and have him or her stop the vehicle). However, it is important to remember the general assumption that the driver will not be required to be alert and attentive to the driving task (National Automated Highway System Consortium, 1995), and any strategy for dealing with emergencies—or failures in general—must take this into consideration.

Throughput Analyses

Combined experimental and analytical data were used to investigate the throughput in an automated lane that can be expected under certain conditions. Although this is not strictly a human factors issue, the analyses are based on driver performance (in Experiments 1 to 4). Using the delay time data, maximum traffic flows were determined under the following conditions:

• For Experiments 1 and 2, an analysis was done on the combined data. The gap between vehicles within a string was set to allow a vehicle to leave the automated lane without disrupting traffic flow in that lane, given that the vehicle enters a lane in which the speed limit is 88.6 km/h (55 mph). It was assumed that the vehicle leaving the automated lane was not the last vehicle in a string.

- For Experiments 3 and 4, the gap between strings of vehicles was set to allow a vehicle entering the automated lane to reach the design speed without disrupting traffic flow in the automated lane, given that the entering vehicle is traveling 88.6 km/h (55 mph) when it enters the automated lane.

Assumptions were also made about the number of vehicles in a string, the gap between vehicles within a string and between strings, and the average vehicle length. The results from all three analyses are shown in Table 5.11. By comparison, on an unautomated roadway with vehicles traveling at a conventional 2 s intervehicle spacing and 104.7 km/h (65 mph), and given the same average vehicle length as that assumed for the analyses, the expected traffic flow is 1672 vehicles/hr. Note that in the results from Experiments 1 and 2, based on a vehicle leaving the automated lane, the two higher speeds actually lead to a decrease in throughput with the AHS. In all cases, the lower the speed—or, more important, the smaller the differential in speed between the unautomated and automated lanes—the greater the improvement in throughput. In the best case, given a fully automated transfer of control from the driver to the AHS (Experiment 4) and a speed differential of only 16.1 km/h (10 mph), there is an increase in throughput of over 400%. Thus, under the assumptions made for these analyses and given the scenario used in these experiments, the smaller the speed differential between the unautomated and automated lanes, the better the expected throughput will be in the automated lane.

TABLE 5.11
Throughput Analyses

	Throughput in Automated Lane, Vehicles/hr		
Speed in Automated Lane, km/h (mph)	Experiments 1 and 2	Experiment 3	Experiment 4[a]
104.7 (65)	2088	7426	8528
128.8 (80)	1379	3546	4130
153.0 (95)	635	1816	2111

[a]The throughputs are higher in the Experiment 4 analyses than in the Experiment 3 analyses because in the former a steeper acceleration profile was used for the entering vehicle and a fully automated method of control transfer was used.

Summary of Findings on Human Factors Issues

Table 5.12 summarizes the findings regarding the human factors issues posed at the beginning of the chapter.

Future Research Issues

To say that we have just scratched the surface in addressing the human factors issues associated with the automated highway system would be a major understatement. Among the issues that still must be addressed, either experimentally, analytically, or using a combination of the two approaches, are the following:

- What effect does travel in an automated lane have on a driver's reaction to unexpected events (e.g., another vehicle cutting in sharply ahead of the driver's vehicle) when the driver is again in manual control of the vehicle?
- What are the effects on manual driving of an extended period of automated travel, say 2 hr?
- What test should be given to a driver to ensure that he or she is ready and able to resume control of the vehicle following automated travel? Will the same test be needed after a relatively short period of automated travel as after a relatively long period? What constitutes "relatively short" and "relatively long" travel?
- How can we ensure that the driver is alert when it is necessary (e.g., so that the driver test for readiness to resume manual control can be started)? What method(s) will work for a hearing-impaired driver, for a driver who is asleep, or for a driver working in the back of his or her minivan? Note that there are two possibilities here: One can try to maintain driver alertness throughout the period of automated travel (e.g., by having the driver look for infrequent events), or one can allow the driver to do as he or she wishes until his or her attention is needed, and then try to capture it by presenting some stimulus.
- What is the best method for merging a vehicle into high-speed (say 65 mph), high-density traffic—in terms of the gap between strings of vehicles or between individual vehicles where there are no strings, and in terms of the acceleration profile of the merge—so that it is acceptable to the driver? One needs to investigate this issue from the perspective of both the driver of the entering vehicle and of the driver of the vehicle immediately behind the entering vehicle.

TABLE 5.12
Summary of Findings on Human-Factors Issues

Issue	*Finding*
Will drivers accept small gaps?	• On average, drivers would have preferred larger gaps than the ones used in our experiments. The smallest gaps were 1.8 m (6.0 ft), 2.2 m (7.3 ft), and 2.7 m (8.7 ft) at 104.7 km/h (65 mph), 128.8 km/h (80 mph), and 153.0 km/h (95 mph), respectively. Gaps as small as 1 m (3.3 ft) at design speeds of 200 km/h (124 mph) have been proposed for the AHS.
Will drivers accept high speeds?	• On average, drivers were not uncomfortable at any speed used in our experiments, with the highest being 153.0 km/h (95 mph).
What will be the effect on drivers when another vehicle enters the automated lane ahead of them?	• On average, drivers became uncomfortable when another vehicle entered the lane ahead of them and the distance between the two vehicles closed. Time to collision may have triggered the decrease in comfort.
How should control be transferred from the driver to the AHS?	• If the vehicle must enter the automated lane from an adjacent unautomated lane, the AHS should take full control of the vehicle while it is still in the unautomated lane.
How should control be transferred from the AHS to the driver?	• Simply giving control back to the driver will not be acceptable, at least at the highest speed used in our experiments (153.0 km/h [95 mph]).
What effects will automated travel have on manual driving?	• If drivers are given control of their vehicles at relatively high speeds (128.8 and 153.0 km/h [80 and 95 mph] in our experiment) in the automated lane, there may be a carryover effect on speed--drivers may drive much faster than the posted speed limit (88.6 km/h [55 mph] in our experiment) when they enter the unautomated lane.
What role can the driver be expected to play when a failure occurs?	• With a hard-over steering failure, it will not be reasonable to expect the driver to provide any meaningful steering response.

- How is a driver's situation awareness during manual driving affected by automated travel, where situation awareness consists of taking into consideration the behavior of vehicles other than those immediately ahead and to either side of both the driver's vehicle, and of nearby nonvehicles?
- What effect does time of day have on manual driving following a period of automated travel? Of interest would be comparisons of dawn versus daylight versus dusk versus night and comparisons of different situations in which the automated travel begins in one time of day condition and ends in another (e.g., the trip starts in daylight, and the transfer of control from the AHS to the driver is at dusk).
- How does travel in an automated lane affect the manual driving behavior (e.g., gap acceptance in a lane change) of younger drivers (e.g., ages 16 to 20 years), of other drivers?
- What meaningful, safe role can a driver play under failure conditions other than a hard-over steering failure?
- For the psychological comfort of the driver, what are the allowable limits within which the gap between vehicles in a string must be maintained? One may need to look at this under conditions in which the vehicle accelerations are smooth and those in which they are jerky, depending on assumptions about the actual controls capabilities.
- What is the optimal speed differential between the automated and unautomated lanes, given that vehicles must travel between the two lanes, to maximize throughput on the AHS as a whole?
- What is the best way, considering both safety and efficiency, to drive several vehicles from a string of vehicles out of an automated lane at the same exit?
- Where the automated lane is not segregated from the other lanes, what is the safest and most efficient method for transferring a string of vehicles from the automated lane on one freeway to the automated lane on another freeway?
- For the psychological comfort of the driver, what lane widths are acceptable when there are barriers separating lanes? If there is to be manual traffic on one side of the barrier, then the issue needs to be investigated regarding both the driver under automated control and the driver under manual control.
- For the psychological comfort of the driver, in a nonsegregated highway scenario with no barriers between lanes, what lane widths are acceptable? The issue needs to be investigated regarding the driver under both automated and manual control.

ACKNOWLEDGMENTS

Work reported in this chapter was sponsored by the Federal Highway Administration under Contract No. DTFH61-92-C-00100. The authors wish to express their thanks for the support and guidance provided by the contracting officer's technical representative, Ms. Elizabeth Alicandri, of the Federal Highway Administration's Turner-Fairbank Highway Research Center.

REFERENCES

Bloomfield, J. R., Buck, J. R., Carroll, S. A., Booth, M. S., Romano, R. A., McGehee, D. V., & North, R. A. (1994). *Human factors aspects of the transfer of control from the Automated Highway System to the driver.* Revised Working Paper (Contract No. DTFH61-92-C-00100; FHWA-RD-94-114). McLean, VA: Turner-Fairbank Highway Research Center, Federal Highway Administration.

Bloomfield, J. R., Buck, J. R., Christensen, J. M., & Yenamandra, A. (1994). *Human factors aspects of the transfer of control from the driver to the Automated Highway System.* Revised Working Paper (Contract No. DTFH61-92-C-00100; FHWA-RD-94-173). McLean, VA: Turner-Fairbank Highway Research Center, Federal Highway Administration.

Bloomfield, J. R., Christensen, J. M., Carroll, S. A., & Watson, G. S. (1995). *The driver's response to decreasing vehicle separations during transitions into the automated lane.* Draft Working Paper (Contract No. DTFH61-92-C-00100; FHWA-RD-95-107). McLean, VA: Turner-Fairbank Highway Research Center, Federal Highway Administration.

Bloomfield, J. R., Christensen, J. M., Peterson, A. D., Kjaer, J. M., & Gault, A. (1995). *Human factors aspects of transferring control from the driver to the Automated Highway System with varying degrees of automation.* Revised Working Paper (Contract No. DTFH61-92-C-00100; FHWA-RD-95-108). McLean, VA: Turner-Fairbank Highway Research Center, Federal Highway Administration.

DeMers, R., Frazzini, R., Funk, H., Meisner, J., Plocher, T., Case, A., & Zhang, W. B. (1994). *Function, mechanization, reliability, and safety for AHS health management precursor system analysis.* Working Paper (Contract No. DTFH61-93-C-00197). McLean, VA: Office of Safety and Traffic Operations R&D, Federal Highway Administration.

Kuhl, J. G., Evans, D. F., Papelis, Y. E., Romano, R. A., & Watson, G. S. (1995). The Iowa Driving Simulator: An immersive environment for driving-related research and development. *IEEE Computer, 28,* 35–41.

Kuhl, J. G., & Papelis, Y. E. (1993). A real-time software architecture for an operator-in-the-loop simulator. *Proceedings of the Workshop on Parallel and Distributed Real-Time Systems* (pp. 117–126). Los Alamitos, CA: IEEE CS Press.

Lings, S. (1991). Assessing driving capability: A method for individual testing. *Applied Ergonomics, 22*(2), 75–84.

National Automated Highway System Consortium. (1995). *Automated Highway System (AHS). System objectives and characteristics.* Troy, MI: Author.

Plocher, T. A., & DeMers, R. E. (1994). *Human factors design of automated highway systems: Driver performance requirements.* Revised Working Paper (Contract No. DTFH61-92-C-00100). McLean, VA: Turner-Fairbank Highway Research Center, Federal Highway Administration.

Tsao, H. S. J., Hall, R. W., Shladover, S. E., Plocher, T. A., & Levitan, L. (1993). *Human factors design of automated highway systems: First generation scenarios.* Final Working Paper (Contract No. DTFH61-92-C-00100; FHWA/RD-93/123). McLean, VA: Turner-Fairbank Highway Research Center, Federal Highway Administration.

6

THE ADVANCED TRAFFIC MANAGEMENT CENTER

Michael J. Kelly
Dennis J. Folds
Georgia Institute of Technology

The half dozen or so men and women are seated at their low, futuristic-appearing consoles in the dimly illuminated control room. Heavy blinds on the windows are closed to block out the first rays of emerging daylight. The bright colors of the situation displays on the myriad of computer monitors are reflected in the faces of the computer operators. Images crawl slowly across dozens of closed-circuit television monitors. A huge projection television in the front of the room displays a familiar-appearing spiderweb of lines. Right now this web is mostly green, but if you watch closely, you can see line segments, one-by-one and minute-by-minute, changing to yellow. Two people are talking on their telephones about component failures and maintenance needs; another is keying data into the computer; a fourth is intently consulting a color-coded weather radar display; the rest are sitting back watching—relaxed but alert. It is the start of morning rush hour in the advanced traffic management center.

The intelligent transportation systems (ITS) era is introducing high technology to traffic management. Advanced traffic management systems (ATMSs) incorporate large numbers of highly capable traffic flow and roadway condition sensors to provide the basis for enhanced ATMS operator situation awareness. They incorporate data fusion and information-processing equipment to predigest the large volumes of newly available data into usable information. Some advanced systems may employ automation of routine decisions and actions to help control the operators' workload and standardize the responses seen by drivers. Finally, ATMSs provide new technologies for controlling traffic and for communicating information to vehicles and drivers.

The traffic management center (TMC) is the primary ganglion in this network of data sensing, information processing, communications, and control devices. From the TMC, human operators and central computers monitor the traffic status and carry out any measures necessary to promote smooth traffic flow. A complex, closed-loop information and control system is the result (see Fig. 6.1).

In the past, traffic control centers have been responsible for relatively low-technology functions such as traffic signal control and maintenance. Typically, the older centers have been relatively independent from other traffic-related services and have had relatively few communication and coordination requirements. Arguably, the greatest impetus toward ATMS evolution came from Los Angeles' high-technology response to the challenges of traffic management for the 1984 Olympic Games.

As the new technologies (including automated support systems) are introduced, the role of the operators is expanding. An older center might be staffed by one or two operators capable of controlling and troubleshooting traffic signals. In the ITS-era TMC, the operators might have to (a) control traffic signals and ramp meters in real time, using memorized algorithms; (b) remotely control a suite of dozens of closed circuit television (CCTV)

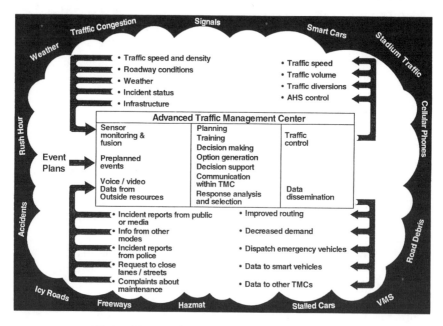

FIG. 6.1. The typical advanced traffic management system.

cameras and monitor their images; (c) identify roadway incidents and co-ordinate emergency responses; (d) interact with complex, computer-based support systems; and (e) transmit commands via various in-vehicle, road-side, and broadcast devices to control traffic and influence driver behavior.

TRAFFIC MANAGEMENT CENTER OBJECTIVES

Folds et al. (1993) conducted a series of structured interviews with TMC designers and operators who were nominated as "visionaries" by their peers in the transportation community. The goal was to develop a comprehensive description of the capabilities and operations of an ideal ITS-era TMC. The visionaries agreed that the overall mission of the TMC is "to facilitate the safe movement of people and goods, with minimal delay, throughout the roadway system."

To perform this mission, the TMC must pursue the following system objectives:

1. *Maximize the available capacity of area-wide roadway systems.* The first objective of the TMC is to create the maximum effective capacity for existing roadways. By distributing the traffic load spatially and temporally, the TMC seeks to minimize congestion and delay, thereby increasing effective through-put.

2. *Minimize the impact of roadway incidents (accidents, stalls, debris).* Roadway incidents, particularly during high demand times, have a significant impact on travel times and create a threat to public safety. Reducing the effects of these incidents may involve efforts on two fronts: reducing the likelihood of incidents occurring and minimizing the delays associated with incidents that do occur.

3. *Assist in the provision of emergency services.* Active involvement of the TMC may facilitate the provision of emergency services. Interaction with emergency service providers may include incident detection and verifica-tion, incident notification, coordination of responses where multiple services are needed, and modification of system parameters to improve speed or accuracy of emergency response. TMC support for emergency services is not limited to roadway incidents. It may be needed in any situation in which the roadway is used by emergency responders (e.g., creating ambulance paths by clearing highways between a disaster site and trauma hospitals).

4. *Contribute to the regulation of demand.* The presence of maintenance activities, special events, or major incidents may create conditions in which the overall demand exceeds the available capacity, no matter how efficiently the traffic flow is managed. Addressing this problem requires persuading

drivers to reschedule trips, reroute trips, or take alternative modes of transportation. Regulation of demand may be short or long term, reactive or proactive. It involves cooperative efforts with the media and with other transportation agencies.

5. *Create and maintain public confidence in the TMC.* To fulfill objectives 1 through 4, the TMC must be perceived by the public as reliable and competent. The TMC must be seen as providing accurate and timely information to the public. Public relations efforts may be conducted to reinforce positive public perception of the TMC.

FUNCTIONS OF THE TRAFFIC MANAGEMENT CENTER

In support of these objectives, ITS-class TMCs must meet all, or some subset, of the operational requirements listed for the data-sensing and input requirements, TMC performance requirements, and TMC output requirements listed in Table 6.1.

TABLE 6.1
TMC Operational Requirements

Data Requirements	*Decision-Making Requirements*
• Current traffic conditions	• Predicted traffic conditions
• Current roadway conditions	• Maintenance requirements and plans
• Incident location and severity	• Traffic management tactics
• Emergency vehicle status	• Incident management tactics
• Rail traffic conflict status	• Emergency response tactics
• Current/expected weather	• Demand regulation tactics
• Overall transportation demand	• Public confidence support tactics
• Indicators of public confidence	
Output Requirements	*TMC Support Requirements*
• Control vehicle motion	• Collect and store traffic data
• Control vehicle speeds	• Collect and store incident data
• Control vehicle access	• Collect road toll data
• Influence vehicle routes	• Train TMC staff
• Influence trip planning	• Plan future responses and activities
• Influence travel mode selection	• Provide administrative support
• Dispatch maintenance providers	
• Cause dispatch of emergency providers	

It should be noted that, at this level, all of these operational requirements are free of technology specification. Current traffic conditions, for example, may be sensed by the traditional inductive loop detector arrays, closed circuit television cameras (color or monotone), image processing systems, radar, ultrasound, or even by a human standing on an overpass with a handheld counter. Route selection may be influenced by broadcast radio, highway advisory radio, variable message signs, in-vehicle information displays, or other means. Some TMCs profess a design goal of complete automation; others are more satisfied with a labor-intensive design. The automation philosophy obviously has a major impact on the methods selected to meet the requirements. During TMC design, a detailed function allocation procedure should be used to determine how each requirement will be met by this specific TMC.

CURRENT AND ITS TECHNOLOGY

The organic assets of the TMC (i.e., assets directly owned or controlled by the TMC) will include sensor systems, data processing systems, and effector systems. In addition to its own sensor and effector systems, the TMC will obtain data from other sources (e.g., citizen telephone calls) and provide output through other channels (e.g., radio traffic reporters), thereby gaining access to assets of other organizations. Collectively, these assets will form the infrastructure of the TMC. Each type of asset is discussed in the following paragraphs.

Data Input

The ATMS sensors must provide data in real time. The sensor array should be self-sufficient in that all of the data necessary for managing traffic in real time should be available on demand. The TMC will not be dependent on outside sources (such as helicopter traffic reports) that may provide sporadic or inaccurate information, although it may use such information when it is available. The following are the major types of information sources that may be used by the TMC:

Roadway sensors are used to identify traffic congestion, including patterns of congestion that may suggest the presence of an accident or other incident. The sensors must be able to measure traffic speed and traffic volume. In key areas of the roadway system, the sensors must be capable of providing these data by lane and by vehicle size. The most common current technology consists of inductive wire loops embedded in the pavement. The loop detectors identify the passage of automobiles (or other large ferrous objects) by reporting variations in the electrical current within the loop. By using a pair of the loops a known distance apart, it is possible to compute the speed of each individual vehicle. Other traffic detection technologies such as radar, ultrasound, and visual image processing are being tested. As the technology of visual image processing advances, it is replacing the inductive loops in some applications.

Visual sensors, which usually consist of networks of closed-circuit television (CCTV) cameras and monitors, are used for confirming traffic conditions, verifying incident location, and determining incident severity. Typically, these cameras will not be intended as the primary source of information about traffic flow, but will be intended as secondary sources to support incident management and check the status of signals and signs. However, the designer's intentions notwithstanding, many operators will adopt the CCTV system as their primary information source (Kelly, Gerth, & Whaley, 1995.) At least one TMC contracts with individuals who live or work in areas with frequent accidents to phone in detailed reports of accidents they see.

Although older installations have generally used black-and-white CCTV because of its better performance under low illumination, color CCTV is thought by many designers to be more effective during daylight hours, and it is preferred by operators. Camera selection, pointing, and zooming is usually done by operators using joysticks and button boxes. More recently, automated aiming and zooming on sites of suspected incidents have become available, relieving the operator of this manual control function. Results of recent research (Folds et al., 1995) suggest that automated camera aiming is inefficient unless it is combined with manual control, allowing the operators to make fine adjustments in the cameras' slewing angle.

Specially instrumented probe vehicles will provide data about current traffic conditions. These "smart" vehicles will include participating commercial fleet vehicles, government vehicles, public transportation vehicles, and some private vehicles. Several demonstrations of this technology have been performed in the United States and in Europe. Typically, the location of each probe vehicle fleet member is determined by a global positioning system (GPS) receiver and continuously reported to the TMC. The amount of time required to move between any two points can be used to calculate average vehicle speed and to estimate the level of congestion. Accuracy of probe vehicle data is somewhat questionable because of variability in the GPS derived position accuracy (an expected error of tens of meters), difficulty in receiving GPS satellite signals in an "urban canyon" where signals may be blocked or reflected by buildings, and by any faulty assumptions in the tracking software. During a test drive for one probe vehicle demonstration, the authors were amused to find that their smart vehicle had reported heavy congestion to the TMC while it was being driven slowly around a sparsely populated block looking for a parking space.

Road condition and weather condition sensors monitor precipitation and road surface conditions to predict the effects of precipitation on driving conditions. These sensors will also monitor the friction coefficients of dry pavements and provide early detection of worn pavements that need resurfacing. Although not yet widely used, sensors now exist that can measure and transmit data on pavement temperature and moisture, air temperature, and salt content of pavement moisture. Some of these sensors will be maintained at fixed locations such as on bridges that are likely to freeze. Others may be mobile sensors, perhaps attached to maintenance vehicles.

System monitoring devices identify sensor and effector maintenance needs such as burned out bulbs in traffic signals and malfunctioning loop detectors. These technologies exist today, but emerging technologies will increase the reliability of malfunction reports. One TMC manager reported that complaints from citizens now provide the quickest and most reliable source of malfunction data.

Other data sources will be used to provide outside information to the TMC. To complement its own sensor assets, the TMC will obtain data from a number of external sources, some of which are described in the following paragraphs.

Voice communications with the traveling public will be maintained via telephone (including cellular phones and roadside emergency phones) and citizen band (CB) radios. These data sources will be used primarily to obtain reports of incidents and system malfunctions. There are great differences of opinion among TMC designers concerning the value and reliability of information from this source.

Voice communications with public safety authorities will be maintained over radio channels used by police, fire, and ambulance services as well as over dedicated telephones. These data will be used to obtain reports of incidents, to verify incident location and severity, and to help coordinate incident response.

Computer data links will be maintained with public safety agencies, commercial railroad companies, commercial vehicle fleet operators, and roadway maintenance providers. These will include dispatch databases of police, fire, and ambulance services. Any dispatch of a vehicle by these agencies will be entered into a shared database as part of the act of dispatching, and will be immediately available to the TMC. It can be expected that emergency vehicles will adopt GPS-based tracking systems so that dispatchers will know their locations at all times. These data may also be available in the TMC for use in incident responses.

National Weather Service data will provide weather forecasts and Doppler radar displays of current weather. Inclement weather is the most significant factor in traffic flow. The importance of Doppler radar, which can precisely iocate storm cells, has just begun to be recognized by traffic managers. A

Doppler radar map, superimposed over the roadway network map, will allow traffic managers to more effectively understand, predict, and manage weather-related congestion.

Output

Effector assets will provide the means for the TMC to have an impact on traffic conditions. The following will be the primary effectors owned and operated by the TMC:

Traffic signals will remain a primary means for traffic control in the next decade or two. Using control systems now available or under development, the TMC will have sophisticated control over the timing patterns for traffic signals. Timing patterns may be adjusted to optimize the local flow of traffic through an intersection controlled by a given signal, and groups of signals in an area can be coordinated manually or automatically to optimize flow in accordance with current traffic conditions.

Ramp metering in some form will become an increasingly important tool for controlling entry onto limited access roadways in the near future. By spacing the vehicles entering the freeway, TMCs can reduce congestion and accidents downstream from the on-ramps. With the incorporation of ITS technology, freeway ramp metering will be coordinated with nearby traffic signals on the arterial network to distribute the overall system load.

Variable message signs (VMS, also known as changeable messages signs, or CMS) will be used to distribute information relevant to the local traffic. They will present information on expected travel times and delays as well as incidents ahead, and in some situations will suggest alternate routes. A VMS at a major railroad crossing might provide an estimated time until clearance when traffic is stopped for a train. Various degrees of automation are being explored in the selection and posting of VMS messages.

Highway advisory radio (HAR) systems, consisting of networks of low-powered transmitters, each with a limited coverage region, will broadcast messages that supplement messages displayed on a VMS. This system will be available over the conventional AM/FM radios in most vehicles. It allows localized advisories to be received by traffic in the affected area. These recorded messages also may be heard over the telephone.

Other Output Channels. In addition to its organic outlets, the TMC will also send data to other organizations that, in turn, will transmit information to the public. The major output channels will be as follows:

Commercial radio and TV traffic advisories will continue to disseminate traffic information in the near future. The TMC will provide data for use in preparing these advisories. Commercial radio and TV will also be used in major advertising campaigns to advise the public of major construction or maintenance projects, and of traffic conditions associated with special events.

Cable TV traffic channels will relay TMC messages and data about current traffic conditions. They will also carry information from public transportation authorities. The traffic channels will be shown at information kiosks at airports, truck stops, hotels, and other places where travelers are making trip and route decisions, as well as in homes and offices.

Traffic bulletin board services will be accessible by computer over telephone dial-up. Several major cities, for example, already provide real-time traffic reports on the InterNet. This service typically will contain data on current traffic conditions throughout the metropolitan area. Traffic volumes and rates will be reported by route, and incident locations posted upon verification. Third-party vendors are expected to create software that will take this data and create tailored traffic advisories and route-planning assistance for users. Major users will include commercial vehicle operators, but many private individuals will also use the service. The service will be accessible from homes and dispatch centers (for use in trip planning), and also from smart vehicles equipped with intelligent systems that use the data for optimized navigation.

Print media, especially local newspapers, will be used to convey information about construction and maintenance projects that will impact traffic, traffic management for special events in the area, and general information on the operations of the TMC. The TMC will support preparation of print media features concerning such topics as defensive driving and the use of public transportation systems.

Information Processing and Decision Making. The most important functions of the TMC are to interpret the incoming information, recognize indications of current and future problems in traffic flow, select response tactics, carry out those tactics, and then reevaluate their effectiveness. In current TMCs, these functions that require higher levels of cognition are largely the responsibility of the human operators. Some systems have made limited attempts to automate some of these functions by modeling expert knowledge. Increases in automation to support the operators' decision making, as well as to make and implement decisions on routine tasks, has become a major technical goal of TMC designers.

There are a number of data-processing assets that will be needed to perform the TMC's decision making and information throughput. The following breakout is not intended to imply that each asset will be implemented in computer software, nor that separate software products will be created for each asset. Many of these data-processing requirements are now fulfilled by human operators according to their experience and the heuristics they have been taught.

Adaptive traffic control systems (ATCS) will take data from the roadway sensors and use it to control and coordinate traffic signal timing and ramp metering. The ATCS will maintain optimal flow by automatically responding to current and predicted variations in traffic. A key feature of the future ATCS will be the integration of control over surface street and freeway traffic. The algorithms used by ATCS will be robust, and will automatically accommodate unusual traffic patterns, such as those associated with special events.

Predictive traffic modeling systems (PTMS) will use data from the roadway sensors, origin–destination (O-D) data from vehicles currently in the roadway system, and data about incidents to predict traffic flow a few minutes into the future (say 5 to 30 minutes). This is done so that preemptive actions can be taken well before an actual problem arises. PTMS will predict problems in traffic flow by using a baseline of current traffic conditions and signal timing plans, taking into account O-D data, known incidents, and traffic advisories currently in effect. Systems similar to the PTMS are under development, but this function is normally now performed by the operator, based on accepted procedures and past experience.

The PTMS model will contain expected baseline levels of traffic, which will be time and day dependent and adjusted for weather conditions. Many of the specific components of this baseline will be provided by historical data recorded by the TMC. The PTMS will also predict the response of travelers to the implementation of traffic controls by the TMC. For instance, it will predict the percentage of people who will take an alternate route when it is suggested on a VMS, and use this prediction in its forecast of future traffic conditions.

Incident detection and location systems (IDLS) will receive data from roadway sensors and external data links. Its purpose will be to detect and verify the presence of incidents on the roadway system, and to determine the exact location of the incident. In some cases, the first indications of an incident will be abnormalities in traffic flow in the immediate area. In other cases, especially when traffic volumes are light, the first indications may come from other sources, such as calls over cellular phones that subsequently result in an entry into a dispatch database.

Incidents, as defined by the IDLS, will include not only traffic problems, but also system malfunctions such as traffic signal outages. Upon detection of a probable incident and determination of its apparent location, incident verification is attempted. The IDLS will provide a preliminary estimate of incident severity (perhaps simply a yellow or red icon) based on the data available. Existing automated incident-detection algorithms are based on speed and volume data from loop detectors. Performance of these algorithms has not yet proven adequate in terms of the percentage of incidents detected, the false alarm rate, and the time required to detect them. Recent research has found that the time lag between incident occurrence and its detection is the most crucial of these three interdependent measures and should be optimized by the algorithms (Folds, Mitta, Fain, Beers, & Stocks, 1995.)

Response advisory systems (RAS) will help determine the appropriate response to an incident. Responses could be doing nothing, informing motorists of a slight delay, rerouting some traffic and, in extreme cases, closing a major freeway and rerouting all traffic. In the case of rerouting, the RAS will determine the best alternate route(s). The RAS will also monitor the clearance of an incident and help determine when the operation should return to normal. For example, the RAS will determine when to clear an advisory message from a VMS. RAS capabilities, based on expert systems techniques, are in use to post VMS message patterns in several centers.

Information dissemination systems (IDS) will interface with the communication links to response organizations such as police, ambulance services, fire department, and towing services, and to the data services supplied to the mass media, the Traffic Channel, and the bulletin board service. The IDS will be capable of posting messages on VMSs and creating voice messages for broadcast on highway advisory radio.

HUMAN FACTORS ISSUES IN TMC EVOLUTION

During the coming years, dozens of traffic control centers and outdated TMCs will upgrade their computer suites and software systems, increase their levels of automation, and modernize other equipment and procedures. In some centers, this will be a revolutionary process in which the older technology will be discarded and replaced. In others, the ITS era will arrive one step at a time perhaps with the introduction of a VMS network now and a new support system in a couple of years. A third form of system evolution is development of an ITS-class traffic management system for a small geographic area initially and then a continuous extension of the boundaries of the controlled area and the functions of the operators.

Kelly, Gerth, and West (1994) and Kelly, Gerth, and Whaley (1995) reported visiting a number of the more progressive traffic control centers in North America and Europe. The purpose was to explore the design methods (including human factors contributions), the technologies and procedures used, the lessons learned during initial operations, and the expansion plans. During design of the centers, as well as their equipment and furnishings, typically little attention was given to human factors issues. Only one of the two dozen centers visited had been designed under a user-centered process involving existing operators. During that European center's design, a complete top-down function and task analysis was performed based on extensive interviews with operators. Center floorplan and workstation mockups were constructed and repeatedly refined using the operators' recommendations. Finally, custom computer display screens were developed to meet the operators' documented information needs.

Some other centers used less formal methods of integrating user requirements. One center, for example, polled its operators to estimate the number of CCTV monitors they could effectively manage, then based its design on those estimates. A small-scale mockup of a proposed control room was constructed, and operators were encouraged to experiment with different console and equipment layouts using this model.

Another center, highly dependent on VMS messages for incident and congestion management, became justifiably concerned that the semantic content of its messages be understood by the public. Focus groups were recruited from among local drivers to discuss their interpretations of various candidate messages. The message philosophy and suite of messages actually implemented for use was based on the results of those focus groups.

In general, however, human factors engineering issues had little impact on the TMC design process. Basic ergonomic and environmental factors such as illumination, noise, and air quality are well understood by architects and are seldom a major problem. More complex workstation, display, and job design issues related to the perceptual and cognitive aspects of human operator performance were rarely addressed.

Who Is on Duty in the TMC?

One crucial question for designers and operators of the newly designed TMC involves selection of operators. The qualifications of the TMC staff must be directly related to the design and function allocation philosophy. If the operators' functions are relatively repetitive, predictable, and noncritical, operators with a lower level of qualifications may be acceptable. If unique problems are frequent, rapid reaction is needed, or criticality is high, operators with higher levels of training and experience are indicated. One commonly used approach is to supplement operators of lower skill with more highly capable supervisory personnel who can assume control over complex problems.

The minimum qualification for a TMC operator would seem to be good verbal skills, a degree of computer literacy, and good reasoning skills (common sense). This, indeed, represents the standard of several centers here and abroad (at least one as a result of a formal job study). A few centers employ part-time students as operators, usually under the supervision of a manager or senior operator. Centers active in incident management generally require more highly qualified operators than do signal control centers or those that see their role as congestion management. At some centers, operators come from a technical staff of traffic engineers or computer scientists who have other assigned technical duties or who may be given special projects. One European center was totally staffed by police personnel; others had police liaisons on duty to handle interactions with officers in the field. A small number of centers require college engineering or technical degrees.

User-Centered Design

Designing a complex human–machine system like the TMC from a user-centered standpoint must go far beyond the more traditional human factors role of assessing individual display and control components and their arrangement in the workstation. Human factors inputs to the design process should have a significant impact on the high-level design philosophy. In particular, the rationale underlying the allocation of functions among operators and machines must be driven largely by human factors considerations.

Early approaches to function allocation from the human factors perspective divided functions between operator and machine according to which could best perform the function. The "humans are better at" and "machines are better at" (HABA-MABA) approach serves well for simple tools that do not employ extensive feedback loops between operator and system. The results of this approach are far too simplistic, though, for today's complex systems because it does not provide a way to share the allocation of functions performed jointly by humans and machines. New approaches to function allocation define the operators' type of interaction with the machine system. A long-term, evolutionary goal of some traffic management systems is total automation; the goal of others is to remain operator intensive. Each philosophy has significant design implications for the TMC.

As a crucial part of TMC configuration trade studies, the role of the human operator in the system must be defined. Each of the selected functions may be allocated to a human operator, an automated system, or some combination of the two. The function allocation becomes a basis for a detailed specification of what the TMC hardware and software systems are expected to do and what the human operator is expected to do in interacting with the system. A detailed analysis of the operators' tasks and activities can provide a major benefit both in center design and in the selection and training of the operators. Important job aspects identified during task analyses should be documented in a concise, accessible, well-indexed procedures manual. As the structure and architecture of the TMC are clarified, it becomes possible to define and design the operators' interfaces with the workstation items, with other staff members, and with outside agencies. When there are existing operators, their expertise is invaluable to help refine system design.

During replacement of a European traffic signal facility, a comparison was made between the old center and the new center that replaced it (Kelly et al., 1995). The old center had many desirable features, from the operators' perspective, that were missing in the new center. The job, for example, required close coordination between a radio operator and a traffic signal operator. In the old center, the two sat side-by-side and communicated to a large extent by gestures and body language. In the new center, by contrast, operators are assigned to consoles according to function, and these two operators are now separated by approximately 10 meters. In the old center, the two dozen CCTV monitors were arranged in two rows slightly below the operators' level field of gaze (the recommended position for monitors). In the new center, they are arranged on floor-to-ceiling racks.

Rapid Prototyping of Displays During TMC Design

As the large amounts of incoming data are fused into meaningful chunks, it is usually necessary to provide summaries in graphical format to support the operators' situational awareness. The use of maps, graphs, color coding, and blink-pattern coding can allow the operator to assimilate large amounts of data and provide a warning when something needs immediate attention. Designing effective displays for this information can be a challenge. Numerous rapid-prototyping software packages are now available by which replicas of candidate display screens can be programmed off-line and displayed for operator critique or performance testing. The screens can be easily refined until they are acceptable and then, with some of the packages, display software can be produced for rehosting to the real TMC. Such prototyping can eliminate display characteristics that might cause human performance difficulties before they become part of the TMC.

In some TMC displays, for example, 10 or more conditions had to be discriminated on the basis of color coding alone. One interface used a 12-color scale to indicate system status. The varied elements of the system had different numbers of descriptive levels. One subsystem had perhaps three levels and used the top three colors; another may have used all twelve available colors. As a consequence, the color that meant "worst possible" in the three-level subsystem was encoded as "pretty good" in the 12-level system. User testing of a screen prototype most likely would have discovered and corrected the confusion that this coding created.

User-centered TMC design, however, goes far beyond information display design. Issues may involve integrating displays across several monitors or defining the most efficient job descriptions and procedures for a group of operators. In such cases, prototype testing in a full-scale TMC simulator, using realistic traffic management scenarios, can provide important experience and information. Figure 6.2 shows such a simulator at the Georgia Institute of Technology that is used for conducting human factors studies to support TMC design (Folds, Kelly, & Mitta, 1994).

Automation of TMC Functions

An important aspect of user-centered design is that system automation is designed to assist and support the operator. The operator must always remain "in the loop" and fully aware of what the automation is doing, why it is doing it, what it is going to do next, and modes of intervention if the automated system fails. The operator must have, at a minimum, the ability

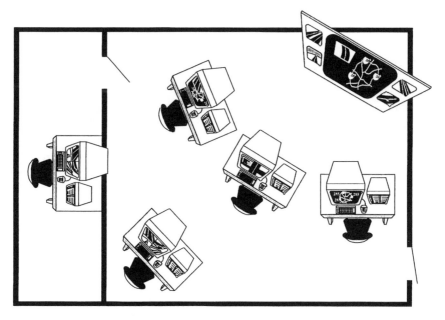

FIG. 6.2. Artist's concept of the ATMS human factors research simulator.

to switch off the automated system to prevent a future problem. At a highly automated facility, Kelly et al. (1995) interviewed two experienced operators about the automated systems and intervention modes. One operator maintained a high degree of situation awareness and demonstrated how he frequently intervened in automated decisions to fine-tune the ATMS responses. The second operator explained that it is not possible to override the automated system and that, even if it were possible, it should never be attempted.

As the next example shows, automation can increase the operators' workload and induce frustration. At one center using standard signal control software, all signal timing plans are operated on time-of-day schedules in which there are automatic changes in the amount of green and red time at 11:00 a.m. and 6:00 p.m. Operators can manually override the system and customize the timing plans for individual intersections. Beginning at about 10:30 a.m., the operator was observed to make extensive timing plan alterations in response to congestion from a major special event. At 11:00 a.m., without warning, the automated system wiped out the operator's work and inserted the preprogrammed mid-day timing plan. Later that same day, the operator's manually programmed timing plans (prepared for evening special events) were wiped out when the normal evening timing plans came on at 6:00 p.m. There was a software "switch" that could disable this automatic timing change, had the operator remembered to use it, but there should also have been a warning to the operator before the automated system took control.

VMS Automation Philosophies

VMS systems can help to support the drivers' situation awareness by providing information about the conditions ahead. Drivers may be influenced to alter their routes in response to delays or to decrease their speed and become more alert in situations of minor congestion and delays. TMC operators who control a network of these signs need to be continually aware of the message content to ensure that it is appropriate and up-to-date. There are frequent stories about temporary signs that were left on because the operators forgot to remove them. At least two approaches would ameliorate this problem. First, signs containing temporary messages should be indicated by special color coding (or flash pattern coding) on the map display of system resources. Second, an aging time might be applied to different messages, and the operator would be cued at the appropriate time to review the continuing need for the messages.

VMS operation is often the most highly automated aspect of TMC activities. Numerous automation philosophies are used, each assigning different functions and levels of responsibility to the operators, computers, and machines.

Manual Input. At a few centers, VMS signs are manually keyed into the system. In most such cases, the user interface presents a template on the computer screen that the message must fit. Also, there are usually strict limits on the vocabulary that may be used. One center has installed a working model of the actual VMS sign so that the manually produced signs can be refined, using variable letter spacing, before it is actually displayed. In one observed instance, the operator was unable to produce a suitable sign for the situation (police requesting a "stronger recommendation" that traffic exit at a given street) because of the vocabulary restrictions.

Manual Input With Computer Assent. In systems that have complex response plans involving VMS or lane-control signs, there are many rules concerning the content of signs. For example, in lane-control systems speed limits may not decrease by more than 20 km/h on consecutive signs or adjacent freeway lanes. Operators using manual control frequently request an inappropriate pattern of messages. Decision aids are reaching the market that reject these inappropriate patterns, explain why they are inappropriate, and suggest acceptable alternatives. Operators may accept the suggested pattern, try a new manual input, or at their own risk, insist on installing the rejected pattern.

Operators Select From Computer-Generated Menu. Some other centers provide the operators with a comprehensive menu of sign messages that may be displayed. The operator selects a sign from the menu, and inserts the command that it be displayed.

Computer Determines Response Plan and Operator Assents. In more highly automated centers, the system becomes aware of congestion or incidents through sensor data or operator input. Using internal response logic, an appropriate response plan is developed by the automated system and reviewed by a central computer. The response plan, including appropriate messages on a series of signs, is presented to the operator for approval or revision. Only after approval by the operator is the set of messages displayed. Human operators have been shown to be relatively poor monitors of automated response plans and tend to accept a high percentage of inappropriate messages suggested by the computer.

Computer Determines and Carries Out Response Plan. In the most highly automated centers, the computer becomes aware of traffic congestion from the traffic sensors. An appropriate response plan is developed, reviewed and implemented without interaction with operators. As the plan is being implemented, the operators are informed of the changes on a display of computer actions and may manually override the new plan if necessary. Such overrides very rarely occur.

Approaches to Response-Plan Automation

Expert Knowledge Base. Several of the newer centers employ a comprehensively preplanned set of responses (VMS message patterns) to potential incidents. At the COMPASS center on Highway 401 in Canada, for example, a team of traffic management experts divided the freeway through Toronto into hundreds of unique zones based on location, distance upstream or downstream from interchanges, roadway geometry, locations of VMSs, and other similar factors. Appropriate response plans were then developed for various incident scenarios within each unique response zone.

When no incidents are present, the VMS displays congestion information in response to loop detector data. If no congestion is present, other messages including navigation (e.g., name and distance of next exit) or public safety (e.g., drinking driver admonitions) are displayed. When COMPASS operators detect an incident, they call up an incident report screen on their workstation CRT and, according to prompts, key in data defining the location code and type of incident. The automated incident response system searches its file of incident templates until it finds one matching the operator's report. Predetermined sign patterns are displayed to the operator who consents (usually) or makes editorial changes (rarely). The automated response is limited to sign displays. The operators must coordinate incident reporting and incident response of other agencies by voice.

Automated Lane Control. Lane-control systems, such as the European Motorway Control and Signaling System (MCSS) in the Netherlands, employ a relatively dense system of sensors and changeable signs, as well as a broadly distributed computer network to control freeway traffic. Approximately every half kilometer there is an array of double loop detectors, a computer outstation, and a gantry over the road with independent advisory signs for each traffic lane. In hazardous areas, loop detectors may be packed as closely as every 50 meters to improve the efficiency of incident detection.

The loop detector arrays detect changes in traffic speed that signal the beginning of a congested situation. Data are sent to the outstation computer which, using specific algorithms, determines whether drivers should be alerted. The outstation computer automatically composes an appropriate pattern of messages advising drivers to slow to specific speeds (multiples of 20 km/h), to change lanes, or that the lane is closed. The outstation computer then requests permission from the central computer to post the instructions. The outstation computers upstream and downstream are then notified of the changes and will calculate and post on their signs any changes indicated by the initial changes in the adjacent sign. Various rules govern allowed message patterns. Adjacent lanes, for example, cannot have posted speeds that vary by more than 20 km/h. Also, speeds must be stepped down in 20 km/h increments, For example, a 60 km/h sign must be preceded by an 80 km/h sign which, in turn, must be preceded by a 100 km/h sign.

The system is capable of operating in a fully automated manner, although small staffs of human operators are on hand in the centers to (occasionally) override the computers, to input relevant data such as maintenance closures of lanes, and to take over in case of unplanned circumstances (e.g., a major airplane crash near the Amsterdam roadway in 1992). The facility incorporates both visual and auditory alarms to alert the operators when the computers detect a need for human intervention.

On initial installation in the early 1980s, the system operated at a lower level of automation. The lone operator was required to consent to every sign change. Sign changes were found to occur at a rate of about 200 per hour, generating an impossibly high workload. All signing decisions based on loop detector data are now made by the computer and displayed for the operators' information—after they have been implemented. The operators' primary function is to order changes in the signs (e.g., to close lanes for work zones). The computer, however, does not completely trust the operator and makes a check on the legality of every operator input. Occasionally, the computer and the operator will disagree on the appropriate signal pattern. In such cases of disagreement, the signs will display the more restrictive of the message sets. Operators maintain radio contact with roadway maintainers and with a fleet of "help trucks" that patrol the freeway looking for vehicles in distress.

Design for Situation Awareness

Maintaining awareness of the traffic situation on a complex roadway network provides a significant challenge to the TMC operator. With the large amount of information and communication technology available in the ITS-era TMC, it is possible to provide operators with everything they need to paint a detailed picture of the dynamic traffic situation. The perceptual and cognitive limitations of the human operators, however, allow them to attend to only a relatively small portion of this picture at a given time.

The data available to the operator can be aggregated at various levels of granularity to meet the operator's immediate needs. For example, the binary status of a single loop detector may be important during maintenance troubleshooting. The volume and speed of traffic passing that single loop detector may be of interest for incident detection functions. In contrast, the volume and speed of vehicles passing a series of loop detectors is of interest during most traffic monitoring. A signal from a support system that no indication of anomalous traffic flows are seen in the network may be sufficient information when the operator is busy with other duties.

For each different operator function, the ideal is to present all the necessary information—but no more. New graphical user interfaces to computer systems allow significant tailoring of data granularity, display content, and display techniques. Use of these tools can promote improved situation awareness in the TMC if the displays are designed to meet validated user information needs.

For example, in one TMC, multiple levels of "zoom" were desired for the main map display of the TMC area. Five levels, ranging from the entire geographic region down to the individual intersection, were implemented. Experienced operators reported that only two levels of zoom, a local region (perhaps a mile square) and the individual intersections, were actually used.

There may be significant differences between operators concerning the amount of detail they want to see on the screens. Inexperienced operators, for example, may want to have street names displayed, whereas experienced operators have these memorized. One form of dynamic screen redesign, a "clutter–declutter" capability, allows operators to select the amount of screen detail by toggling through, say, three predefined levels of detail.

Privacy Issues

The sensors and information-processing equipment available in the ITS-era TMC provide unprecedented opportunities for invading the privacy of individual drivers and other citizens. It is feasible, for example, using GPS satellite and cellular phone technology, to track an individual driver from trip origin to trip destination, including any stops for errands. It also would be possible to calculate each driver's speed along the route. All this information could be archived for later use.

A database of origins and destinations (O-D) for every driver would be very useful in predicting congestion (e.g., if many drivers were converging on the same destination). However, such a database could be misused by police to automate the issuance of speeding citations. It also could be used by divorce attorneys to provide evidence of a driver's visits to a paramour. For these reasons, several TMC managers expressed the opinion that they want nothing to do with collecting or storing such archived O-D data.

The TMC's CCTV cameras with their high magnification zoom capabilities and (future) application of night-vision technology provide another opportunity for privacy invasion. Several jurisdictions have already faced the dilemma of operators focusing their cameras on windows and swimming pool decks of roadside apartment buildings. Metal masks mounted around the cameras now limit the CCTV field of view. This solution serves both as a deterrent to peeking and as visible evidence to the public that their privacy has been restored.

A more serious dilemma occurs when privacy is invaded on the roadway itself. What, for example, is the proper TMC response when operators observe a crime taking place on the highway shoulder? Should the incident be taped? Should the police be called? Should a close-up view of the incident be taped including license numbers and faces of perpetrators and victims? Do the answers depend on whether there is a police presence on the TMC staff? These thorny questions are now being addressed by some of the pioneering TMCs.

CONCLUSIONS

Traffic on today's network of streets and highways has reached capacity in many urban areas, resulting in increased accidents and delays. The cost of this congestion, including accidents, injuries, and productivity loss because of delays has been estimated at $170 billion per year (Constantino, 1993). Advanced traffic management systems, using networks of state-of-the-art sensing, communication, and information-processing systems have been shown to increase roadway capacity and reduce the number and severity of accidents.

Of approximately two dozen progressive TMCs visited by the authors, only a small fraction employed formal human factors engineering practices in the design of systems and tasks. Only one center was largely designed by ergonomists, and the full user-centered design process was successfully employed. A few centers use local university ergonomics consultants, and many seek inputs from their existing operators. Usually, however, designs are simply evolutions of the architects' previous work, including previous solutions and previous mistakes.

Substantial amounts of automation are becoming possible in the TMC, and some systems are even capable of managing normal traffic flow without human intervention. Full-service TMCs, however, that coordinate responses to roadway incidents and manage traffic around roadway work zones, will continue to need human operators far into the future. ITS systems must be designed to support, not replace nor peripheralize, the human operator.

ATMS users (operators, maintainers, drivers) and their specific needs must be considered at the beginning of the design process for new and evolving TMCs and their support systems. Designers and vendors must realize that there is more to human factors engineering than providing the operators with a colorful graphic user interface. Numerous publications are present or forthcoming that will provide guidance in important human factors issues and ways of incorporating human factors engineering into the forefront of the design process.

HUMAN FACTORS GUIDELINES

Numerous published human factors guidelines are available or in the process of being written. These may be consulted for detailed human factors engineering recommendations for hardware, software, console, and control room design for the TMC. Some of these are provided in the following list:

ANSI/HFS 100-1988 is the standard for human factors design of video display terminals.

NUREG-0700 is the Nuclear Regulatory Commission guideline for control room design. A major revision of this older document was released in draft form for comment in 1995.

Guidelines for Designing User Interface Software was written by MITRE Corporation for the U. S. Air Force.

Human Factors Handbook for Advanced Traffic Management Center Design was prepared for the Federal Highway Administration by Georgia Tech Research Institute to provide detailed human factors guidelines and promote consideration of other TMC design issues.

REFERENCES

Constantino, J. (1993). Statement of Dr. James Constantino. *Technology policy: Surface transport infrastructure R&D: Hearings before the Subcommittee on Technology, Environment, and Aviation of the Committee on Science, Space and Technology, U.S. House of Representatives* (pp. 6–19). Washington, DC: U.S. Government Printing Office.

Folds, D. J., Beers, T. M., Stocks, D. R., Coon, V. E., Fain, W. B., & Mitta,D. A. (1995). A comparison of four interfaces for selecting and controlling remote cameras. *Proceedings of the Human Factors and Ergonomics Society 39th Annual Meeting.* San Diego: Human Factors and Ergonomics Society, pp. 1137–1141.

Folds, D. J., Brooks, J. L., Stocks, D. R., Fain, W. B., Courtney, T. K. & Blankenship, S. M. (1993). *Functional definition of an ideal traffic management system.* Atlanta: Georgia Tech Research Institute.

Folds, D. J., Kelly, M. J., & Mitta, D. A. (1994). Human factors experimentation in the IVHS TMC simulator. In ERTICO (Eds.), *Towards an intelligent transportation system: Proceedings of the First World Congress on Applications of Transport Telematics and Intelligent Vehicle-Highway Systems* (pp. 1709–1716). London: Artech House.

Folds, D. J., Mitta, D. A., Fain, W. B., Beers, T. M., & Stocks, D. R. (1995). *Toward guidelines for the design of incident detection support systems.* Atlanta: Georgia Tech Research Institute.

Kelly, M. J., Gerth, J. M., & West, P. D. (1994). *Comparable systems analysis: Evaluation of ten command centers as potential study sites* (Tech. Rep. FHWA RD 93 158). Washington, DC: Federal Highway Administration.

Kelly, M. J., Gerth, J. M., & Whaley, C. J. (1995). *Comparable systems analysis: Design and operation of advanced control centers* (Tech. Rep. FHWA-RD-94-147). Washington, DC: Federal Highway Administration.

7

MODELING DRIVER DECISION MAKING: A REVIEW OF METHODOLOGICAL ALTERNATIVES

Fred Mannering
University of Washington

Research in the general area of Intelligent Transportation Systems (ITS) has given rise to important questions regarding drivers' decision making, which has always been fundamental to understanding vehicle traffic flow and to developing models for predicting urban traffic congestion. However, because of the complexities involved in modeling and understanding drivers' decision making, standard urban transportation modeling has often focused on a more aggregate level: viewing traffic flow on highways as an observational unit to be studied. This has led to an entire body of literature that considers traffic flow to be roughly analogous to fluid flow. This literature applies principles of fluid flow, such as shock-wave analysis (May, 1990), to the modeling of traffic flow. Such an approach is an attempt to replicate the product of individual driver decision making and has been useful in many applications. However, in the presence of rapidly changing technology, such as that offered by ITS, the focus of research must be directed toward the primary decision-making unit—the driver—because the standard fluid-flow analogy is not likely to apply in an environment of vehicles containing possibly different levels of this technology in a single traffic stream.

To be sure, there has been a considerable amount of research on driver decision making and its ultimate impact on traffic flow. This research has been conducted in a variety of disciplines including human factors, transportation engineering, and accident analysis. However, models of driver decision making have yet to be integrated successfully into a more general model that is capable of predicting traffic flow in urban areas. Such a predictive model is clearly important to the successful implementation of many ITS technologies that seek to mitigate the impact of traffic congestion. In light of this, it is important that studies of driver decision making continue, but with the intent of leading us closer to more realistic and accurate predictive models of traffic flow.

The objective of this chapter is to review methodological alternatives for modeling driver decision making, and to give examples of how driver decisions particularly relevant to ITS can be approached. The methodological techniques chosen for presentation all have the capability of being integrated into a more general predictive model of urban traffic flow. Thus, the focus of this chapter is not just on the analysis of driver decision making, but also on the prediction of driver decision making in the presence of a changing environment of technology and information.

In reviewing methodological alternatives for modeling driver decision making, this chapter seeks to provide enough technical background to familiarize the reader with the basic concepts of the approach along with some of the more advanced concepts. Emphasis is placed on the appropriate application of methodological approaches, and examples of inappropriate applications are given to clarify the limitations of the methodology. In all cases, references are provided to allow the reader to further investigate the intricacies of the methodological approaches. The reader will note that many of the references are drawn from the econometric literature. It is the author's belief that the techniques used by economists to analyze behavior offer great potential for analyzing driver decision making.

The chapter's methodological review is organized according to the type of driver decision making being modeled. The first type considered is that which produces count data. This is followed by driver decision making that results in continuous-continuous, discrete, discrete-continuous, duration, and ordered data. After the presentation of methodological alternatives, estimation software is discussed.

COUNT DATA

Count data, which is data that assumes only non-negative integer values, is encountered frequently in the modeling of driver decision making. Traditional examples include the number of driver route changes and trip departure changes per week. In terms of future applications, count data will be generated in evaluating drivers' frequency-of-use of ITS technologies (e.g., the number of times per week that advanced guidance systems and traffic information are used to change routes, departure times, or both). Clearly, count data are commonly encountered in many aspects of driver decision making. In spite of this, surprisingly few researchers actually model count data with an appropriate statistical methodology. The most common mistake is to model count data as a continuous variable by applying standard least-squares regression methods. This is not strictly correct because regression models can give predicted values that are not integers and that can, in some instances, be negative. These limitations make standard regression analysis inappropriate for modeling count data.

Count data can be properly modeled by using a number of methods, the most popular of which are the Poisson and negative binomial regression models. Poisson regression is the more popular of the two, and it can be applied to a wide range of driver decision-making count data. To demonstrate the application of a Poisson regression, consider a driver's decision to use an in-vehicle navigation system. Assume that this navigation system provides the driver with real-time traffic information on selected routes, including expected route travel times. Further assume that we are interested in modeling the number of times this navigation system is used by the driver during a week. In a Poisson model, the probability of driver i using the navigation system n_i times per week (where n_i is a non-negative integer) is

$$P(n_i) = \frac{\exp(-\lambda_i)\lambda_i^{n_i}}{n_i!}, \tag{1}$$

where $P(n_i)$ is the probability of driver i using the navigation system n_i times per week, and λ_i is the Poisson parameter for driver i, which is equal to driver i's expected number of system uses per week (i.e., $E[n_i]$). Poisson regressions are fitted to data by specifying the Poisson parameter λ_i to be a function of explanatory variables which, in this case, could include driver age, income, marital status, traffic congestion faced during typical trips, prior experience with the ITS technology, and so on. This is done by specifying the Poisson parameter as

$$\ln\lambda_i = \beta\mathbf{X}_i, \tag{2}$$

where $\mathbf{X}_i$ is a vector of explanatory variables and β is a vector of estimable parameters. The Poisson regression specified in equations 1 and 2 is estimable by standard maximum likelihood methods (Greene, 1992) with the likelihood function,

$$L(\beta) = \prod_i \frac{\exp[-\exp(\beta\mathbf{X}_i)][\exp(\beta\mathbf{X}_i)]^{n_i}}{n_i!}. \tag{3}$$

Poisson regression is a powerful analysis tool, but it can be used inappropriately if its limitations are not fully understood. There are two common analysis errors. The first is failure to recognize the possibility of truncated data. For example, if we are modeling the number of times (per week) an in-vehicle navigation system is used on the morning commute to work, during

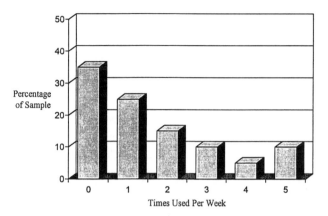

FIG. 7.1. Example of right-truncated count data: Number of times per week in-vehicle navigation systems are used on the commute to work.

weekdays, the data are right-truncated at 5, which is the maximum number of uses in any given week (see Fig. 7.1). Estimating a Poisson regression model without accounting for this truncation will result in biased estimates of the parameter vector β, and erroneous inferences will be drawn. Fortunately, the Poisson model can be adapted easily to account for such truncation. The right-truncated Poisson model is written as (see Johnson & Kotz, 1969)

$$P(n_i) = [\lambda_i^{n_i}/n_i!] / \left[\sum_{m_i=0}^{r} (\lambda_i^{m_i}/m_i!) \right], \tag{4}$$

where $P(n_i)$ is the probability of commuter i using the system n_i times per week; λ_i is the Poisson parameter for commuter i; m_i is the number of uses per week; and r is the right truncation (in this case, 5 times per week). A driver decision-making example of a right-truncated Poisson regression is provided by Mannering and Hamed (1990) in their study of weekly departure delays.

The second common analysis error relates to the property of the Poisson distribution that restricts the mean and variance of the distribution to being equal (i.e., $E[n_i] = \text{var}[n_i]$). If this equality does not hold, the data are said to be overdispersed, and the parameter vector will be biased if corrective measures are not taken. The solution to this is to add an error term, ζ, to equation 2 such that

$$\log \lambda_i = \beta X_i + \zeta_i, \tag{5}$$

where $\exp(\zeta_i)$ is gamma-distributed with a mean of one and a variance of γ. The addition of this term produces a negative binomial model that allows the mean to differ from the variance such that

$$\text{var}[n_i] = E[n_i][1 + \gamma E[n_i]] , \tag{6}$$

where γ is an additional estimable parameter. If γ is not significantly different from zero, then the data are not overdispersed, and the simple Poisson model is appropriate.

Applications of Poisson and negative binomial models to driver decision making have been comparatively rare (Mannering, 1989; Mannering & Hamed, 1990), whereas applications in other fields have been abundant. Sources that deal with important aspects of these models include Cameron and Trivedi (1986), Greene (1993), Hausman, Hall, and Griliches (1984), Lee (1986), and Maddala (1983).

CONTINUOUS-CONTINUOUS DATA (CONTINUOUS EQUATION SYSTEMS)

Continuous-continuous data refers to a system of interrelated equations in which both dependent variables (i.e., left-hand side variables) are continuous. An example in driver decision making would be the time spent using two technologies in a car (e.g., a map screen on the dashboard and a heads-up display [HUD]). Although it is tempting to model these as independent regressions, it is important to recognize their interdependence. That is, more time spent using HUD means less time spent using the dashboard map screen and vice versa. In equation form this can be written as

$$u_h = \beta_h X_h + u_d + \upsilon_h , \tag{7}$$

$$u_d = \beta_d X_d + u_h + \upsilon_d , \tag{8}$$

where u_h and u_d are the time spent using heads-up and dashboard displays, respectively; X is a vector of explanatory variables (e.g., driver's age, gender, technology preferences, etc.); βs are vectors of estimable parameters and υs are error terms. It can be shown (Pindyck & Rubinfeld, 1981) that estimation of equations 7 and 8 individually, using standard least-squares regression will result in biased estimates of the parameter vector β because a right-hand side "explanatory" variable (e.g., u_d in equation 7) changes when the left-hand side variable changes (e.g., u_h in equation 7). This violates a fundamental least-squares regression assumption that right-hand side variables are fixed in the presence of changing values of the left-hand side variables (Theil, 1971).

Statistically, there are a number of solutions to the problem of nonfixed right-hand side variables in regression modeling. These include single-equation methods such as indirect least squares (ILS), instrumental variables (IV), two-stage least squares (2SLS), and system-equation methods such as three-stage least squares (3SLS) and full information maximum likelihood (FIML). These methods are described in detail in numerous sources, including Greene (1993), Kmenta (1986), Pindyck and Rubinfeld (1981), and Theil (1971). Single-equation methods estimate equations individually, typically using standard least squares, but take corrective measures to account for the nonfixed right-hand side variable. The most common correction (instrumental variables) is to regress the nonfixed right-hand side variable against all available exogenous (fixed) variables and to use the regression-predicted values of the nonfixed variable as the right-hand side variable instead of the variable's actual value.

System approaches differ from the single-equation methods in that equations are estimated simultaneously and possible correlations between error terms are taken into account (i.e., v_d and v_h in equations 7 and 8). The result is that system-equation approaches generally produce more precise parameter estimates (i.e., smaller variance, higher t statistics).

Continuous-continuous data are often encountered in driver decision making. A common example is drivers' use (in miles per year) of individual vehicles when they have a choice of more than one (i.e., drivers in multivehicle households). This problem has been studied using three-stage least squares (Greene & Hu, 1984; Mannering, 1983; and Hensher, Smith, Milthorpe, & Barnard, 1992). Other examples include drivers' travel time to an activity (e.g., shopping) and the length of time the driver is involved in that activity (Hamed & Mannering, 1993).

Although continuous-continuous data are frequently encountered, they are often not recognized as such. This has led to the application of standard regression models when corrective single-equation or system-equation techniques are warranted. Because of the potential bias in parameter estimates, great care must be taken when applying standard regression methods. The researcher must actively consider the possibility that continuous-continuous data, and not simply continuous data, are present.

DISCRETE DATA

Discrete data can describe a binary choice (e.g., a yes or no answer) or the choice of a specific alternative. A driver decision-making example of a specific alternative choice could be the choice of ITS technologies or the choice of routes on a morning commute. The most common assumption in dealing with discrete data is that individual decision makers choose the alternative that provides the most utility (i.e., utility maximization from microeconomic theory). Thus, the probability of individual n choosing alternative i from the set of alternatives I is

$$P_n(i) = P(U_{in} \geq U_{In}) \supset I , \tag{9}$$

where P denotes probability and U_{in} is the utility provided by alternative i to individual n. To estimate this probability, the utility function must be specified. This is usually done in a linear form such that

$$U_{in} = \beta X_{in} + \varepsilon_{in} , \qquad (10)$$

where X_{in} is a vector of measurable characteristics that define utility (e.g., age, gender, price of alternative i, other characteristics of alternative i, etc.); β is a vector of estimable parameters; and ε_{in} is an error term that accounts for unobserved factors influencing an individual's utility. The term βX_{in} in this equation is said to be the observable portion of utility because the vector X_{in} contains measurable characteristic variables (e.g., age of individual n, price of alternative i, etc.), and ε_{in} is the unobserved portion. Before continuing, it is important to note that equation 10 is not really utility maximization in the classic microeconomic sense because prices and income can be arguments in the utility function and not part of the income constraint (i.e., utility is usually maximized subject to an income constraint). The economic term for equation 10 is "indirect utility," and it has many of the same interpretations as standard or "direct" utility. The reader is referred to McFadden (1981) for a complete discussion of this distinction.

Given equations 9 and 10, the following can be written as

$$P_n(i) = P(\beta X_{in} + \varepsilon_{in} \geq \beta X_{In} + \varepsilon_{In}) \supset I \qquad (11)$$

or,

$$P_n(i) = P(\beta X_{in} - \beta X_{In} \geq \varepsilon_{In} - \varepsilon_{in}) \supset I . \qquad (12)$$

With equation 11 an estimable discrete choice model can be derived by assuming a distributional form for the error term. A natural choice would assume that this error term is normally distributed. If this is done, a probit model results. However, because of computational difficulties, probit models are rarely used in situations when many choice alternatives are present (i.e., I is large in equation 9; Ben-Akiva & Lerman, 1985). A more common approach is to assume that ε_{in}s are generalized extreme value (GEV) distributed. The GEV assumption produces a closed form model that can be readily estimated by standard maximum likelihood methods. It can be shown (McFadden, 1981) that the GEV assumption results in the multinomial logit model,

$$P_n(i) = \exp[\beta X_{in}] / \sum_I \exp[\beta X_{In}] , \qquad (13)$$

where all variables are as previously defined, and the vector β is estimable by maximum likelihood (Ben-Akiva & Lerman, 1985; Train, 1986).

Logit models have been applied to countless discrete choice situations including the choice of transportation mode (Ben-Akiva & Lerman, 1985), choice of automobile type (Hensher, Smith, Milthorpe, & Barnard, 1992; Mannering & Winston, 1985; Train, 1986), and drivers' choice of commuting route (Mannering, Abu-Eisheh, & Arnadottir, 1990). Moreover, logit models have some very desirable properties. Perhaps one of the most useful of these properties is the fact that the denominator of a logit model (e.g., equation 13) gives the expected maximum indirect utility from the choice process being modeled. This has important implications because any change in the characteristic vector $\mathbf{X}$ will either result in an increase or decrease in total expected maximum utility depending on the direction of the characteristic change and whether the corresponding parameter estimate in the vector β is positive or negative. This property can form the basis of evaluating changes in total benefits to decision makers resulting from changes in alternative characteristics as shown by Small and Rosen (1981). Example applications of this property include estimates of benefit changes to commuters resulting from the imposition of high-occupancy vehicle lanes and to automobile purchasers resulting from mandatory automobile safety regulations (Winston & Mannering, 1984).

Although logit models have enjoyed great popularity, some caution must be exercised in their application. The most restrictive property of the multinomial logit model is independence of irrelevant alternatives (IIA). This property arises from the model's derivation in which independence of error terms (ε_{in}s), among choice alternatives, is assumed. The IIA property restricts changes in probabilities resulting from the change in the characteristics (i.e., $\mathbf{X}$ vector) of one alternative to be proportional among all other alternatives. To illustrate this, suppose a driver is deciding to purchase one of three ITS technologies: two are dashboard visual displays, and one is an audio information system. Assume all provide the same information to the driver, but in different forms. A multinomial logit model of these three choices will predict that a price reduction in one of the visual display alternatives will decrease the probabilities of the driver selecting the other visual display and the audio system, equally. A more realistic outcome would have the decrease in the other visual system's probability to be greater than the decrease in the audio system's probability because the two visual display alternatives are more direct substitutes. A formal way of looking at this is that the error terms of the two visual displays are likely to be correlated (which violates the error-term independence assumption made to derive the multinomial logit model) because common unobserved effects may be found in both error terms (e.g., an unobserved general preference among some drivers for visual information).

Formal statistical tests for possible IIA problems are discussed in numerous sources including Small and Hsiao (1985). Fortunately, the GEV error-term assumption allows any IIA problem to be readily addressed in a *nested* logit structure. The idea is to group alternatives with correlated error terms into a nest by estimating a model that includes only individuals choosing the nested alternatives. Because choice probabilities are determined by differences in utilities (both observed and unobserved, see equation 11), shared unobserved effects will cancel out in each nest providing that all alternatives in the nest share unobserved effects. This canceling out will not occur if a nest (group of alternatives) contains two alternatives that share unobserved effects and a third that does not (as was the case in our ITS example). In equation form this nested logit structure is

$$P_n(i_v \mid V) = \exp[U_{ni_v}] / \sum_{l_v} \exp[U_{nl_v}] , \qquad (14)$$

where $P_n(i_v \mid V)$ is the probability of individual, n, selecting alternative i_v from nest V given that an alternative from nest V is chosen and that U_{niV} is the observable portion of utility (i.e., βX). After this model is estimated using maximum likelihood procedures identical to those used to estimate equation 13 (but using only those individuals observed making a choice from the alternative set V), the denominator of this resulting model is used to calculate a *logsum* (also called an *inclusive value*) for all individuals (i.e., even those that were not observed to chose from the alternative set V). This variable is computed by taking the natural logarithm of the denominator of equation 14,

$$h_{nl_v} = \ln\left[\sum_{l_v} \exp[U_{nl_v}] \right]. \qquad (15)$$

This term is then used in an unconditional model that considers the choice among other alternatives and the alternative nest V. This is written as

$$P_n(i) = \exp[U_{ni}] / \left[\left(\sum_{j \neq l_v} [\exp[U_{ni}]] \right) + (\exp[U_{nl_v} + \alpha h_{nl_v}]) \right] \qquad (16)$$

where α is an estimable parameter (i.e., as the parameters in the β vector), and other terms are as previously defined. A visual comparison of a standard multinomial logit model and this nested logit model as applied to our ITS example of the choice among visual and audio technologies is provided in Fig. 7.2.

The resulting estimate of α has important implications. If α is not significantly different from one, the utility function error terms are independent and the nested logit structure is not warranted (i.e., a simple multinomial logit model is appropriate). If α is greater than one or less than zero, the

Example of a Standard Multinomial Logit Model:

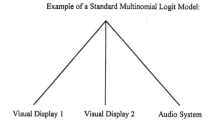

Visual Display 1   Visual Display 2   Audio System

Example of Nested Logit Model:

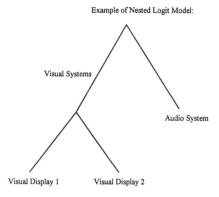

Visual Systems

Audio System

Visual Display 1   Visual Display 2

FIG. 7.2. Examples of standard multinomial logit and nested logit models.

model is inconsistent with utility maximization and must be reformulated (McFadden, 1981).

State Dependence and Heterogeneity in Discrete Choice Models

A potential estimation problem can arise in discrete choice models if information on previous choices is used to explain current choice probabilities. To demonstrate this, consider a driver choosing among three routes on a morning commute to work. On day 1, the driver is observed taking route 2. In modeling the driver's choice on day 2, it is tempting to use information observed from the previous day (i.e., that route 2 was taken) as an independent variable (in the X vector). Behaviorally, this may make sense because it could be capturing important habitual behavior. Such habitual behavior is called *state dependence* (Heckman, 1981). Unfortunately, the inclusion of such a state variable may also be picking up residual heterogeneity, which would lead one to observe spurious state dependence. To see this, suppose that the error term includes unobserved characteristics (i.e., unobserved heterogeneity) that are commonly present among drivers selecting specific routes. If a variable indicating previous route selection is included, it is unclear whether the estimated coefficient of this variable is capturing true habitual behavior (state dependence) or is simply picking up some mean commonalty in drivers' error terms (unobserved heterogeneity). This is an important distinction because the presence or absence of habitual behavior could lead one to draw significantly different behavioral conclusions.

Isolating true state dependence from unobserved heterogeneity is not an easy task. However, a number of statistical tests can be conducted as discussed in Mannering and Winston (1991). Other reference sources for this problem include Griliches (1967) and Heckman (1981).

DISCRETE-CONTINUOUS DATA

As was the case with continuous-continuous data, it is possible to have interrelated variables that are *discrete-continuous*. An example would be drivers' choice of ITS technology (e.g., visual dashboard display or audio system) and the extent to which it is used (e.g., in minutes per week). To see why this must be considered as part of an interrelated equation system, consider an ITS technology usage equation,

$$u_{kn} = \beta_k X_{kn} + v_{kn} \, , \qquad (17)$$

where u_{kn} is the minutes of use of technology k by driver n over some time period; X_{kn} is a vector of driver and technology characteristics; β_k is a vector of estimable parameters; and v_{kn} is an error term. In estimating this equation, information on technology use is only available from those people actually observed using the technology. No information is available for drivers that have not yet purchased the technology or have purchased another competing technology. Because the people observed using a particular technology are not likely to be a random sample of the population, a self-selected bias will result. This is illustrated in Fig. 7.3.

In this figure the usage data for people observed using ITS technology k is given by the '+' values. The '−' indicates the usage of technology k by nontechnology k users had they been observed using technology k. It is clear from line 1 that estimation of equation 17 with only observed '+' values will produce biased βs when compared to the true unbiased line (line 2). This is referred to as a selectivity bias problem (Mannering, 1986).

Statistically, this selectivity problem is corrected by realizing that the problem arises because of a correlation between discrete and continuous error terms. Thus, if the indirect utility from the choice of technology k is

$$U_{kn} = \phi_k Z_{kn} + \varepsilon_{kn} \, , \qquad (18)$$

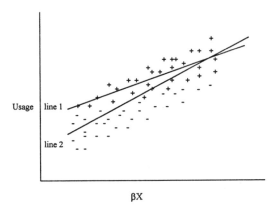

FIG. 7.3. An illustration of self-selected bias in discrete-continuous data.

where ϕ_k is a vector of estimable parameters and $\mathbf{Z}_{kn}$ is a vector of individual and technology characteristics, then $E(\varepsilon_{kn}v_{kn}) \neq 0$. Thus unobserved factors that tend to increase usage may also tend to increase (or decrease) the likelihood of a specific alternative being selected.

To correct the selectivity bias problem, note that equation 17 can be written as

$$E(u_{kn}\,|\,k) = \beta_k\mathbf{X}_{kn} + E(v_{kn}\,|\,k) \, , \tag{19}$$

where $E(u_{kn}\,|\,k)$ is the use of the technology conditioned on the choice of technology k, $E(v_{kn}\,|\,k)$ is the conditional error term, and other terms are as previously defined. This equation will produce estimates of β that are corrected for possible selectivity bias because the conditional expectation of v_{kn} is accounted for. All that is needed is a closed form solution for $E(v_{kn}\,|\,k)$. Such a solution has been derived by Hay (1980) and Dubin and McFadden (1984) by their assuming that ε_{kn} is GEV-distributed (i.e., a logit discrete choice model) and is

$$E(v_{kn}\,|\,k) = (-1)^{k+1}(\sigma 6\rho_k)/\pi^2 \left[(1/k)\sum_{j \neq k}^{K} [(P_j \ln P_j)/(1 - P_k + \ln P_k)] \right] \tag{20}$$

where σ^2 is the unconditional variance of v; ρ_k is the correlation between v and the resulting logistic error term (i.e., the subtraction of GEV error terms); K is the total number of discrete alternatives; P_j is the probability of selecting alternative j; and P_k is the probability of selecting alternative k. Although this equation looks cumbersome, its use to correct selectivity bias is straightforward. The estimation procedure is as follows:

1. Estimate a multinomial logit model to predict the probabilities of individuals selecting alternatives, k.
2. Use the logit-predicted alternative probabilities to compute the portion of equation 20 in large brackets (i.e.,[.]) for each individual.
3. Use the values computed in step 2 to estimate equation 19 using standard least-squares regression methods. The term $\sigma 6 \rho_k / \pi^2$ becomes a single estimable parameter.

The above discrete-continuous correction is referred to as the bias correction approach. Other corrective measures include instrumental variables, expected value, and full information. The reader is referred to Mannering and Hensher (1987) for a review of alternative correction methods and model structures.

Discrete-continuos problems are frequently encountered in driver decision making. Previously explored examples include drivers' choice of vehicle and the extent of its use (Hensher & Milthorpe, 1987; Mannering & Winston, 1985; Train, 1986), drivers' choice of route and travel speed (Mannering, Abu-Eisheh, & Arnadottir, 1990), and drivers' choice of activity (e.g., shopping, social, etc.) and activity duration (Damm & Lerman, 1981; Hamed & Mannering, 1993). It is important that researchers studying driver decision making carefully consider the possibility of selectivity bias and the implied discrete-continuous model structure.

DURATION DATA

Duration data generally refers to data measuring the time until the occurrence of an event. Driver decision-making examples of such data include the amount of traffic congestion delay needed to induce a change in drivers' route or departure time choice, the length of time between trips, the length of time drivers spend in specific activities, and the length of time until drivers' adoption or purchase of an ITS technology. Duration data are continuous and can, in most cases, be modeled using standard least-squares regression. However, the use of models based on hazard functions can often provide more insights into the underlying duration problem (Hensher & Mannering, 1994; Kiefer, 1988).

For years, hazard-based duration models have been extensively applied in biostatistics (Fleming & Harrington, 1990; Kalbfleisch & Prentice, 1980) and economics (Kiefer, 1988). However, their application to driver decision making and transportation problems in general has been limited, although the popularity of the approach is increasing.

The concept of hazard-based duration models is to focus on the conditional probability of a duration ending at some time t, given that the duration has continued (or survived) until time t. For example, suppose an in-vehicle navigation system is made available at time t. Hazard-based duration models account for the possibility that the probability of an individual driver buying the system at time t, t + 1 and t + 2 could be different. As the time since the system's introduction passes, those drivers that have not yet bought the system could have purchase probabilities that are either increasing, decreasing, or remaining constant over time. This change in probabilities over time is ideally suited to hazard–function analyses.

Mathematically, the hazard function can be expressed in terms of a cumulative distribution function, $F(t)$, and a corresponding density function, $f(t)$. The cumulative distribution is written as

$$F(t) = P[T < t] , \tag{21}$$

where P denotes the probability, T is a random time variable, and t is some specified time. In the case of the time until purchase of a new ITS navigation system, equation 21 gives the probability of trying the new system before some transpired time, t.

The corresponding density function (the first derivative of the cumulative distribution with respect to time) is

$$f(t) = dF(t)/dt , \tag{22}$$

and the hazard function is

$$h(t) = f(t)/[1 - F(t)] , \tag{23}$$

where $h(t)$ is the conditional probability that an event will occur between time t and t + dt, given that the event has not occurred up to time t. As stated earlier, the hazard, $h(t)$ gives the rate at which events (such as trying a new in-vehicle navigation system) are occurring at time t, given that the event has not occurred up to time t.

Another important construct in hazard-based models is the survivor function. The survivor function gives the probability that a duration will be greater than or equal to some specified time, t. The survivor function is written as

$$S(t) = P[T \geq t] , \tag{24}$$

and therefore is related to the cumulative distribution function by

$$S(t) = 1 - F(t) , \tag{25}$$

and to the hazard function by

$$h(t) = f(t)/S(t) . \tag{26}$$

Graphically, hazard, density, cumulative distribution, and survivor functions are illustrated in Fig 7.4. This figure provides a visual perspective of the preceding equations.

Turning specifically to the hazard function, its slope has important implications because it reflects the possibility that the probability of ending a duration may be dependent on the length of the duration. That is, the likelihood of purchasing an ITS technology is a function of the length of time that has transpired since the system's introduction in which the driver did not purchase the system. This is referred to as duration dependence, and the first derivative of the hazard function with respect to time (i.e., the slope of the hazard function) provides this information.

To illustrate this, consider the four hazard functions shown in Fig 7.5. In this figure, the first hazard function, $h_1(t)$, has $dh_1(t)/dt < 0$ for all t. This is a hazard that is monotonically decreasing in duration, implying that the longer individuals go without purchasing the technology, the less likely they are to purchase soon. The second hazard function is nonmonotonic and has $dh_2(t)/dt > 0$ and $dh_2(t)/dt < 0$ depending on the length of duration t. In this case the purchase probabilities increase or decrease in duration. The third hazard function has $dh_3(t)/dt > 0$ for all t and is monotonically increasing in duration. This implies that the longer individuals go without purchasing the technology the more likely they are to purchase soon. Finally, the fourth hazard function has $dh_4(t)/dt = 0$, which means that technology purchase probabilities are independent of duration and no duration dependence exists.

Information relating to duration dependence, as derived from the first derivative of the hazard function with respect to time, can provide important insights into the duration process being modeled. However, there are clearly

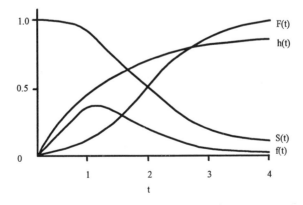

FIG. 7.4. Illustration of hazard ($h(t)$), density ($f(t)$), cumulative distribution ($F(t)$), and survivor functions ($S(t)$).

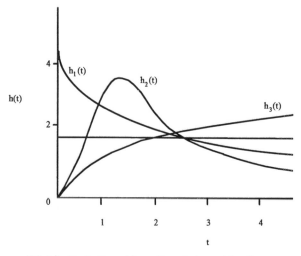

FIG. 7.5. Illustration of four alternate hazard functions.

important determinants of duration (e.g., socioeconomic characteristics) that must be accounted for in the modeling approach as well. These determinants are included in hazard-based models using two alternate methods: proportional hazards and accelerated lifetime.

Proportional-hazards models operate on the assumption that determinants (i.e., factors that affect duration) act multipicatively on some underlying hazard function. This underlying (or baseline) hazard function is denoted $h_0(t)$, and this is the hazard function assuming that all elements of a determinant vector, X, are zero. The manner in which determinants are assumed to act on the baseline hazard is usually specified as the function $\exp(\beta)$, where β is a vector of estimable parameters. Therefore, the hazard rate with determinants, $h(t|X)$, is given by the equation,

$$h(t|X) = h_0(t)\exp(\beta X) . \tag{27}$$

This proportional-hazards approach is illustrated in Fig 7.6.

An alternate approach of incorporating determinants in hazard-based models is the accelerated lifetime model. This model assumes that the determinants rescale time directly (i.e., accelerate time) in a baseline survivor function, which is the survivor function when all determinants are zero. Assuming that the determinants act in the form $\exp(\beta X)$, as was the case for the proportional-hazards model, the accelerated lifetime model can be written as

$$S(t|X) = S_0[t\exp(\beta X)] , \tag{28}$$

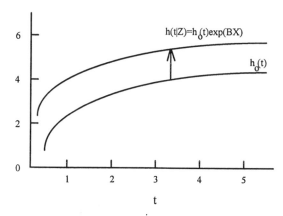

FIG. 7.6. Illustration of the proportional-hazards model.

and it follows that this model can be written in terms of hazard functions as

$$h(t \,|\, \mathbf{X}) = h_o[t \exp(\beta \mathbf{X})] \exp(\beta \mathbf{X}) . \qquad (29)$$

Accelerated lifetime models have, along with proportional-hazards models, enjoyed fairly widespread use (see Kalbfleisch & Prentice, 1980). The selection of accelerated lifetime or proportional-hazards models is often determined on the basis of distributional assumptions (i.e., the assumed distribution of durations). Commonly used distribution assumptions are discussed below.

Distributional Alternatives

Two general approaches to implementing the hazard-based model are possible. One is to assume a distribution of duration (e.g., Weibull, exponential, etc.), and the other is to apply a generalized approach that does not require a distributional assumption. The former approach is called "fully parametric" because a distributional assumption is being made for the hazard along with an assumption on the functional form with which determinants act in the model (i.e., the $\exp(\beta \mathbf{X})$ used in the previous section). The latter approach is semiparametric because only the determinant functional form, $\exp(\beta \mathbf{X})$, is specified.

Fully parametric models can be estimated in proportional-hazard or accelerated lifetime forms, and a variety of duration-distribution alternatives are available including gamma, exponential, Weibull, log-logistic, and log-normal. The choice of any one of these alternatives can be justified on theoretical grounds, and each has important implications relating to the shape of its underlying hazard functions. Three common distributions (exponential, Weibull, and log-logistic) are summarized next.

The exponential distribution is the simplest to apply and interpret. With parameter $\lambda > 0$, the exponential density function is

$$f(t) = \lambda \exp(-\lambda t) \tag{30}$$

with hazard,

$$h(t) = \lambda . \tag{31}$$

Equation 31 implies that this distribution's hazard is constant, and thus the probability of exiting a duration is independent of the time length of the duration. This is a fairly restrictive assumption because the exponential distribution does not allow any sort of duration dependence to be captured.

The Weibull distribution is a more generalized form of the exponential in that it allows for positive-duration dependence (hazard is monotonic increasing in duration), negative-duration dependence (hazard is monotonic decreasing in duration) or no duration dependence (hazard is constant in duration). With parameters $\lambda > 0$ and $C > 0$, the Weibull distribution has density function,

$$f(t) = \lambda C (\lambda t)^{C-1} \exp[-(\lambda t)^C] \tag{32}$$

with hazard,

$$h(t) = \lambda C (\lambda t)^{C-1} . \tag{33}$$

Note with equation 33, if the Weibull parameter C is greater than one, the hazard is monotone increasing in duration; if C is less than one, it is monotone decreasing in duration; and if C equals one, the hazard is constant in duration and reduces to the exponential distribution's hazard (i.e., $h(t) = \lambda$). Because the Weibull distribution is a more generalized form of the exponential distribution, it provides a more flexible means of capturing duration dependence, but it is still limited because of the restriction it places on the hazard (i.e., that it is monotonic). In many applications, a nonmonotonic hazard may be theoretically justified.

The log-logistic distribution allows for nonmonotonic hazard functions and is often used as an approximation of the more computationally cumbersome log-normal distribution. The log-logistic with parameters $\lambda > 0$ and $C > 0$ has the density function,

$$f(t) = \lambda C(\lambda t)^{C-1}[1+(\lambda t)^{C}]^{-2} \qquad (34)$$

and hazard function,

$$h(t) = [\lambda C(\lambda t)^{C-1}]/[1+(\lambda t)^{C}] . \qquad (35)$$

Note that the log-logistic's hazard is identical to the Weibull's except for the denominator. Equation 35 shows that if $C < 1$, then the hazard is monotone decreasing; if $C = 1$, then the hazard is monotone decreasing from parameter λ; and if $C > 1$, then the hazard increases from zero to a maximum at time, $t = (C-1)^{1/C}/\lambda$, and decreases toward zero thereafter.

The alternative to assuming a distribution of the hazard is to use a nonparametric approach for modeling the hazard. This is convenient when little or no knowledge of the functional form of the hazard is available. Such an approach was developed by Cox (1972) and is based on the proportional-hazards approach. The Cox proportional-hazards model is semiparametric because $\exp(\beta X)$ is still used as the functional form of the determinants. The model is based on the ratio of hazards so that the probability of an individual, i, exiting a duration at time t_i, given that at least one traveler exits at time t_i, is given as

$$\exp(\beta \mathbf{X})_i / \sum_{j \in R_i} \exp(\beta \mathbf{X}_j) , \qquad (36)$$

where R_i denotes the set of individuals with durations greater than or equal to t_i.

Proportional hazards and accelerated lifetime models are readily estimated using standard maximum likelihood estimation procedures. The reader is referred to Kalbfleisch and Prentice (1980) for details.

Hazard-Based Duration Models: Concerns and Limitations

There are a number of factors to consider before applying hazard-based duration models. First, there is the likelihood of censored data, which occurs when individual observations are not observed to have completed durations. In our example, it would be drivers that have not yet purchased the new in-vehicle navigation system by the time the data was collected. This censoring differs from the truncation discussed earlier in the Poisson model because with truncated data no information is available on observations that result in data that goes beyond the truncation. For censored data, information is available on observations (i.e., drivers) even though their duration goes beyond the censored time. In hazard-based models, censoring is handled with comparative ease, and only slight modifications to the likelihood function are required. The reader is referred to Kalbfleisch and Prentice (1980) for more information relating to censoring.

Second, heterogeneity, discussed earlier in discrete choice analysis, is also a problem in hazard-based models. This is because the assumption implicitly made in proportional-hazards models is that the survivor function (see equation 29) is homogeneous over the population being studied. As such, all of the variation in durations is assumed to be captured by the determinant vector $\mathbf{X}$. A problem arises when some unobserved factors (i.e., not included in $\mathbf{X}$) influence durations. This, as in the discrete-model case, is unobserved heterogeneity and can result in a major specification error that can lead one to draw erroneous inferences on the shape of the hazard function and determinant coefficient estimates (Gourieroux, Monfort, & Trognon, 1984; Heckman & Singer, 1984). Ignoring heterogeneity is equivalent to leaving out an important determinant in the $\exp(\beta\mathbf{X})$ function. Fortunately, a number of corrections have been developed to explicitly account for heterogeneity. The most common is to include a heterogeneity term designed to capture unobserved effects across the population and to work with the conditional duration–density function. With a heterogeneity term, w, having a distribution over the population, g(w), along with a conditional duration–density function, f(t | w), the unconditional duration density function can be determined from

$$f(t) = \int f(t\,|\,w)g(w)dw \ . \tag{37}$$

With this formulation, hazard models can be derived using procedures identical to those used in the derivation of the nonheterogeneity hazards models.

The problem in operationalizing such a heterogeneity model is that a distribution of heterogeneity in the population must be selected. There is seldom any theoretical justification for selecting one distribution over another, and the economics and marketing literature is strewn with papers that have used a wide variety of heterogeneity distributions. However, the gamma distribution (Greene, 1992; Gupta, 1991; Hui, 1990) is the most popular. The selection of a heterogeneity distribution must not be taken lightly. The consequences of incorrectly specifying g(w) are potentially severe and can result in inconsistent estimates as demonstrated both theoretically and empirically by Heckman and Singer (1984). Fortunately, from the perspective of choosing among many possible distributions, it has been shown (Kiefer, 1988) that if a correctly specified duration distribution is used, the coefficient estimate results are not highly sensitive to alternate distributional assumptions of heterogeneity.

The third concern relates to the presence of a state dependence variable. An example of such a variable would be the duration until a driver's purchase of a previously introduced navigation technology. This can cause problems in the presence of unobserved heterogeneity (see earlier discussion on discrete models). Again, this is because the coefficient of the state variable may be capturing spurious correlation caused by unobserved heterogeneity. Heckman and Borjas (1980) and Kim and Mannering (1996) give a thorough discussion of this point.

Finally, there is the possibility of time-varying determinants (i.e., in the **X** vector). If **X** changes over the duration time being studied (e.g., the price of the new in-vehicle navigation system), parameter estimates may be biased. Time-varying determinants are difficult to account for in duration models (Greene, 1992), and if they are truly important, using discrete instead of continuous time–duration data may be warranted. The discrete time approach segments the duration time into discrete intervals and models the binary choice to purchase or not to purchase in each discrete time period. Because this is a discrete choice, the logit model discussed earlier is appropriate. Examples of the discrete time approach are given in Abbot (1985), Green and Symons (1983), Ingram and Kleinman (1989).

ORDERED DATA

Some data on driver decision making are in the form of ordered responses. This type of data would result from survey questions that ask drivers to give responses such as "never," "sometimes," "frequently," or by rating an item on some scale (e.g., from 1 to 10). This type of data either comes in an integer form or can be translated into an integer form. For example, "never," "sometimes," "frequently" can be coded as 1, 2, and 3, respectively. This type of data is well suited to ordered probability models (Greene, 1993).

Ordered probability models define an unobserved variable, z, that is used as a basis for modeling the ordinal ranking of the data. This unobserved variable is specified as

$$z = \beta \mathbf{X} + \varepsilon, \tag{38}$$

where **X** is a vector of characteristics determining individuals' choice of ranking category; β is a vector of estimable parameters; and is a random disturbance. Using this equation, observed ordinal rankings, y, are defined as

$$
\begin{aligned}
y &= 0 && \text{if } z \leq \mu_0 \\
&= 1 && \text{if } \mu_0 < z \leq \mu_1 \\
&= 2 && \text{if } \mu_1 < z \leq \mu_2 \\
&= \ldots \\
&= j && \text{if } z \geq \mu_{j-1}
\end{aligned}
\tag{39}
$$

where μs are estimable parameters that define y, which corresponds to integer rankings.

If the disturbance term in equation 38 is assumed to be standard normal (with mean = 0 and variance = 1), then an ordered probit model results, and if the disturbance is assumed to be standard logistic, then an ordered logit model results. Unlike the case of discrete choice models, the ordered logit model does not have a significant computational advantage over the ordered probit. The choice of one model over the other is often made purely on theoretical grounds.

Ordered probability models encounter many of the same specification problems as standard regression models. See Greene (1993) for additional information on such models.

OTHER SPECIFICATION PROBLEMS

In addition to the specification problems discussed in previous sections, all models must be checked for the more common, well-known specification errors. These specification errors are discussed in detail in Ben-Akiva and Lerman (1985), Pindyck and Rubinfeld (1981), and Theil (1971). The reader is encouraged to consult these sources for additional information on specification error. A summary of common specification errors include the following:

Omitted Variables
If an important variable is omitted from the X vector, the model is likely to be misspecified, and a biased estimate of the parameter vector β will result (i.e., estimated values of β will differ from true values).

Irrelevant Variables
If an irrelevant variable is included in the model, a loss in statistical efficiency will result. The efficiency loss means higher standard errors, and lower t statistics for estimates of the parameter vector β will result. Thus for any given sample size, we will be less certain about estimates of β.

Nonlinearities
The assumption often made is that βX is linear (i.e., $\beta_1 X_1 + \beta_2 X_2 + ...$). If this is not the case and the model is estimated as if it were, the estimate of the parameter vector β will be biased.

Nonzero Error Term Means
Many of the models presented in this paper are derived under the assumption that the mean of the error term is zero. If this is not the case and an intercept is not included in the β vector (i.e., a constant), then β will be biased. If a constant is included, then β will be unbiased, but the

parameter estimate of the constant will be biased. Because of this, it is always a good idea to include a constant in the parameter vector.

Errors in Variables

If either the modeled variable or the variables in **X** are measured with error, estimation will almost always produce biased estimates of β.

Nonfixed Explanatory Variables

The problem of nonfixed explanatory variables was discussed under the continuous-continuous data section of this chapter. However, nonfixed variables can be a problem in almost all model types and can result in a biased estimate of β. Corrective procedures such as instrumental variables (discussed earlier) can be used in most applications.

Errors in Distributional Assumptions

Models are derived by making an assumption as to the distribution of error terms (e.g., normal, GEV, Weibull, etc.). If the assumption is not correct, there will be at least a loss in efficiency in estimating β (i.e., higher standard errors).

Heteroskedasticity

Model derivations typically assume that the variance of the error terms are equal from one observation to the next. If this is not the case, a loss in estimation efficiency will result (i.e., higher standard errors).

Serial Correlation

Model derivations also assume that there is no error-term correlation among observations. Such correlation could result if more than one observation is used from a single individual. A loss in efficiency will result if such correlation exists.

Multicollinearity

If the variables comprising **X** are highly correlated, standard errors of β estimates will become large. However, estimates of β will still be unbiased. As a result, multicollinearity usually is not the severe specification problem that many researchers make it out to be.

Truncated-Censored Data

Although discussed for the Poisson regression and hazard-based duration models, truncated-censored data can be present in all model types. If such restrictions are present and ignored, estimates of β will be biased.

ESTIMATION SOFTWARE

The models discussed in this chapter can be estimated using a number of software packages, each of which has its own strengths and weaknesses. It is important to realize that no one package has preprogrammed commands to handle all of the model types discussed herein. Thus, it is often necessary for the researcher to learn to use programming features, available in most packages, that allow the user to specify estimation functions for virtually any model. As a rule, however, it is recommended that the researcher become familiar with two or three packages to ensure that most required model estimations are covered by preprogrammed commands, and that full advantage is taken of each package's strengths. This second point is important because packages tend to be more efficient in some estimation areas and less efficient in others. For example, some packages may be ideally suited for the analysis of discrete data but less well suited for the analysis of count data. Thus the researcher may want to use a different package when analyzing count data.

Turning to specific software packages, an overview is presented for six packages: SAS, SPSS, Gauss, TSP, SST, and Limdep. Although this list is by no means comprehensive, it does define a representative range of packages. All of these packages run in an MS-DOS environment and, in most cases, on other operating systems as well. With the caveat that these packages are being continually updated and improved, a general overview of their capabilities with regard to the models presented in this chapter is presented next.

SAS

SAS is a large-scale package with a huge international following. SAS has excellent data manipulation capabilities and is generally well documented. Unfortunately, SAS has a primary focus on methodologies that appeals to statisticians, physical scientists, sociologists, and psychologists. As mentioned in the introduction, the models discussed in this chapter are drawn more from the economics literature and, therefore, SAS is not the most efficient package for applying these techniques. Although the user can use SAS to estimate some of the techniques presented in this chapter, it is not the package to choose if only one package is to be chosen.

SPSS

SPSS is similar to SAS in many ways. It is a very large program that has a strong following among statisticians, physical scientists, and noneconomist social scientists. As with SAS, researchers wishing to apply the methodological approaches presented in this chapter will find SPSS to be a limited package.

Gauss

Gauss is a very capable software package that can handle virtually any modeling methodology. Gauss enjoys a loyal following among economists and is well suited for estimating the models presented in this chapter. The drawback with Gauss is that, for some relatively important techniques, preprogrammed commands do not exist. *Gauss* is flexible and can be easily programmed, but this requires the user to become familiar with *Gauss*'s programming language. This may be viewed as an obstacle to researchers not familiar with FORTRAN or C. Also, some of the preprogrammed command and data manipulation methods seem awkward when compared to those of other packages.

TSP

TSP is a package that is very well suited to some of the techniques presented in this chapter. For example, *TSP* has an impressive array of methods that deal with continuous-continuous data. *TSP*'s limitation is that its set of preprogrammed commands excludes some important methodological approaches (e.g., hazard-based duration models). *TSP* is a good choice to complement another package that has a wider range of preprogrammed commands.

SST

SST is a fairly new package that was first developed in the mid-1980s. It is well suited to many of the methods presented in this chapter. *SST* manipulates data easily and is very efficient in both batch and interactive modes. *SST* arguably has the best preprogrammed commands for discrete data analysis (multinomial logit estimation), and its programming language is both versatile and easily accessible to relatively new computer programmers. However, *SST* is limited in that it does not offer preprogrammed commands for a number of techniques discussed in this chapter (e.g., hazard-based duration models). *SST* is an excellent choice for discrete data analysis, but another package is needed to fill in gaps in preprogrammed commands dealing with duration models, negative binomial regressions, and others.

Limdep

Of all known packages, *Limdep* is the one package that has preprogrammed commands for almost all of the techniques discussed in this chapter. *Limdep*'s range of preprogrammed commands is astonishing: Poisson regression, negative binomial regression, system equation methods, discrete choice models, hazard-based duration models, ordered probability models, and many more. Of all the packages on the market, *Limdep* offers the widest range of preprogrammed capability with regard to statistical techniques used by economists. However, *Limdep* is not a perfect package; some data manipulations and batch operations are cumbersome. Although *Limdep* may be the best choice if only one package is to be chosen, researchers would be advised to consider other packages to supplement *Limdep* for simplicity of use (e.g., *SST* for handling discrete data).

SUMMARY

This chapter overviews a wide variety of modeling techniques that can be used to handle the types of data likely to be encountered in modeling driver decision making. An overview of these techniques along with references that can be used for obtaining additional information is provided in Table 7.1. The techniques in this chapter are very powerful tools for analyzing driver decision making. The intent of this chapter is to give the reader a basic understanding of the various modeling approaches along with some of the likely pitfalls that may be encountered. It is important to select the methodology suited to the decision-making process being modeled, and to carefully search for possible model specification errors. The literature in virtually every field is strewn with papers that have adopted the wrong methodological approach or have made serious model specification errors. Before applying any of the techniques discussed herein, it is recommended that the researcher take the time to read some of the many referenced articles under each methodological heading in Table 7.1. Such reading will help clarify any lingering questions concerning the modeling methods.

REFERENCES

Abbot, R. (1985). Logistic regression in survival analysis. *American Journal of Epidemiology, 121*, 465–471.

Ben-Akiva, M., & Lerman, S. (1985). *Discrete choice analysis: Theory and application to travel demand.* Cambridge, MA: MIT Press.

Cameron, C., & Trivedi, P. (1986). Econometric models based on count data: Comparisons and applications of some estimators and tests. *Journal of Applied Econometrics, 1*, 29–53.

Cox, D. R. (1972). Regression models and life tables. *Journal of the Royal Statistical Society, B34*, 187–220.

Damm, D., & Lerman, S. (1981). A theory of activity scheduling behavior. *Environment and Planning A, 13*, 703–718.

Dubin, J., & McFadden, D. (1984). An econometric analysis of residential electric appliance holdings and consumption. *Econometrica, 52*, 345–362.

Fleming, T., & Harrington, D. (1990). *Counting processes and survival analysis.* New York: Wiley.

Green, D., & Hu, P. (1984). The influence of the price of gasoline on vehicle use in multivehicle households. *Transportation Research Record, 988*, 19–23.

Green, M., & Symons, M. (1983). A comparison of the logistic risk function and the proportional hazards model in prospective epidemiologic studies. *Journal of Chronic Diseases, 36*, 715–724.

Greene, W. (1992). *LIMDEP 6.0, econometric software.* Bellport, NY: Econometric Software, Inc.

Greene, W. (1993). *Econometric analysis.* New York: Macmillan.

TABLE 7.1
Summary of DataTypes, Model Alternatives, and References

Data Type	Model Alternatives	References
Count data	Poisson regression Negative binomial regression	Johnson & Kotz (1969) Maddala (1983) Hausman, Hall, & Griliches (1984) Lee (1986) Cameron & Trivedi (1986) Mannering (1989) Mannering & Hamed (1990) Greene (1982, 1993) Poch & Mannering (1995) Shankar, Mannering, & Barfield (1995)
Continuous-dontinuous data	Indirect least squares Instrument variables Two-stage least squares Three-stage least squares Full-information maximum likelihood	Theil (1971) Pindyck & Rubinfeld (1981) Green & Sh (1984) Hensher (1985) Kmenta (1986) Greene (1992, 1993) Mannering (1983) Hamed & Mannering (1993)
Discrete data	Multinomial logit Multinomial probit	McFadden (1981) Small & Rosen (1981) Ben-Akiva & Lerman (1985) Maddala (1983) Small & Hsaio (1984) Winston & Mannering (1984) Mannering & Winston (1985, 1991) Train (1986) Hensher, Smith, Milthorpe, & Bernard (1992) Mannering & Grodsky (1995)

TABLE 7.1 (continued)

Data Type	Model Alternatives	References
Discrete-continuous data	Selectivity correction term Instrumental variables Expected value Full-informtion maximum likelihood	Hay (1980) Damm & Lerman (1981) Maddala (1983) Dubin & McFadden (1984) Mannering & Winston (1985 Mannering (1986) Train (1985) Hensher & Milthorpe (1987) Mannering & Hensher (1987) Mannering, Abu-Eisheh, & Arnadottir (1990) Hamed & Mannering (1993)
Duration data	Proportional hazards Accelerated lifetime	Heckman & Borjas (1980) Kalbfleisch & Prentice (1980) Green & Symons (1983) Gourieroux, Monfort, & Trognon (1984) Heckman & Singer (1984) Abbot (1985) Kiefer (1983) Ingram & Kleinman (1989) Hui (1990) Mannering & Hamad (1990) Fleming & Harrington (1990) Gupta (1991) Greene (1992, 1993) Mannering (1993) Hensher & Mannering (1994) Paselk & Mannering (1994) Kim & Manrering (1995)
Ordered data	Ordered logit Ordered probit	Greene (1992, 1993) Mannering, Kim, Barfield, & Ng (1994)

Gourieroux, C., Monfort, A., & Trognon, A. (1984). Pseudo maximum likelihood methods: Theory. *Econometrica, 52,* 681–700.

Griliches, Z. (1967). A note on serial correlation bias in estimates of distributed lags. *Econometrica, 29,* 65–73.

Gupta, S. (1991). Stochastic models of interpurchase time with time-dependent covariates. *Journal of Marketing Research, 18,* 1–15.

Hamed, M., & Mannering, F. (1993). Modeling travelers' postwork activity involvement: Toward a new methodology. *Transportation Science, 17,* 381–394.

Hausman, J., Hall, B., & Griliches, Z. (1984). Econometric models for count data with an application to the patents-R & D relationship. *Econometrica, 52,* 909–938.

Hay, J. (1980). *Occupational choice and occupational earnings.* Doctoral dissertation, New Haven, CT: Yale University.

Heckman, J. (1981). Statistical models for discrete panel data. In C. Manski & D. McFadden (Eds.), *Structural analysis of discrete data with econometric applications* (pp. 179–195). Cambridge, MA: MIT Press.

Heckman, J., & Borjas, G. (1980). Does unemployment cause future unemployment? Definitions, questions and answers from a continuous time model of heterogeneity and state dependence. *Econometrica, 47,* 247–283.

Heckman, J., & Singer, B. (1984). A method for minimizing the impact of distributional assumptions in econometric models for duration data. *Econometrica, 52,* 271–320.

Hensher, D. (1985). An econometric model of vehicle use in the household sector. *Transportation Research, 19B,* 303–313.

Hensher, D. A., & Mannering, F. (1994). Hazard-based duration models and their application to transportation analysis. *Transport Reviews, 14,* 63–82.

Hensher, D., & Milthorpe, F. (1987). Selectivity correction in discrete-continuous choice analysis: With empirical evidence for vehicle choice and use. *Regional Science and Urban Economics, 17,* 123–150.

Hensher, D., Smith, N., Milthorpe, F., & Barnard, P. (1992). *Dimensions of automobile demand.* North-Holland, Netherlands: Elsevier.

Hui, W. T. (1990). *Proportional hazard Weibull mixtures.* Working paper, Department of Economics, Australian National University, Canberra.

Ingram, D., & Kleinman, J. (1989). Empirical comparisons of proportional hazards and logistic regression models. *Statistics in Medicine, 8,* 525–538.

Johnson, N., & Kotz, S. (1969). *Distributions in statistics: Discrete distributions.* New York: Wiley.

Kalbfleisch, J., & Prentice, R. (1980). *The statistical analysis of failure time data.* New York: Wiley.

Kiefer, N. (1988). Economic duration data and hazard functions. *Journal of Economic Literature, 26,* 646–679.

Kim, S.-G., & Mannering, F. (1996). Panel data and activity duration models: Econometric alternatives and applications. In T. Golob, L. Long, & R. Kitamura, (Eds.), *Panels for transportation planning: Methods and applications.* Norwell, MA: Kluwer.

Kmenta, J. (1986). *Elements of econometrics.* New York: Macmillan.

Lee, L.-F. (1986). Specification test for Poisson regression models. *International Economic Review, 27,* 689–706.

Maddala, G. (1983). *Limited-dependent and qualitative variables in econometrics.* Cambridge, England: Cambridge University Press.

Mannering, F. (1983). An econometric analysis of vehicle use in multivehicle households. *Transportation Research, 17A,* 183–189.

Mannering, F. (1986). Selectivity bias in models of discrete and continuous choice: An empirical analysis. *Transportation Research Record, 1085,* 58–62.

Mannering, F. (1989). Poisson analysis of commuter flexibility in changing routes and departure times. *Transportation Research, 23B,* 53–60.

Mannering, F. (1993). Male/female driver characteristics and accident risk: Some new evidence. *Accident Analysis and Prevention, 25,* 77–84.

Mannering, F., Abu-Eisheh, S., & Arnadottir, A. (1990). Dynamic traffic equilibrium with discrete/continuous econometric models. *Transportation Science, 24,* 105–116.

Mannering, F., & Grodsky, L. (1995). Statistical analysis of motorcyclists' self-assessed risk. *Accident Analysis and Prevention, 27,* 21–31.

Mannering, F., & Hamed, M., (1990). Occurrence, frequency, and duration of commuters' work-to-home departure delay. *Transportation Research, 24B,* 99–109.

Mannering, F., & Hensher, D. (1987). Discrete/continuous econometric models and their application to transport analysis. *Transport Reviews, 7,* 227–244.

Mannering, F., Kim, S.-G., Barfield, W., & Ng, L. (1994). Statistical analysis of commuters' route, mode, and departure flexibility and the influence of traffic information. *Transportation Research, 2C,* 35–47.

Mannering, F., & Winston, C. (1985). Dynamic empirical analysis of household vehicle ownership and utilization. *Rand Journal of Economics, 16,* 215–236.

Mannering, F., & Winston, C. (1991). Brand loyalty and the decline of American automobile firms. *Brookings Papers on Economic Activity, Microeconomics,* 67–114.

May, A. (1990). *Traffic flow fundamentals.* Englewood Cliffs, NJ: Prentice-Hall.

McFadden, D. (1981). Econometric models of probabilistic choice. In C. Manski & D. McFadden (Eds.), *Structural analysis of discrete data with econometric applications* (pp. 198–272). Cambridge, MA: MIT Press.

Paselk, T., & Mannering, F. (1994). Use of duration models for predicting vehicular delay at U.S./Canadian border crossings. *Transportation, 21,* 249–270.

Poch, M., & Mannering, F. (1996) Negative binomial analysis of intersection accident frequencies. *Journal of Transportation Engineering, 122,* 391–401.

Pindyck, R., & Rubinfeld, D. (1981). *Econometric models and economic forecasts.* New York: McGraw-Hill.

Shankar, V., Mannering, F., & Barfield, W. (1995). Effect of roadway geometrics and environmental factors on rural accident frequencies. *Accident Analysis and Prevention, 27,* 371–389.

Small, K., & Hsaio, C. (1985). Multinomial logit specification tests. *International Economic Review, 26,* 619–627.

Small, K., & Rosen, H. (1981). Applied welfare economics with discrete choice models. *Econometrica, 49,* 105–130.

Theil, H. (1971). *Principles of econometrics.* New York: Wiley.

Train, K. (1986). *Qualitative choice analysis: Theory, econometrics and an application to automobile demand.* Cambridge, MA: MIT Press.

Winston, C., & Mannering, F. (1984). Consumer demand for automobile safety. *American Economic Review, 74,* 316–319.

8

Usability Evaluation for Intelligent Transportation Systems

Jan H. Spyridakis
Ann E. Miller
University of Washington

Woodrow Barfield
Virginia Polytechnic Institute and State University

To a large extent, the success of any information system, such as an Intelligent Transportation System (ITS), will depend on its usability or ability to be easily understood and conveniently employed by a user. Usability is a term with varied meanings, but it generally refers to the ease in which a system or its components can be used. For example, with an ITS, given an in-vehicle display, usability translates into how easy ITS decision making is about whether to take an alternative route, or a particular exit, or whether to delay a trip. The purpose of this chapter is to discuss how "usability analyses" can assist developers in designing ITS, formerly known as Intelligent Vehicle Highway Systems (IVHS). Drawing its theoretical base from human factors, cognitive psychology, technical communication, ethnography, and business, usability evaluation can help ensure a good match between the information provided by an ITS and users' information needs. Specifically, a usability analysis can help match the ITS capabilities to the users' needs, goals, and risk-taking comfort levels. Our basic premise is that for an ITS to be successfully employed and accepted by the public, its design must be based on principles of human factors. One of major themes of the human factors approach is to involve the user early in the design of the system. As shown below, usability analysis is a methodology that, if properly employed, will accomplish the goal of early user involvement in ITS design.

The relevance of usability analysis for ITS design is readily apparent when one considers some of the various components of an ITS. An ITS will consist of advanced technology involving communications, computer displays, and control processes that include the following five functional areas (IVHS America, 1992): (a) Advanced Traveler Information Systems (ATIS), (b) Commercial Vehicle Operations (CVO), (c) Advanced Traffic Management Systems (ATMS), (d) Advanced Vehicle Control Systems (AVCS), and (e) Advanced Public Transportation Systems (APTS). Although the methodology outlined in this chapter can be used to assist developers in the design of any particular component of an ITS, we focus on the ATIS portion, primarily because an ATIS will be heavily used by the general public.

It has been suggested that an ATIS will improve the timeliness, efficiency, and accuracy of driving by providing an integrated system structure consisting of the following components: an in-vehicle routing and navigation system (IRANS), an in-vehicle motorist services information system (IMSIS), an in-vehicle signing information system (ISIS), and an information safety advisory and warning system (IVSAWS). Because each of these subsystems involves the user as a component in the system (who must process information, make decisions, and initiate actions), the usability analysis methodology is particularly relevant. Specifically, because each of the subsystems involves providing the driver with information in the car, making sure the provided information matches what the driver wants or needs is critical to the success of the system. However, discovering what information the driver needs is not an easy task, but must be based on a formal methodology that examines what information is required, where and when it should be delivered, and in what format. Furthermore, a close match between the goals of the ITS and user information needs must be made. Again, this task should be based on a formal methodology (i.e., usability analysis, with users involved in the design process).

Previously, usability analysis methodology was employed to analyze the ease-of-use of various consumer products (e.g., VCRs, computers). The notion of adopting usability analysis to the design of ITS is not a departure from traditional uses of the methodology for product design because many components of ITS are very similar to consumer products that have already benefited from a usability analysis. As an example, consider an in-vehicle display. The information provided by the in-vehicle display must be based on user requirements. Thus a usability analysis must be done to determine those requirements. In addition, the ease of use of interacting with the display must be determined by performing a usability analysis. Finally, the readability of the display under various weather and driving conditions must be determined by a usability analysis. For these reasons, and because much of the literature on usability uses the term "consumer product," in the preceding chapter we found it convenient to discuss the usability of transportation systems just as we would discuss the usability of a consumer product.

According to findings focusing on user acceptance of consumer products, drivers will resist using information products that make their driving task more difficult or complex. For example, if the information provided on alternative routes is too complex or if too many options are offered, that information will likely be ignored or misinterpreted. Unfortunately, developers of traveler information hardware and software have often learned about the importance of usability the hard way, finding that hard-to-use or unfriendly user interfaces have discouraged consumer interaction. It should be noted that, one way or another, usability evaluations are always conducted on consumer products (whether the products are transportation products or not), no matter if such evaluations are done intentionally by the development team before the system is introduced to the public or by default by the user after release. For this reason, we propose that all components of an ITS must undergo a formal usability analysis because such an analysis will ultimately be done anyway. Usability testing is particularly relevant to developers of ITS as the goals of such systems go beyond mere product acceptance (i.e., drivers will actually use the ATIS to assist in the performance of actual tasks).

Although most developers and programmers acknowledge the value of usability testing, many still see it as a hindrance to a timely and orderly product development process. At best, they anticipate time-consuming reprogramming efforts, and at worst, they fear an overhaul of the product from the ground up. These fears are justified when a usability evaluation serves only as a final checkpoint before the product is released to the public. However, when usability evaluations are integrated into the product development process from the beginning, the end result is a better product within the boundaries of the original schedule.

Usability testing can improve existing processes for a development team by including endusers' input, but it does not dictate or depend on a particular product development model. No matter what type of product development model is being used, usability involvement is important during each phase of a product's life cycle. However, usability results can have the largest positive impact during the investigation, definition, and design phases (Skelton, 1992). The *investigation* phase starts when a potential market is recognized and the development team investigates whether it is feasible to proceed with building a product (see Table 8.1). For ITS, the potential market is tremendous considering the number of automobiles on the road. The *definition* phase begins with the output of the investigation phase—a recognized need for product development—and ends with the input to the design phase—a complete description of the intended product. For ITS, recognized needs include the alleviation of congestion and the improvement of safety on our roadways. The *design* phase, the next step in the product development process, occurs when the product team takes the

TABLE 8.1
Phases of Usability Evaluations

Phases	Characteristics
Investigation	Recognize market and determine feasibility
Definition	Describe intended product
Design	Create specifications for product
Construction	Build product
Function testing	Determine whether product works as designed
Product release	Release product to market

definition of product features, including the "look and feel," and creates specifications for an actual product.

When usability has been fully employed during the first three phases of the development process, the primary role for usability evaluation during the *construction* and *function* testing phases is to provide usability input as design changes are proposed. When the product enters the controlled and general *product* release phase, the design is frozen from further changes. Usability information gathered during this phase will be used during the investigative phase of the product's next version. When usability tests are done only at the end of the product development cycle, if the system does not work as expected, there is no way to incorporate findings without major hardware or software changes. Currently, the prevailing thought within the usability community is that usability evaluations should be an iterative process incorporated into all phases of the development cycle (Dieli, 1989; Gould & Lewis, 1985; Poltrock, 1989; Schneiderman, 1987).

This chapter makes a case for usability testing as a development tool that can add value to the development of any ITS by examining usability evaluation methods in general, their application in the development of Traffic Reporter (an ITS created at the University of Washington), and the value of their application.

USABILITY RESEARCH METHODS

Usability research methods rely on many principles of traditional experimental design often used in social, engineering, and physical science research. However, experimental design in usability research is augmented with observation and verbal protocol analysis to uncover the reasons for performance such as errors, not just to study the errors themselves. For ITS, observations focusing on the use of an-in vehicle display can be made by having a member of the design team ride with the driver, by examining a videotape of the driver using an in-vehicle display, or by examining a driver using an in-vehicle display in the laboratory under controlled conditions. The results of such research can then show whether usability problems exist with the ITS design, what the cause of the problem is, and what methods can be employed to build a more effective ITS. Furthermore, for an ITS, usability analysis can assist developers in determining information requirements for ITS subcomponents. This will be a particularly important application of usability analysis for transportation systems because these systems must provide the information that users want and need in order to gain user acceptance. Concerning usability, formulating valid research designs is an art that depends on eliciting thoughtful responses without leading subjects into particular actions or answers. The following discussion summarizes research design principles as they apply to usability research.

1. *Determine what information is needed to further the team's goals and to answer ITS development questions.* Usability researchers should agree on the deliverables required by the development team and the required time frame for securing results from the usability research. The type of information desired and the time frame may affect significant aspects of the research design. Examples of deliverables for ATIS include answers to questions such as these: What, if any, roadside services should be displayed with the in-vehicle technology? Should alternative routes be displayed? If so, how many? How far ahead does a driver need to be informed of congestion? Each of these is a potential deliverable of an ATIS.

2. *Identify the usability method(s) that will most effectively answer the research questions.* The choice of methods must be driven by the questions and constraints of the situation rather than by allowing inertia to tailor the questions to fit a customary research method. An overview of usability research methods is presented in the next section of this chapter.

3. *Determine the number of subjects required and target the research population.* The interdependency of sample size and administration methods, budget, and data analysis techniques influence a research design, and this interdependency cannot be ignored. Sample size also may influence the conclusions available to the researcher and the generalizability of the results. Many usability studies rely on narrative data and descriptive statistics (e.g., means) because of extremely small sample sizes. If the sample is too small, the variance may be quite large, thus decreasing the power of the design and reducing the chance of obtaining significance. Many formulas exist in the literature for determining sample size if the population variance is known, which often is not the case. Other methods for determining sample size are based on the statistical techniques to be used. For example, Ahlgren and Walberg (1975) state that 20 subjects are needed per independent variable if regression analyses are to be used. The choice of sample size involves

trade-offs. A small sample size allows for extensive questioning, personal administration methods, and rapid research, but a larger sample size, while restricting the number of research questions and the administration methods, increases the external validity.

After determining the necessary sample size, the usability team must identify the characteristics of potential subjects that are important to the research questions. Researchers may be quite clear about what information they wish to gather, but they must be certain to identify variables that initially may appear irrelevant yet ultimately may provide the basis for interpreting other data. Specific subject characteristics regarded as demographic or subject descriptor information may actually become critical variables for drawing inferences and comparing results across studies. In any research, subject classification variables such as income distribution may correlate with other variables being examined and help the researcher to interpret data patterns as well as allow other researchers to compare their findings.

For ITS, the issue of sample size and that of targeting the end-user population in the selection of the sample will be a difficult problem given the diversity of drivers. For example, the usability team needs to consider private drivers, commercial operators, and dispatchers as potential end-users of ATIS. Furthermore, within each category there are several subpopulations. For example, there are long- and short-distance commercial drivers, those who deliver hazardous materials, those who drive emergency vehicles, and so on. Finally, different subject ages and genders for each category of drivers must be represented in the sample selection. If not, the results of the usability analysis may apply only to a small subset of the driving population.

Finally, it is important to design a process for generating the sample. Although a random sample is preferred, some designs may require random selection from selected subpopulations (i.e., stratified sampling) or direct recruitment of subjects who typify end-users. For example, to identify a driver sample consisting of 400 subjects for a telephone interview, a 1987 Los Angeles survey team developed a list of 2,000 valid telephone numbers from a random list of computer-generated numbers for a given set of area codes (Shirazi, Anderson, & Stresney, 1988). In contrast, Houston's 1981 driver questionnaire identified its sampling frame for a mail questionnaire from a license plate study of a specific freeway section (Huchingson et al., 1984). If the survey's sampling frame is to accurately reflect the driver population in question, the sample must be generated in a manner that creates a sample representative of that population. For this reason, many researchers use random sampling, realizing that attempts to stratify samples may create sampling biases.

4. *Examine the ITS using basic usability design rules to identify gross usability problems.* For example, with an in-vehicle map display, if green is used to

code "stop" rather than "go," then our population stereotype for color coding is not met. Also the researchers should review the product from the research population point of view, noting additional usability issues to include in the research, such as prerequisite background knowledge.

5. *Determine the content of the individual questions or tasks and refine the wording so the questions or directions are not leading or ambiguous, and do not use obscure terminology.* The researchers should carefully check for questions or tasks that can be eliminated because they do not have immediate value at the current phase of product development. Additional questions, however tempting, may lead to nonresponse or subject fatigue, compromising the findings.

6. *Determine the response form to each question (fixed-alternative or open-ended) or task (observed behavior or verbal protocol).* Usability specialists often employ several research methods in concert, such as laboratory testing with a pre- or posttest questionnaire, to gather a variety of data. With the questionnaires for transportation data, the response form will determine whether the data are purely quantitative, qualitative, or narrative, and will influence what type of statistical analysis to perform. In addition, a common technique used in laboratory usability testing is a verbal protocol in which subjects verbalize as they perform tasks. Verbal protocols are used to uncover subjects' strategies, problem-solving processes, task sequences, expectations, concerns, and perceptions of the product. This technique, however, diminishes the accuracy of timed performance data because of variations in reading speed and the subjects' ability to solve problems while verbalizing.

7. *Sequence the questions or tasks in a logical order, often using global questions or basic tasks at the beginning, critical questions or tasks in the middle, and difficult or sensitive questions or tasks in the later part to ensure that the majority of the issues are addressed.*

8. *Determine the specific characteristics of how the questions or tasks will be delivered within the chosen usability research method.* These might include the format of written questionnaires and materials, whether the interview or focus group respondents are told the actual company name, and the location of the research (mailed, in a laboratory, or at a field site).

9. *Review the proposed research design with the product development team to ensure that their questions will be addressed and that any concerns are answered before the research is undertaken.* The usability researcher should revise the design if necessary.

10. *Pretest to identify questions and tasks that are unclear or leading.* The researcher should revise again to correct any problems.

11. *Conduct the research and collect data.*

12. *Analyze the results after the data have been collected.* Researchers use a wide variety of statistical analysis techniques: descriptive or inferential.

However, in many motorist surveys, few inferential statistics are used, often because the data do not lend themselves to inferential statistics or the samples are too small. Unstructured interviews with open-ended questions or field studies create varied responses that may be difficult to quantify. Furthermore, many of the responses to interviews and written questionnaires represent a nominal scale of measurement that violates the assumptions of many inferential statistics. Thus, many motorist surveys report only raw data or response frequencies, though occasionally correlations or Chi square analyses are conducted. More quantitative data of the appropriate numerical scale (ratio or interval) lend themselves to the use of inferential statistics that can identify significance of survey findings, and as a result, can increase generalizability of the findings to specific populations.

USABILITY EVALUATION METHODS

During the different phases of the development cycle, stereotypic questions will arise about the features of the ATIS and its users; each of these questions will best be answered by a particular usability evaluation technique. Questions about who might use the ATIS, what they would do with it, what the culture and environment are like in the place where they would use it, and what other sources of traffic information are currently used often are asked during the investigative phase of the development cycle. These types of questions are best addressed with field study, questionnaires, interviews, focus groups, and laboratory testing of previous versions or the ATIS. Questions about the proposed features, the model (e.g., a desktop), or the overall "look and feel" of the ATIS can be addressed with walkthroughs or laboratory testing of prototypes. These questions often arise during the ATIS's definition phase. Questions about the implementation of the features or about the documentation can be addressed using laboratory testing and walkthroughs with high-fidelity prototypes, usually during the design phase of the development cycle.

Field Study: Ethnography, Contextual Inquiry, and Cognitive Work Analysis

Field study methods include ethnography, contextual inquiry, and cognitive work analysis. Field study methods allow researchers to gather information about end-users and their tasks in the users' environments. This environment may be a home, an individual's office, or a cubicle in a company, but for an ATIS it will most likely be an automobile. Contact with end-users in their environments is important for learning about them as they interact with their tools to reach their goals. It is often the only way to study the context in which complex products are used in modern technical and social environments. The findings from the observed data of field studies provide a description of the tasks, goals, environment, and culture of the individuals who will use an ATIS, but they do not generally provide quantifiable data that can be subjected to statistical analysis. Field studies can also provide representative tasks and scenarios for use during other types of usability evaluations (e.g., laboratory testing; Kvavik, Fafchamps, Jones, & Karimi, 1992; Mirel, 1987; Schneiderman, 1990; Sullivan & Spilka, 1992).

Field study methods are commonly used during the investigative phase for a proposed ATIS to provide information on the nature of the users, their goals, tasks, and culture; the way end-users adapt from old to new technology; the product's integration into the existing workflow; the long-term ease-of-learning; and the evolution of the tasks over time to include new processes while still accomplishing the same goal.

Field studies take researchers out of their own environments, which may act as unconscious information filters that serve to validate the researchers' own values, perspectives, and identity, and instead place the researchers squarely in the middle of the end-users' worlds so the most appropriate tool for the task can be identified. Conversely, it is very difficult to uncover specific implementation issues during field study, because the interaction between end-users and usability researchers is too random to ensure exploring any one facet of the interface. The most powerful and accurate usability input will come from using a variety of methods, beginning with field study during the investigative phase of the ATIS development process and continuing with other methods as the development process continues.

Ethnographic Field Study. One field study method that usability researchers have adapted is ethnography, traditionally used by researchers in anthropology to study people in other cultures. Ethnographic field study is an intensive, albeit laborious and time-consuming, way of gathering information about work environments and people, particularly in situations where the culture determines the acceptance or rejection of a product, for example an in-vehicle map display. The method gives researchers tools to enter an environment without elaborate preconceptions and allows the data to evolve as they observe and participate in the target end-users' lives (Agar, 1986; Monk, Nardi, Gilbert, Mantei, & McCarthy, 1993).

Contextual Inquiry. Contextual inquiry is a technique that involves the study of end-users through focused interaction in the environment in which they will use an ATIS. The researcher observes typical users of a system as they perform their daily routines and interviews them about their tasks, other systems they use, and their perceptions of the systems and their environments. This technique is well-suited for usability field research because it is built on interaction focused on a particular aspect of the tasks or particular tools used to perform the tasks (Good, 1989; Holtzblatt & Jones, 1992).

In contextual inquiry sessions, the researcher observes end-users as they work, interjecting questions about the reasons for their actions and about their perceptions of why things work as they do. The art of contextual inquiry depends on allowing end-users to explore areas of primary concern for them while remaining focused on the information needed to make generalizations about the direction the product should take.

Cognitive Work Analysis. Cognitive work analysis is a method for studying the many interrelated factors and analyzing modern, complex environments from several diverse perspectives. It gives the researcher a framework for collecting data and analyzing user tasks, the task environment, the interaction of people in a task process, the history of changes to the process, the information processing and decision-making patterns, the culture of the individual and society at large, and interactions among these factors (Rasmussen, Pejtersen, & Schmidt, 1990).

Cognitive work analysis will allow the usability researcher to predict how the introduction of an ATIS will affect, not only the tasks it is intended to aid, but also changes in behavior as end-users and society adapt to it. Although the ultimate goals of the users may remain constant, the introduction of new technology forces end-users to adapt in various ways to meet their original goals. This adaptive process, called goal-following, helps usability researchers to excavate through the layers of adaptation to find the original goals so that a product can be built that will meet those needs efficiently.

Questionnaires: Printed Surveys, Interviews, and Focus Groups

Questionnaires are designed to gather information through structured interaction with specific people; they vary in the depth of the interaction and the delivery method. Whether questionnaires are delivered in print through surveys or orally though interviews or focus groups, they rely on a specific, predetermined list of questions that respondents will answer.

Printed Surveys. Researchers frequently use printed surveys to obtain information from large samples because this method places the fewest demands on personnel and budgets. Furthermore, printed surveys or questionnaires allow for respondents' anonymity and a geographically widespread sample. The amount of information that can be gathered with printed surveys is limited because respondents are usually unwilling to fill out lengthy forms. Moreover, printed surveys work best when questions have closed or fixed alternative responses. The form of the response and the format of a written survey are often important for the respondent's ease of filling it out, and also for the researchers' ease of processing the completed questionnaires. Typically, with usability research, printed surveys are used to collect subject demographic data, and oral methods are used to obtain data regarding the current product.

Interviews. Interviews or orally delivered questionnaires can be conducted in person or over the telephone and with one person or several people at a time. As interviews are labor intensive and hence costly, they tend to use smaller sample sizes than printed surveys. The form of the interview will depend on the information required, the access to the target respondents, and the amount of interaction required between the respondents and the interviewer. Interview methods are particularly useful when the development questions have subjective answers that cannot be answered easily or cost-effectively through other methods. They can also be used in conjunction with other methods, such as after a laboratory test, to determine what respondents liked, did not like, and their overall reaction to a product.

A traditional interview is short, and conducted in person, with only one person at a time. Interview questions tend to be open-ended, encouraging respondents to supply more detailed information than they normally would be willing to write down on a questionnaire. A number of authors have suggested that as demands placed on respondents are reduced, the quality of responses increases (Sharp & Franke, 1983). Not only do interviews have higher response rates than other survey methods, but this method also allows the interviewer to clarify questions when respondents are confused (Babbie, 1989). Additionally, in-person interviews allow a wider set of responses to items in the survey as well as positive reinforcement to participants for their participation, both at the outset and at the conclusion of the interview. Successful interviews achieve a balance between gathering answers to the predetermined list of questions, systematically within the time allowed, and allowing respondents to completely deliver their thoughts (Metzler, 1989).

Focus Groups. Focus group research uses the same process as printed questionnaires and interviews to develop research questions. Focus groups are useful when the researcher wants to elicit controversy or agreement about an issue from the respondents. In fact, their hallmark is "the explicit use of the group interaction to produce data and insights that would be less accessible without the interaction found in a group" (Morgan, 1988, p. 12). They are well suited to explore peoples' attitudes about a product or process, their thoughts about the task or process, and constraints that may impede the use of certain technologies.

Walkthroughs

A walkthrough is a valuable usability tool during the definition phase of the ATIS development cycle for validating prototype features, the logic employed to access the features, and the overall "look and feel" of the proposed design. A walkthrough uses a prototype that can be in virtually any stage of refinement with a person operating the prototype as well as with one or several people giving feedback (Karat, Campbell, & Fiegel, 1992). Also available for usability research is a cognitive walkthrough technique, which is a mental exercise used by a product development team to collectively critique a design (or portions of a design) from the target end-users' point of view and identify gross problems that can be resolved before they are discovered during other forms of usability research (see Lewis & Polson, 1992, for more information).

In a usability walkthrough, the person operating the prototype uses a prepared list of task goals, such as checking traffic conditions, to guide a discussion about the features, the methods of accessing the features, the logic of the process, and possible design flaws. A group of end-users observe as the researcher shows how the tasks would be completed, interjecting when they feel that the task, the feature, or the implementation is inappropriate for them or their situation. The end-users also are invited to discuss their preferences and design ideas.

Because the purpose of a walkthrough is to gather information about the feature set without interference from a non- or only partially working interface, it is often desirable for the researcher directing the walkthrough to operate the ATIS or prototype. The researcher can then operate it in any way necessary, showing only the portions of interest, without needing to provide elaborate instructions to the subjects, wasting time, or losing focus in the manipulation.

Laboratory Testing

Laboratory testing is the most common of all the usability evaluation methodologies. By using this method, usability researchers can control the environment, the tasks and their context, the effects of learning, the interaction with the subject, and many other factors that may confound the results. The resulting data are also more precise because of the ability to collect data on elapsed time, performance, and deviations from the expected path. Because rigorous controls can be placed on the research, laboratory testing more closely resembles true experimental design than any of the other methods previously discussed. Laboratory testing can be used effectively throughout the development cycle to evaluate the existing version of a product, prototypes, competitors products, and documentation, as well as to conduct basic research (Desurvire, Kondziela, & Atwood, 1992; Helander, 1988; Muckler, 1992; Nilsen, Jong, Olson, Biolsi, Rueter, & Mutter, 1993; Spyridakis, 1992; Wright & Monk, 1991).

Although laboratory usability tests will differ depending on the questions being addressed and the attributes of the ATIS being considered, each will use basic experimental design principles. Laboratory testing uses a variety of data collection methods, such as observation, keystroke capture, verbal protocol, stimulated recall, and self-report.

Laboratory testing is a cost-effective way to gather large amounts of data quickly, where the controlled process is the important factor, not the financial investment in space and equipment. The power of laboratory testing comes from controlling as many variables as possible and understanding the effects of those beyond control. When usability testing is done correctly, the results are concrete and objective.

Three usability methods were used during different phases of the Traffic Reporter project: printed questionnaires to find large-scale demographic and commute pattern information, interviews to gather additional information about commuter habits and perceptions, and iterative laboratory testing with successive versions of the software to find usability problems with the design. The rest of this chapter elaborates on the application of these methods.

UNIVERSITY OF WASHINGTON–WSDOT ITS—TRAFFIC REPORTER

Traffic Reporter (TR) is a PC-based, graphic, ITS developed at the University of Washington for the Washington State Department of Transportation (WSDOT; see Fig. 8.1). The system seeks to improve traffic flow by influencing commuter behavior and decisions concerning alternative routes, departure times, and transportation modes. It is a real-time system in that it delivers up-to-the-minute information about traffic conditions. A mainframe computer located at the Washington State Department of Transportation Traffic Systems Management Center sends data to TR from detectors buried in the pavement of freeway lanes. The mainframe provides one-minute data summaries, and TR's graphical map updates its display with map colors representing freeway speeds, and a trip information window displays estimated travel speeds and times.

Usability testing of TR has been ongoing throughout the development of the system. At this time, three usability evaluations have been conducted during three phases of TR's development: questionnaires and in-person interviews when TR was in the investigation stage; laboratory testing when a TR prototype had been designed; and most recently, laboratory testing

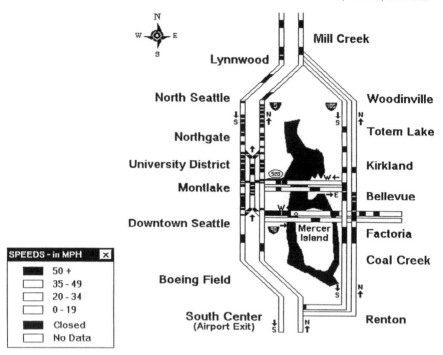

FIG. 8.1. Representation of "Traffic Reporter," a real-time traffic information system developed at the University of Washington.

during construction, just before TR was scheduled to become available for limited public use.

Investigation Phase: Printed Questionnaires

While TR was in the investigation stage, researchers distributed 9,000 on-road questionnaires to motorists commuting from the North into downtown Seattle on Interstate 5 (I-5), the main north–south corridor in and out of the city. The questionnaires focused on demographics, commuter driving patterns, and route and time flexibility (Spyridakis, Barfield, Conquest, Haselkorn, & Isakson, 1991). Profiles of commuters' use of traveler information for making driving decisions emerged from the questionnaires.

Initial On-Road Survey Respondents. The first steps in the project were to identify the survey population, the sampling frame, and the actual sample. The goal in selecting the initial population was to identify a large number of Washington State freeway commuters who experience traffic congestion and who have access to various forms of traffic information. In response to this goal, two Seattle corridors were initially specified for three reasons: (a) these Seattle corridors have extremely large traffic volumes that would allow for obtaining a sizable sample; (b) motorists who use these corridors during commuter rush hours frequently experience traffic congestion; and (c) these corridors contain various message delivery systems. The northern I-5 corridor was finally selected because it contained a variety of motorist information delivery media, including two on-road traffic message delivery systems: variable message signs (VMS) and highway advisory radio (HAR). Based on the corridor selection, the population was narrowed to include all motorists who use I-5 to commute from the North into the greater Seattle downtown area.

From this population, a sample frame was selected that would provide a sufficient number of motorists to represent the population, as well as motorists who have access to various forms of traffic information and who experience traffic congestion. To be included in the sample frame, motorists had to be the driver of a commuting vehicle; be traveling south on some portion of the northern I-5 corridor to downtown Seattle during peak morning hours (5:45 a.m.–8:45 a.m.); and use this corridor at least once a week.

We determined that a minimum sample size of 500 would provide findings with sufficiently small standard errors to accurately represent the driving population for the selected highway corridor. However, a secondary concern was to identify respondents who would be willing to participate in follow-up, in-person interviews and usability laboratory tests. Therefore, it was necessary to start with a much larger initial sample size. Assuming that 25% of the total initial survey respondents were likely to volunteer for the follow-up evaluations, we realized that a minimum of about 2,000 respondents to the initial surveys would be needed to ensure the success of the follow-up interviews and usability tests. Furthermore, with a minimal expected response rate of 20% for the mail-in survey, the desired motorist sample grew to almost 10,000, making this the largest motorist survey of its kind.

When this survey was repeated 1 year later for commuters traveling north on the southern I-5 corridor, using similar parameters for defining the population, we reduced the survey sample to 5,000, because we had learned from the northern corridor survey that the response rate of Seattle commuters to this survey was around 40%.

On-Road Survey Administration. In deciding what survey distribution method would be most effective for the on-road survey, we evaluated five distribution approaches. All methods were weighed against the following criteria: time frame for accessing motorists, potential response rate or bias, accuracy of contacting the desired sample frame, and time differential between time of commute and receipt of survey. We wanted to identify a method that would allow us to access motorists easily and without much delay, obtain a high response rate, obtain a representative sample, and reduce the time between motorists' commute and receipt of the survey. We evaluated five distribution methods: (a) geographic selection by business location, (b) identification by license plate, (c) geographic selection by home address or telephone, (d) general distribution through the media, and (e) on-road solicitation. Although somewhat labor intensive, the on-road solici-tation method was selected because it offered an in-person distribution method during commute time, a timely distribution method, and accurate targeting of the correct sample frame. (Details about the administration method can be found in Spyridakis et al., 1991.)

Questionnaire Design and Data Analyses. This section explains how the questionnaire was developed and how strategic objectives affected the development of the questionnaire. The content was based on specific infor-mation needs of this study, an extensive review of similar surveys, analysis of human factors issues relevant to driving and commuting, and suggestions from WSDOT personnel. After numerous questions were designed by all relevant parties, the questions were grouped into four categories: (a) char-acteristics of the commute itself, (b) motorist choices and behavior, (c) delivery of traffic information, and (d) descriptive data for driver classifica-tion and identification (voluntary). Through discussion with WSDOT, a pre-test of 25 drivers, and elimination of numerous questions (some of which became part of the in-person follow-up interviews), the final questionnaire was formulated. The layout and physical form of the questionnaire were determined by the distribution–return methodology: The survey was de-signed to be a three-fold self-mailer. Space was provided on the interior margins for respondents to add comments, and to facilitate accurate data entry, response boxes or input lines were provided for each question.

Frequencies were calculated for all variables of the total sample; then the sample was separated by gender, and gender differences were assessed with t-tests for interval data and Mann-Whitney U tests for ordinal data. Next, Pearson correlations were applied to interval scaled data and Spearman correlations were applied to all ordinal and a few nominal scaled variables if the coding met the assumptions of the Spearman routine. A factor analysis then was conducted on relevant variables (Spyridakis et al., 1991). Finally, cluster analyses and chi squares within clusters were conducted to identify possible motorist groups and the significant differences between the groups' responses on specific questions (Conquest, Spyridakis, Haselkorn, & Bar-field, 1993).

Results From the On-Road Questionnaire. The questionnaires revealed commuters' preferences for the delivery of traveler information:

- Commuters preferred information delivered at home.
- Commuters least preferred information delivered via computer (commuters in the later in-person interviews were quite open to information delivery via computer).
- Those commuters willing to make driving decisions based on traveler information (77% of all respondents) wanted commute time information and time estimates for alternate routes, and desired feedback to help them verify that they indeed had made the best choice.
- Commuters liked receiving time information in a numerical format and were willing to receive graphic information.
- Commuters wanted credible traveler information.

Through a statistical routine called cluster analysis, the questionnaires revealed four types of commuters in terms of their use of traveler information for making driving decisions:

1. *Route changers* (RC): commuters willing to change route before or after entering the freeway (21%).
2. *Route and time changers* (RTC): commuters willing to change time and route (40%).
3. *Pretrip changers* (PC): commuters willing to change time, route, and travel mode before leaving their residence, but unwilling to change route during their trip (16%).
4. *Nonchangers* (NC): commuters unwilling to change departure time, route, or travel mode (23%).

Whereas the cluster analysis was conducted to identify groups of commuters based on their willingness to adjust their commute in response to traffic information, further analysis was conducted to determine commonalties among commuter responses in the survey. A principal components factor analysis of the correlation matrix from the original data set revealed a five-factor solution across the 62 variables: (a) issues affecting route choice; (b) distance–time information; (c) traffic information, particularly via TV and radio; (d) traffic information, particularly via VMS, HAR, and telephone hot line; and (e) commute attributes and flexibility.

Recommendations for System Design. We concluded from these questionnaires that traveler information could influence four types of commuter choices: departure time, transportation mode, selection of pretrip route, and selection of on-road route. In response to these conclusions, we defined the following goals for our ITS: to tailor pretrip information to the RTC and

PC groups—those commuters most likely to alter departure time and route choice; to develop a system for home delivery of commuter information; and to include a feedback mechanism in any system developed.

The initial on-road survey provided an extensive amount of data on commuter behavior. Cluster analysis further provided a means of conceptualizing the design of the in-person survey, and the principal components factor analysis indicated that additional information was required regarding issues that might affect route choice. The on-road survey also raised a number of interesting questions, which the in-person interviews were designed specifically to probe.

Investigation Phase: In-person Interviews

In-person interviews of Seattle-area commuters probed three broad areas of interest derived from the basis of the analysis of responses to the on-road survey: (a) behavior and decisions of commuters relative to their route choice before departure, (b) behavior and decisions of commuters while driving, and (c) responses of commuters to a set of messages manipulated for order and wording that might be displayed on VMS. All the interviews were conducted at the University of Washington.

Interview Respondents. Subjects were recruited from the initial survey respondents who indicated a willingness to participate in an in-depth study. They were randomly selected from each of the four clusters identified in the initial survey, in numbers proportional to the original cluster sizes. Of 120 subjects recruited, 96 participated (80% participation rate).

Interview Administration. In selecting an interview method, we considered a number of issues: demands placed on the subjects, reliability, response mode, and tone and presentation. The in-person format, rather than a telephone interview, was selected on the basis of a literature review and after consideration of the issues surrounding the types of questions to be asked.

Two steps were taken to control for threats to internal validity, a common control issue with this method. First, interviewers were trained by one researcher and were required to meet specific criteria before interviewing commuters. Second, interviewers worked from a written questionnaire that specified all interviewer prompts and provided categories for recording commuter responses. Subjects were interviewed individually. They were informed that participation was voluntary and that they could take a break or terminate the interview at any time.

Interview Design. The areas of interest to the development team were converted into three sets of research questions. The first group explored the environment in which commuters receive and use pretrip information by questioning the respondents about the amount of time they have before departure, and the number and type of demands they experienced during that time. Also included in this first set were questions probing the respondents' access to and use of traffic information before departure.

The second set of questions probed commuters' behavior and decisions during the commute between home and work. Commuters were asked to describe orally one primary and two alternative routes (if known and used) and to trace those routes on detailed street maps. Additional questions probed sources of information used either to confirm or refute decisions to alter routes. The final items in this set of questions probed commuters' perceptions of flexibility in arrival time, penalties for arriving late, and stress of using alternative routes. Commuters also were asked to estimate the number of times they arrived late for work each month because of traffic conditions.

The third set of questions probed commuters' responses to two sets of message manipulations that might be displayed on variable message signs. The first set of manipulations involved two variables known to affect task performance: task instruction (to use a specific alternate route or a generic instruction to take another route) and message order (suggestion for route or reason for traffic problem) (Dixon, 1987; Kulhavy, Schmid, & Walker, 1973; Yekovich & Kulhavy, 1976). The order of messages was randomized so that the task instruction appeared either before or after the reason (the description of the traffic problem). A second manipulation involved the type of reason (generic or specific) presented in the message and the presence or absence of a suggested response to the traffic situation. Messages that contained specific reasons presented a specific description of the traffic problem (e.g., accident at Mercer Street exit); messages that presented generic reasons gave a more general description (e.g., accident ahead). Messages were manipulated to present either a suggested response to the traffic situation or a general statement (e.g., expect delays).

Data obtained from the in-person interviews were analyzed in a manner similar to the data analysis completed for the initial on-road survey. All statistical analyses were conducted using a standard statistical software package (SYSTAT). Responses were examined for patterns across the entire sample and for patterns across clusters and gender. Finally, a principal components factor analysis was conducted to reveal commonalties of responses (Wenger, Spyridakis, Haselkorn, Barfield, & Conquest, 1990).

Results From In-person Interviews. These interviews provided information about commuters behavior and their decisions before they depart and en route, and their responses to variable message sign information. On the whole, commuters are somewhat receptive to traffic information delivered before departure. They reported that the period before departure is not stressful and that they have a relatively small number of tasks to accomplish. Their low rate of modification to route, mode, and departure time may indicate that commuters may receive the traffic information passively and may not find it credible. This second inference is supported by comments to this effect received from commuters on the initial on-road survey. The low rate of route modification may be caused as well by temporal delay between receipt of the information and decision because the majority of commuters reported receiving their first traffic information more than one hour before departure.

Commuters have a high degree of knowledge about their primary and alternative routes. A majority make use of alternative routes at some time, and nearly half of the alternative routes include some portion of I-5 (the primary route used by commuters into downtown). Commuters make a small number of adjustments to their primary route, mainly on the basis of their observations of traffic conditions. However, it appears that commuters decide to use an alternative route on the basis of traffic reports received either at home or in the car. They receive little feedback regarding their choice to use an alternative route and, when received, this feedback is delayed. Finally, commuters are not burdened with tasks other than commuting to their workplace, and approximately one third of all commuters experience high levels of stress en route, with the perceived level of stress increasing if they use an alternative route.

Commuters were more likely to correctly interpret VMS messages when they are presented with a specific rather than a generic task. Furthermore, they were more likely to interpret the message correctly when the reason was presented before the task. Interestingly, a pattern in total contrast to the one just described was observed for the probability of commuters changing route in response to the message. Commuters indicated that they would be more likely to change their route when the message presented a generic (rather than specific) task and when the task (rather than the reason) was presented first. Finally, commuters indicated that they would be most likely to change their route in response to a message if the message presented a generic reason and did not present the task.

Recommendations for System Design. The in-person interviews produced a picture of an extremely complex commuting population, but one with definable needs, thus suggesting important implications for the design of motorist information systems This description of the behavior and decision of commuters before departure and en route, as well as their responses to the message manipulations, has a number of implications for the design of information systems. The results reinforce the notion that demonstrating system credibility may be a significant issue. Furthermore, the results indicate that commuters may have time to use an interactive graphic traffic information system before departure, one that would demand some active engagement. Although commuters do rely on traffic reports, they require that the information be more current and specific than the information currently available and that the accuracy of information be verified through feedback. Increasing currency and specificity might well increase the probability that commuters would choose an alternative route (as opposed to merely making minor adjustments to the primary route). Incorporating feedback mechanisms into on-road systems for delivering information might also encourage commuters to choose alternative routes.

When delivering information, designers should note that task information appears to be of secondary importance to commuters. Moreover, commuters prefer generic reasons. This finding may indicate that commuters wish only to know that a traffic problem exists and then to tailor their response to their specific commuting goals (Dudek, Weaver, Hatcher, & Richards, 1978). These findings also may be medium dependent in that the observed pattern of commuter responses might not be observed if, for example, the messages were delivered by radio. Furthermore, they can be generalized only to information delivered en route, not to information delivered before departure.

Design Phase: Laboratory Test

Usability testing of the TR prototype conducted at the University of Washington between September 1990 and February 1991 focused on the subjects' interaction with the system, the interface design, and the likelihood of changing commuting behavior based on the current system. The test subjects used the TR prototype to complete stereotypical commuter tasks, and after the tasks they commented on the interface during a posttest interview (Crosby, Spyridakis, Ramey, Haselkorn, & Barfield, 1993).

Laboratory Test Subjects. Three different user groups with somewhat different needs were assessed during this study: commuters, media personnel, and WSDOT traffic managers. The largest group of test subjects was selected from the three commuter groups that had completed the investigative phase questionnaire indicating they would change their commuting behavior. Based on questionnaire responses, subjects were randomly selected for the usability test, then equal numbers of subjects were assigned to each commuter group: route changers (RC), pretrip changers (PC), and route and time changers (RTC). Additionally, we tested both those commuters who stated they would and those who stated they would not like to see a computer system developed for delivering traveler information.

TV and radio media personnel, as well as WSDOT traffic managers, were also tested (using a slightly different test methodology) because they were to be the initial users of TR. These subjects consisted of three volunteers from the Seattle news media: a radio traffic reporter, a radio news manager, and a radio–TV reporter (accompanied by a radio reporter–pilot). The WSDOT subjects were six traffic managers asked by the Department of Transportation to assist in our evaluation.

Laboratory Test Design. The tasks selected for the laboratory test were based on earlier usability test results suggesting that commuters need commute time estimates and information in order to alter their commute choices. The commuter groups were given research scenarios in which they faced time constraints or heavy traffic, as well as tasks to perform with TR in which they obtained mean speeds of freeway traffic, received speed and time estimates for a specific trip, and selected the most efficient exit and entrance ramps. Similar information was obtained from media personnel but with a focus on traffic reporting. Finally, from WSDOT traffic managers, we elicited subjective evaluations of TR.

The version of TR used during the study displayed a map of I-5 (from the King-Snohomish County boundary to the North to slightly south of downtown Seattle), which changed color in small segments to reflect the traffic speed on that portion of the freeway. The displayed freeway grid was edged by exit and entrance ramp labels. Users could interact with the system to obtain information about the travel times and average speeds between these ramps as well as detailed information about specific sections of the freeway.

Laboratory Test Administration. To begin the test sessions, subjects were given a series of introductory materials: (a) a consent form; (b) a subject profile questionnaire to confirm that subjects belonged to the anticipated commuter and computer groups; (c) a scripted introduction stating the purpose of the usability test and of the system; (d) a scripted demonstration of the main screen, the mouse, the four functions of the system, and the help screen; (e) a tutorial that taught basic operations and four system functions (zoom, time and speed information between two ramps, information about specific entrance ramps, and information about specific exit ramps); and (f) a diversionary task to counteract the immediate memory effects of the tutorial and to provide an opportunity for subjects to practice verbal protocols. Media personnel were not given subject profile questionnaires or diversionary tasks. The WSDOT traffic managers were given a brief introductory statement concerning the purpose of the system, and a demonstration of the basic features of TR and the manipulations required to access them.

The commuter and media groups were then asked to perform, using a task list, sterotypical commuter tasks using TR. Each task required subjects to perform one of the four system functions being tested. The test administrator and the datakeeper recorded the order of the functions that subjects used as they interacted with the system, and the number of times subjects accessed the Help screen. The datakeeper also timed the subjects as they performed the tasks—from the moment they finished reading until they finished the task. Periodically, subjects were asked qualitative questions, on a 5-point scale, about functions they had just used. These questions asked about certain aspects of the interface, such as the clarity of the information or the usefulness of various dialogs, as well as subjects' likelihood of changing commuting route or time based on the information.

WSDOT traffic managers were tested in a group situation and were asked to explore the system, comment on it, and discuss it as a group. They were encouraged to comment about their concerns for TR and suggest additions or improvements.

The next part of the test, a questionnaire (using a 5-point Likert scale), examined the system overall, the interface, commuter behavior based on TR information, TR information versus other available forms of traveler information, additions or changes to the system, the meaning of arrows on the main screen, preference for delivery of TR information, and use of the system. A closing open-ended interview consisted of questions about commuters' thoughts on expansion of the system, different color codings or ways to show traffic flow, and other issues related to commuting and the Seattle freeway system in general. The media personnel were asked about ways to improve the representations of speed and traffic flow, their thoughts on expanding the system, and other issues related to commuting and commuter behavior.

The results were analyzed for scaled, timed, ranked, and subjective information gathered during the posttest interview. Although we applied both parametric and nonparametric statistical procedures to the data, and these statistics supported our conclusions, we did not report these statistics because of their instability given the small cell sizes. Rather, we reported descriptive statistics: medians, minimums, maximums, and frequency.

Results From Laboratory Test—Commuter Group. The laboratory test revealed that the interface was, overall, understandable to the commuter group (the RC, RTC, and PC groups had median scores of 43 to 44, with 30 being somewhat understandable and 50 being very understandable). The medians for computer preference groups were similar. Although the subjects reported that the interface was understandable, they had problems with several specific areas of the design:

- All commuters used virtually twice as many mouse actions as necessary to complete the tasks. This excess number of mouse clicks was the result of confusion about single- versus double-clicking and confusion between left and right mouse buttons.
- The colors chosen to indicate the range of speeds were acceptable, except for purple to indicate 20 to 34 mph traffic. Subjects also wanted an indicator for stopped, or stop and go traffic.
- Arrows to indicate exit or entrance-only ramps were not understandable.

The subjects suggested features that they would require in TR if they were to use the product in making commuting decisions. These additions to the feature set include the cause of delays (normal heavy traffic, accident, or repairs); the overall traffic conditions throughout the area; arterial information; the estimated duration of traffic congestion; and the estimated time savings by the change of routes.

All subjects said they would use TR if it were readily available to them but voiced concerns about the viability of the system for their personal use. These concerns included the cost to use a system such as TR, their actual ability to connect to TR hardware or software, and the credibility of the information displayed on the system (e.g., malfunctioning highway sensors).

This test also provided information about the subjects' potential for changing their commuting behaviors based on having TR available for their use. Changes in commuting behavior are reflected by a willingness to change departure time, route, or mode of transportation. The results of the research in these areas show that, given adequate information, subjects were willing to change their departure time based on the information provided by TR, but subjects were less willing to change their mode of transportation.

Results From Laboratory Test—Media Personnel Group. The media personnel felt that TR would be a welcome addition to the sources of information they already use in identifying potential traffic problems. They also cited specific issues with the current design of the system:

- The color code used to indicate speeds was good, except for purple.
- Inconsistency in the orientation of listed information with the map metaphor caused confusion.
- They wanted TR to report more understandable information, such as comparisons about the time it would take commuters to travel alternative routes. They also wanted comparisons of the current traffic situations with historical averages in weather, time of day or year, road conditions, and traffic-flow patterns with an indicator to highlight areas of unusual congestion.

- More specific information about the causes of traffic problems was needed.
- Verification of the accuracy of displayed information was requested.
- A larger highway area, with major arterials, was needed to give a confident picture of the commuting situation.
- Media personnel who report traffic information are referred to in their industry as traffic reporters, a title they found conflicting with the system name of Traffic Reporter (TR).
- They often use touchscreen technology in performing their jobs and would want to have TR implemented with that technology.

Results From Laboratory Test—WSDOT Traffic Manager Group. The traffic manager group had many of the same usability problems with TR as the commuters and media personnel:

- Purple used as a speed indicator, and no color indicator for stop and go traffic was problematic.
- The validity of the speed and time information displayed by the system was questioned.
- No information about waits at entrance ramps was a major drawback, because they felt much of a commuter's time was spent waiting to enter the highway.
- They wanted to see the inclusion of some type of arterial information.
- They felt that the TR screen design would not work well for TV or CATV because of the small and detailed information, such as ramp labels.
- They expressed logistical and ethical concerns about access, distribution, and cost of the system to the general public.

Recommendations for System Design. We concluded from the usability tests with the three groups that the concept of TR, the basic functionality, and the "look and feel" of the design were sound. The usability problems identified by the tests were not so severe as to jeopardize the project, and many of the issues could be addressed with changes to the existing interface. In response to this research we recommended interface design changes to address each of the usability problems identified by the study.

In our recommendations, we also addressed the implementation of TR, based on concerns raised by the test subjects, specifically that TR initially be developed for the use of TV and radio media, that a dedicated radio station be encouraged to use the system, and that continued development of a PC-based system be geared toward commuters who would be willing to pay for additional information.

Construction Phase: Laboratory Testing

Based on the results of the design-phase usability tests, the TR screen was redesigned and converted to a touchscreen that could be operated either with traditional mouse motions or by pressing on the screen with the user's fingertip. This version of TR was suitable for both interactive and noninteractive software versions. Other changes included replacing the four-color scheme for representing traffic speeds (green, yellow, purple, and red) with a three-color scheme (green, yellow, red), and using flashing red to indicate stop and go traffic. Area labels were restructured to provide submenus of relevant on- and off-ramps, which appeared on the main screen in the prototype. By pressing on area labels, commuters could select specific trips and gain access to trip speed and travel time information. Beyond changes to the interface stemming from usability test results, user comments led to expansion of the geography in the prototype's display. The intent of the usability tests during the construction phase was to determine whether the changes to the system effectively increased the likelihood that it would be easily understood and useful for making commuting decisions. This test was conducted by usability specialists at the University of Washington during the summer of 1992.

Laboratory Test Subjects. Once again, test participants were randomly selected from respondents to the on-road questionnaires. Two members of the research team, who had no previous experience with TR, pretested the test materials, and 16 commuters participated in the actual tests.

Laboratory Test Design. This phase of the usability tests focused on commuters' interaction with the touch screen, their interpretations of the main TR screen, their use and interpretation of the trip information window, their use of the zoom feature, the usefulness of printed guides for TR use, and their overall impression of TR for long-term use.

An IBM computer with a touch screen presented TR for this iteration of usability testing. Each test consisted of two sections: (a) participants interacting with the system, using a verbal protocol, and answering a set of prepared questions; and (b) a researcher then interviewing subjects about the design and function of TR.

Laboratory Test Administration. Subjects were provided with introductory materials similar to those provided during the design-phase test, and were then asked to complete tasks following a task list. Prior to using TR, subjects were asked their initial choice of screen operation, either mouse or finger, and told that they could change during the course of the test if they so desired. During the test sessions, data were recorded about subjects' performance and perceptions of the system and interface. To ensure that data were recorded accurately, that no questions were skipped inadvertently, and that answers were complete and relevant to each question's intent, two researchers (an administrator and a datakeeper) conducted each test session. The test sessions were video- and audiotaped, and the subjects' responses were also recorded by the datakeeper on a keyed response sheet.

Laboratory Test Results. This test revealed that the changes implemented in the system as a result of the previous test were effective for increasing the usability of TR. The change made to the color coding was the most dramatic improvement: Many subjects easily transferred their model of traditional traffic light colors to TR's colored speed indicators. This test also revealed problems that had not been uncovered during previous testing, and problems with the touchscreen implementation:

- Subjects had difficulty manipulating the touchscreen to select information and often had to press the screen more than once to activate the selection. They also had problems selecting information about a particular entrance or exit ramp when highlighting the ramp, obscuring ramp information with their hand, and selecting a ramp from an area list by using a dragging motion.
- Subjects had difficulty identifying the direction of traffic flow, north- and southbound, as well as east- and westbound.
- Subjects had problems identifying and interpreting express lane information, such as the direction of traffic flow, without assistance.
- Subjects preferred a chart format to a graph format for the presentation of specific trip information; they also preferred that trip information be ordered from the least to the most time-consuming options.
- Subjects expressed a preference for the spacing and sequencing of direction and highway descriptors (e.g., south on I-5, rather than south (tab) on (tab) I-5, I-5 south, or south I-5).
- Subjects identified travel time as the most valuable information.
- Subjects preferred a gray background over white or blue backgrounds.
- The zoom feature to obtain exact speeds within a color segment was not useful as designed.
- Ramp reference lists organized by ramp name were accessed more often than lists organized by area labels.

Recommendations for System Design. In response to these findings we recommended increasing the sensitivity of the touchscreen, increasing the size of the labels used in the interface, and reorganizing the screen to eliminate the need for scrolling. Also because of subjects' difficulties in manipulating the touchscreen with the finger, we recommended changes to the way that information was accessed from the touchscreen, such as having the user touch an area label to access a list of ramps, and displaying the list until a specific ramp was selected with a subsequent touch.

We also recommended the addition of labels to indicate geographic direction (N, S, E, and W) and the direction of open express lanes, and a gray indicator to designate closed areas of the express lanes. Further improvements included changing the background color of the trip information window to gray and placing travel time as the first item in the trip window display. We noted that documentation should be included with TR in the form of brief instructions, and ramp reference guides should be organized alphabetically by ramp (instead of area). We further recommended that the zoom feature be eliminated during the trial period.

CONCLUSION

Usability research is crucial to the ease by which people can use a system, the application of the system in their daily lives, and the acceptance the system will have as an information source. To have the most impact on the usability of a system, research should be incorporated into the early phases of the development process and continue as iterative testing during the remainder of the development process.

Usability research can contribute to the development process in five ways: It can (a) inform the design process about the people who ultimately will use the system in their daily lives, including their tasks, goals, background, environment, and culture; (b) validate assumptions about users' preferences in design and requirements for documentation; (c) confirm good design decisions so that developers can guard against inadvertently changing those decisions and so can capitalize on those ideas in future developments; (d) expose unexpected weaknesses in the user interface of which the developers may be unaware of because of their familiarity with the product; and (e) resolve uncertainties when there are differing ideas within the team about the direction of the product or a feature.

REFERENCES

Agar, M. H. (1986). *Qualitative research methods series, Vol. 2. Speaking of ethnography.* Newbury Park, CA: Sage.

Ahlgren, A., & Walberg, H. J. (1975). Generalized regression analysis. In D. J. Amick & H. J. Walberg (Eds.), *Introductory multivariate analysis* (pp. 8–52). Berkeley, CA: McCutchan.

Babbie, E. R. (1992). *Survey research methods.* Belmont, CA: Wadsworth.

Barfield, W., Conquest, L., Spyridakis, J., & Haselkorn, M. (1989). Information requirements for real-time motorist information systems, *Vehicle, Navigation and Information Systems*, Toronto, Canada, 101–104.

Barfield, W., Haselkorn, M., Spyridakis, J., & Conquest, L. (1989). Commuter behavior and decision making: Designing motorist information systems. *Proceedings of the 32nd Annual Human Factors Society Meeting*, Denver, CO, 611–614.

Conquest, L., Spyridakis, J., Haselkorn, M., & Barfield, W. (1993). The effect of motorist information on commuter behavior: Classification of drivers into commuter groups, *Transportation Research: Vol. 1C (No. 2)*, 183–201.

Crosby, P., Spyridakis, J. H., Ramey, J., Haselkorn, M., & Barfield, W. (1993). A primer on usability testing for developers of traveler information systems, *Transportation Research: Vol. 1C (No. 2)*, 143–157.

Desurvire, H., Kondziela, J., & Atwood, M. E. (1992). What is gained and lost when using methods other than empirical testing. *ACM SIGCHI '92 Conference Posters and Short Talks*, 125–126.

Dieli, M. (1989). The usability process: Working with iterative design principles. *IEEE Transactions on Professional Communication, 32*, 272–278.

Dixon, P. (1987). The processing of organizational and component step information in written directions. *Journal of Memory and Language, 26*, 24–35.

Dudek, C.L., Weaver, G. D., Hatcher, D. R., & Richards, S. H. (1978). Field evaluation of messages for real-time diversion of freeway traffic for special events. *Transportation Research Record 682*, TRB, National Research Council, Washington, DC, 37–45.

Good, M. (Ed.). (1989). Seven experiences with contextual field research. *SIGCHI Bulletin, 20*(4), 25–32.

Gould, J. D., & Lewis, C. (1985). Designing for usability: Key principles and what designers think. *Communications of the ACM, 28*(3), 300–311.

Helander, M. (Ed.). (1988). *Handbook of human-computer interaction* (pp. 791–928). New York: Elsevier.

Holtzblatt, K., & Jones, S. (1992, May). Contextual design: Using contextual inquiry for system development. Tutorial presented at *ACM SIGCHI '92 Conference*, Monterey, CA.

Huchingson, R. D., Whaley, J. R., & Huddleston, N. D. (1984). Delay messages and delay tolerance at Houston work zones. *Urban Traffic, Parking and System Management, Transportation research record 957*, Transportation Research Board, National Research Council, Washington, DC.

IVHS America. (1992). *Strategic Plan for Intelligent Vehicle-Highway Systems in the United States*, Washington DC: Author.

Karat, C.-M., Campbell, R., & Fiegel, T. (1992). Comparison of empirical testing and walkthrough methods in user interface evaluation. *Proceedings of the ACM CHI '92 Conference*, 397–404.

Kulhavy, R. W. , Schmid, R.F., & Walker, C. H. (1973). Temporal organization in prose. *American Educational Research Journal, 14*(2), 115–123.

Kvavik, K. H., Fafchamps, D., Jones, S., & Karimi, S. (1992). Field research in product development. *SIGCHI Bulletin, 24*(1), 22–27.

Lewis, C., & Polson, P. G. (1992, May). Cognitive walkthroughs: A method for theory-based evaluation of user interfaces. Tutorial presented at *ACM SIGCHI '92 Conference*, Monterey, CA.

Metzler, K. (1989). *Creative interviewing* (2nd ed.). Englewood Cliffs, NJ: Prentice-Hall.

Mirel, B. (1987). Designing field research in technical communication: Usability testing for in-house user documentation. *Journal of Technical Writing and Communication, 4*, 347–354.

Monk, A., Nardi, B., Gilbert, N., Mantei, M., & McCarthy, J. (1993). Mixing oil and water? Experimental psychology in the study of computer-mediated communication. *Proceedings of the ACM INTERCHI '93 Conference*, 3–6.

Morgan, D. L. (1988). *Qualitative research methods series: Vol. 16. Focus groups as qualitative research.* Newbury Park, CA: Sage.

Muckler, F. A. (1992). Selecting performance measures: "Objective" versus "subjective" measurement. *Human Factors, 34*(4), 441–455.

Nilsen, E., Jong, H., Olson, J. S., Biolsi, K. Rueter, H., & Mutter, S. (1993). The growth of software skill: A longitudinal look at learning & performance. *Proceedings of the ACM INTERCHI '93 Conference*, 149–156.

Poltrock, S. E. (1989). Innovation in user interface development: Obstacles and opportunities. *CHI '89 Proceedings, ACM SIGCHI*, 191–195.

Rasmussen, J., Pejtersen, A. M., & Schmidt, K. (1990, September). *Taxonomy for cognitive work analysis*. Available from Jens Rasmussen, Riso National Laboratory, DK-4000 Roskilde, Denmark.

Schneiderman, B. (1990). Human values and the future of technology: A declaration of empowerment. *Proceedings of the ACM SIGCAS Conference*, 1–5.

Sharp, L. M., & Franke, J. (1983). Respondent burden: A test of some common assumptions. *Public Opinion Quarterly, 47*(1), 36–53.

Shirazi, E., Anderson, S., & Stresney, J. (1988). Commuters' attitudes toward traffic information systems and route diversion. *Commuter Transportation Services, Inc.*, Los Angeles, CA.

Skelton, T. M. (1992). Testing the usability of usability testing. *Technical Communication, 39*(3), 343–359.

Spyridakis, J. H. (1992). Conducting research in technical communication: The application of true experimental designs. *Technical Communication, 39*(3), 607–624.

Spyridakis, J. H., Barfield, W., Conquest, L., Haselkorn, M., & Isakson, C. (1991). Surveying commuter behavior: Designing motorist information systems. *Transportation Research, 25A*(1), 17–30.

Sullivan, P., & Spilka, R. (1992). Qualitative research in technical communication: Issues of value, identity, and use. *Technical Communication, 39*(3), 592–606.

Wenger, M., Spyridakis, J., Haselkorn, M., Barfield, W., & Conquest, L. (1990). Motorist behavior and the design of motorist information systems. *Annals of Transportation Research Board* (No. 1281), 159–167.

Wright, P. C., & Monk, A. F. (1991). A cost-effective evaluation method for use by designers. *International Journal of Man-Machine Studies, 35*, 891–912.

Yekovich, F. R., & Kulhavy, R. W. (1976). Structural and contextual effects in the organization of prose. *Journal of Educational Psychology, 68*(5), 626–635.

9

HUMAN FACTORS PARTICIPATION IN LARGE-SCALE INTELLIGENT TRANSPORTATION SYSTEM DESIGN AND EVALUATION

Rebecca N. Fleischman
General Motors Research and Development Center

Thomas A. Dingus
Virginia Polytechnic Institute and State University

There is historical precedence for the participation and role of human factors in the design and evaluation of large-scale systems. Primary examples include the human factors specialists who have had significant roles in the testing of a wide variety of space and military systems dating back to the late 1940s. The evaluation methodologies conducted for these systems were driven by standards that required verification of the safety and efficiency with which a set of required tasks could be performed. This verification included equipment design, logistics support, training, personnel selection, time-line adherence, procedures, and personnel safety among others.

Although there is long-ranging experience in the human factors community on which to draw for effective design and evaluation of Intelligent Transportation Systems (ITS), this experience has been underutilized. Historically, there have been no established standards for the evaluation of ITS. Rather, ITS systems, evaluation plans, and deployments have been developed and conducted by separate contractors as part of transportation demonstration programs. The system designers and evaluation contractors typically have not emphasized the role of humans in these systems. Although the experience of early tests has been used somewhat, the ITS evaluations have differed substantially in their scope, objectives, and methodologies, making transfer of learning a challenge. To improve this situation, the Federal Highway

Administration (FHA) sponsored the development of the MITRE guidelines for conducting evaluations of ITS, which include some human factors elements (Bolczak, 1993). However, an ongoing effort is necessary to involve human factors professionals and behavioral scientists early enough in the development and deployment of ITS to influence system usability, safety, and acceptance so that the ultimate ITS goals can be achieved.

The purpose of this chapter is to provide some evidence and insight into the potential for human factors contributions in the design and system-level evaluation of ITS. This guidance is presented, when possible, through an exemplar large-scale evaluation recently completed: TravTek (travel technology). TravTek is, to date, the most comprehensive ITS large-scale evaluation (LSE) in terms of both its human engineering in system design and its integrative, programmatic evaluation approach. The TravTek program provided researchers with a rich environment for learning in a real-world laboratory, for addressing speculation and hypothesis with data, and for developing and validating models. The TravTek operational field test is described as a system, a driver–system interface design, and a research and evaluation program. The studies that TravTek comprised are reviewed, and the most relevant results from a human factors standpoint are summarized. The authors participated in both the driver–system interface design and the behavioral research.

HUMAN FACTORS AND THE ITS LARGE-SCALE EVALUATIONS

Large-Scale Evaluations (LSEs) are conducted to demonstrate the technical feasibility and viability of system concepts prior to their full deployment. Traditionally, they have been utilized primarily as technology demonstrations, with no systematized attempt to study the role of the human in the person–machine system. With the advent of numerous LSEs, human factors practitioners and researchers have an opportunity to make major contributions to forthcoming ITS systems. Whereas design engineers demonstrate and assess the technical viability of the system, human factors personnel can provide insight into the safety and usability of the system at a micro-driver system level and into the social–technical interactions that arise from putting technological innovations into the public domain. Some of the issues concerning human and societal factors in ITS systems are these:

- safety
- usability
- driver acceptance
- impact on local communities

- institutional barriers
- privacy for users
- fairness to those who do not purchase an in-vehicle system
- willingness to pay for private as well as public components of ITS

Human engineering should be given prominence in systems development to ensure that sufficient attention is paid to the user interface and to concerns such as safety, usability, and public acceptance. Another critical aspect is the inclusion of human factors in the design of coherent research models for evaluating the system under test. The influence of large-scale evaluations can range beyond a mere proof of concept. The outcome of an LSE may provide a basis for policy making. Therefore, it is essential to frame the ITS system as a person–machine system, and it is fruitful to cast the LSE environment as a real-world laboratory. It also would be productive to see the process of design, evaluation, and redesign as extending into the early deployment of systems, perhaps throughout the lifetimes of newly deployed technologies.

In the ITS domain to the present, Advanced Traveler Information System (ATIS) technology has tended to receive the most human factors attention. However, other types of ITS such as Advanced Traffic Management Systems (ATMS) represent less obvious, yet important behavioral issues.

SPECIAL OPPORTUNITIES FOR HUMAN FACTORS

The ITS large-scale evaluation environment affords opportunities of special interest to human factors professionals and behavioral and social scientists in the areas of design and research. They include the following:

Opportunity to frame system as a person–machine system. The LSE is an occasion to think deeply about the relationship between the transportation infrastructure, hardware, software, data, and the people who use it. It enables transportation engineers, policymakers, behavioral experts, government, industry, and academics to learn from each other, and to work together to solve problems. In particular, there is an opportunity to understand better how the system works and to recast it as a person–machine system.

Often, a system is described as a collection of interconnected physical parts. It is of great benefit to develop functional descriptions of how a system achieves its goals. Human factors experts can be particularly helpful in spearheading discussions of systems, how they work, and how much they rely on human perception, judgment, and behavior to do their work. Often, there is a tendency to make assumptions about behavior and to state hypotheses as fact. The inclusion of human factors professionals on design and evaluation teams help to increase the awareness that many of the potential benefits of ITS may depend not only on the accuracy and timeliness of the information, but on the ability of drivers to extract needed information from displays and to utilize it without significantly diminishing the performance of basic driving tasks. Even benefits such as congestion alleviation can depend on the assumption that when given better information, drivers will behave as the traffic engineers and modelers predict.

Opportunity to learn while doing. The LSE offers the chance to participate in a dress rehearsal, and to learn about the human impact of the technology early enough to influence emerging ITS industries. Human–system interface designs for ATIS present many interesting challenges as well. A vehicle–driver interface design must accommodate a wide range of driver abilities and preferences. The most successful effort will include human factors engineering as an integral part of the conceptualization, design, implementation, and testing cycle, and not just as an afterthought. The opportunities for learning from LSE experience are endless and inevitable.

In particular, the complexity of incorporating an ATIS system into the vehicle environment requires human factors experts to think about vehicle technology as a dynamic integrated system of information, displays, and controls, rather than as more traditional static displays. The person–machine relationship in a highly functional ATIS system is so complex that some human factors issues are only to be resolved adequately—or may only surface—at the time of implementation. Some human factors issues in current ATIS development are closely intertwined with technical possibilities and constraints. For example, lane information, freeway exit signs, and other geographical data can be displayed or otherwise used only if they are included in databases. This requires designers who need specific data for implementing a design to communicate and negotiate with people working on ITS architectures and to request that data providers collect the additional data. Thus, human factors involvement in the design-test-redesign cycle may reach as far back as the inventors and producers of technological innovations.

Opportunity to observe human behavior, particularly driving behavior, in situ. The LSE setting is environmentally rich, with much greater situational variety than is found in driving simulations, in terms of realistic traffic and weather conditions, road configurations, and so on. The LSE environment enables the study of higher-order driving behaviors such as way finding, which are difficult to research realistically in simulators or on test tracks because of limitations in the number of choices and cues that can be presented. The LSE is particularly well-suited for evaluating claimed system benefits and for exploring societal issues.

The LSE environment also makes it possible to investigate the utility of ITS functions in realistic situations. Their utility can be better understood both by the operators' subjective evaluations and by the automatic recording of the choices they make, to determine how often and under what circumstances they use a system's features and functions. Will drivers actually use the system and comply with its recommendations? For example, if offered a new route while in transit to a destination, will they accept the reroute? Do drivers perceive a time savings? Do they venture farther into unknown territory if they have a yellow-pages-style directory and route-guidance capabilities?

Opportunity to collect acceptability and system requirement data from users of an actual test system. Because users operate prototype vehicles over an extended period of time, researchers gain access to their reactions to a wide variety of occurrences in context. These data may provide realistic answers to important questions: Is the information presented useful for its stated purpose (i.e., to aid in decision making, way finding, warning, etc.)? Is the information presented in a timely manner; is it intelligible or distracting? What are the consequences if the system fails? Information gained at the test stage can be applied to improving the design before the real system or product is deployed. LSEs offer the opportunity for reliable market research based on the use of actual prototype systems in meaningful situations.

LSEs also make it possible to test and compare different design solutions in the form of different system architectures and different person–system interface designs. This can be fruitful if human engineering is part of the system design, especially if careful thought is given to alternatives for displays and controls and if alternatives can be implemented for testing. In the case of route-guidance information, for example, the virtues of map displays, simplified maneuver-by-maneuver displays, and voice guidance all have been proposed and debated in terms of theoretical supposition (Dingus & Hulse, 1993). In TravTek, all three display options were available to field-study drivers so that display usage could be logged and user ratings collected. The three were also compared in several controlled experiments that studied their potential for distraction and their usefulness for navigation.

Opportunity to validate laboratory research, metrics, and models. A number of measures have been developed by the human factors research community to study driving behavior, interface usability, and the potential for distraction and errors. Data have been collected in laboratory experiments during rapid prototyping and using part-task simulators. There is also a growing use of higher-fidelity driving simulators to test ITS systems. Metrics used in these studies will need continuing validation by over-the-road and other in situ studies. Major design decisions can rest on the results of these studies, including the rejection of classes of displays and the restriction of access to features while the vehicle is in motion.

Specially equipped camera cars have been used in several over-the-road studies of ATIS, warning system displays, and car phones (Dingus et al., 1995; Green, Hoekstra, & Williams, 1993; McGehee, Dingus, & Horowitz, 1994).

Measures of visual attention have included the number of glances to a display, dwell times and performance measures such as errors in extracting information from a display, errors in simulated tasks and actual driving performance degradation such as lane excursions, vehicle speed and steering wheel angle variability, and reduced glances at the mirrors.

It is important to recognize the limitations of "context-free" research (Mitchell, 1989). In situ studies allow us to validate the metrics used in smaller and necessarily more artificial studies. Although we have focused on driving, this would apply to the development and testing of traffic management center (TMC) operator–system interfaces as well. Taking advantage of opportunities to compare the results of system tests from both the laboratory and a real-road network or real control room will lead to a greater understanding of the sensitivity and meaningfulness of our methods. Insight into the utility of information furnished to drivers provides a realistic basis for adjudicating design trade-offs, and may result in the design of person–machine system interfaces that reduce the demands on the user.

Finally, developing predictive operator–system models, traffic models, environmental models, and social models is essential. However, there is no substitute for the collection of empirical data. The LSE is an opportunity to calibrate and validate models; to generate new models and hypotheses.

Opportunity to study the relationship between safety, human factors measures, and design. One of the major ITS objectives is to increase transportation safety. This benefit may be derived from decreased navigational waste, reduced exposure to congestion, routing to a safer road class, collision warning, and so forth, depending on the particular system. Yet the potential for decreased safety is a concern. Safety experts and human factors experts want ITS to be designed for safety, but they may not be speaking the same language or thinking about safety at same level of analysis.

The traffic safety community tends to use measures such as the number, type, and severity of crashes. Crashes are aggregated and analyzed epidemiologically as a function of exposure. These metrics rely heavily on the statistical analysis of large crash databases collected over a period of time. On the other hand, human factors researchers think of safety issues in terms of the potential for distraction from primary task needs, visual attention and processing demands, and cognitive overload. In evaluating the safety of ITS, a great deal is being made of "safety surrogates," microscopic measures that are generally believed to represent the preceding issues. A major challenge is validating these measures as true safety surrogates by demonstrating that differences in particular metrics represent differences in the risk of crashes as understood by the safety community. For example, how much eyes-off-road time (and under what circumstances) will result in an increased probability of crashes? Compared with traditional human factors studies, the LSE affords the chance to collect crash data over a period of time, with a relatively large data set. When used in conjunction with human factors experiments that observe drivers using the same system and the same road network, the relationship between the two levels of analysis can be better understood.

CURRENT STATUS OF ITS LARGE-SCALE EVALUATIONS AND THE HUMAN FACTOR

Several major research and development projects have been planned by the U.S. Department of Transportation. They have a cooperative private sector–public partnership orientation to field testing, yet differ widely in organization, emphasis, and evaluation approach. The LSEs include automated highway systems (AHS), commercial vehicle operations (CVO), and advanced traffic management systems (ATMS), but most of the human factors design effort and research has thus far been applied to ATIS. Most LSE descriptions emphasize the technological aspects of evaluation and not the human factors aspects (FHWA, 1993). The emphasis on technology is also apparent in the National Program Plan for Intelligent Vehicle Highway Systems, although human factors–related issues are also addressed in many cases (ITS AMERICA, 1993).

A few of the large-scale projects include safety evaluation, older driver issues, driver workload assessment, and laboratory and simulator studies as well as over-the-road human factors experiments and other appropriate in-context studies. For the most part, European and North American LSEs that collect user data at all, do so through questionnaires that include questions about the usability of the system. Following are some examples of U.S. LSEs that have included some evaluation of user interactions with the ITS technology under test.

The Pathfinder project was initiated in 1988 and conducted on a 13-mile section of the Santa Monica Freeway in California. It delivered traffic information to drivers in their vehicles using an Etak map system and synthesized voice messages (Mammano and Sumner, 1991). The evaluation included a human factors survey and over-the-road field studies to test trip efficiency and driver behavior. Driver perception of the system and its benefits were positive on the whole. Although there were no human factors experiments, the importance of human factors became clear, and improvements were recommended by the drivers as well as the evaluators (JHK Associates, 1993). Drivers needed to find alternative routes around traffic congestion using the map display, which may have presented a challenge to drivers. Systems that include route selection and guidance may encourage more drivers to divert from congested freeways.

TravTek was in operation in Orlando from March 1992 through March 1993 offering an opportunity to study a system that included route guidance as well as traffic information. It was the most ambitious LSE in terms of its emphasis on human factors, safety, and behavioral research as part of the system evaluation. The TravTek system and evaluation will be described in the following discussion.

SmarTraveler was an LSE conducted in the metropolitan Boston area from 1993 to 1994. This ATIS project provided users with travel information by conventional and cellular phone. A synchronous audiotext system was developed to organize and disseminate highway and transit information. The evaluation included an assessment of system use as well as a user survey (Juster, Wilson, & Wensley, 1994).

Faster and Safer Travel through Routing and Advanced Controls (FAST-TRAC) is currently underway in Oakland County, Michigan. This ATIS and ATMS field test provides real-time turn-by-turn route guidance to drivers whose vehicles are equipped with the ALI-SCOUT as they drive through corridors of roadside infrared beacons. The Autoscope™ video traffic detection system and the Sydney Coordinated Adaptive Traffic System (SCATS) as well as the vehicles will provide the system with traffic data. The FAST-TRAC evaluation plan includes studies of user perceptions and behaviors, human factors, stakeholder analysis, institutional barriers, and traffic modeling, in addition to evaluations of technical performance and the control center (Underwood, 1994). Planned approaches include questionnaires and studies of the ease of use and safety of the ALI-SCOUT such as experiments with an instrumented car and eye camera. Some preliminary findings are reported by Barbaresso (1994).

The Advanced Driver and Vehicle Advisory Navigation Concept (AD-VANCE) is designed to evaluate the performance of a large-scale dynamic route-guidance system in the Chicago area. The field test is being conducted at this writing. In addition to numerous tests of system effectiveness, reliability, and a benefit–cost analysis, the ADVANCE evaluation plan calls for the evaluation of the information products created by the ADVANCE system and a safety evaluation (Koziol, Wagner, & Bolczak, 1994)

Peters (1995), in an overview of human factors evaluations in ITS large-scale evaluations, mentioned the Seattle Wide-Area Information System (SWIFT) and Genesis as well as TravTek. SWIFT is scheduled to begin in 1995 and will provide users with traffic advisories, personal pagers and informational messages. Three types of high-speed data system (HSDS)-capable devices will be used. They are Delco radio receivers, the Seiko Message Watch™, and IBM portable computers. Portable computer users will receive additional information such as a map with current traffic, bus, and ride-share information. SWIFT plans to include customer acceptance as part of the evaluation process (Science Applications International Corporation, 1995a).

Genesis is part of the Minnesota Guidestar Program. Beginning in 1996, it will test personal communication device (PCD) technology as a way to communicate travel information. User acceptance tests and human factors tests are among the system tests planned (Science Applications International Corporation, 1995b).

PROMETHEUS is an acronym for PROgram for European Traffic with Highest Efficiency and Unprecedented Safety. The emphasis of PROMETHEUS has been on systems having a large in-vehicle component to their design. The ultimate goal of the project is to develop technology such that every vehicle has an on-board computer for monitoring vehicle operation, provide the driver with information, and assist with the actual driving task. The initial research phase of PROMETHEUS has been completed, and the current emphasis has shifted to field tests and demonstrations. Ten common European demonstrations have been or are currently being evaluated in the following areas: vision enhancement, emergency systems, proper vehicle operation, commercial fleet management, collision avoidance, test sites for traffic management, cooperative driving, dual-mode route guidance, autonomous intelligent systems, travel information systems, and cruise control.

Human factors-related objectives are becoming more prevalent. Yet most of the effort is focused on user acceptance and marketing surveys, which can be useful, but which do not get to the heart of usability and safety issues. Furthermore, the degree to which the evaluation objectives are successfully integrated as part of the overall project objectives varies greatly from test to test, even within the auspices of very similar ITS large-scale evaluations. This makes the transfer of learning and the building of knowledge difficult. Perhaps the MITRE guidelines (Bolczak, 1993) will help. To date, ITS demonstrations have been heavily driven by a primary objective of "getting the technology on the road." A danger in this approach is that extremely valuable sources of data will go untapped. Furthermore, unless human engineering and safety are integral parts of the system design, potentially useful technology may not gain wide-ranging acceptance, and progress toward ultimate, effective ITS deployment may suffer.

AN EXAMPLE OF THE COMPREHENSIVE HUMAN FACTORS DESIGN AND EVALUATION IN ITS: THE TRAVTEK PROJECT

TravTek was developed to test and evaluate an integrated advanced traveler information system (ATIS) and supporting infrastructure including a traffic management center (TMC). The system was operational from March 1992 through March 1993 in metropolitan Orlando, Florida. However, the planning and design process began in earnest in 1989, and the evaluation process continued through the first half of 1994. TravTek serves as an example of a successful public–private sector partnership, a major ITS human factors effort, and an integrated research and evaluation program.

TravTek public sector partners were the Federal Highway Administration (FHWA), the Florida Department of Transportation, and the City of Orlando. The private sector partners were the American Automobile Association (AAA) and General Motors. All partners participated in the planning and implementation of the TravTek program. This collaboration was not limited to each partner performing its legally agreed-upon part, but extended to ongoing cooperation in solving technical problems and overcoming institutional barriers. This was due, in part, to a mutually concluded set of project objectives and to regular working group meetings that facilitated communication and problem solution.

TravTek stands as the most comprehensive LSE in its approach to research and evaluation. In addition, TravTek expended unusual effort on human engineering and behavioral issues. Human factors was an early and continuing focus in the TravTek vehicle design process, with the formation of a human factors design team that operated from conceptual design through implementation and operations.

Human factors professionals also participated at the TravTek system level, particularly as members of the Evaluation Working Group and the Evaluation Contract Team. They aided systems developers in designing a program that would satisfy the primary objective of the TravTek partners: to field-test a system that was easy to learn, easy to use, safe, and shown to yield a positive experience for AAA, Avis, and GM customers traveling in Orlando. Early human factors intervention enabled TravTek to exceed the scope of the traditional system evaluation and market research approaches. By calling out issues such as driver attention, performance, and safety early in the program design, the human factors personnel had an opportunity to help shape the fundamental system concept and the scope and nature of the evaluation.

TravTek System Description

To communicate some of the more critical details of the TravTek evaluation, it is important, first of all, to explain some details of the system and its objectives. Among its many objectives, TravTek was designed to improve trip efficiency and avoid traffic congestion by providing vehicle navigation, route selection and route guidance, and real-time traffic updates to drivers. TravTek was also meant to increase mobility and security through an in-vehicle Services and Attractions directory and a hands-free cellular telephone. For individual drivers, improved trip times could be achieved through computer-assisted trip planning, reduced navigational waste, and routing around traffic incidents and congested areas of the greater Orlando traffic network. In a fully deployed system, congestion-reduction benefits could be achieved at the traffic network level, even for nonequipped vehicles, if a sufficient percentage of drivers with equipped vehicles use the functions for a suffi-

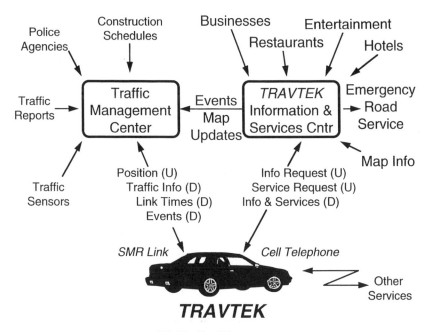

FIG. 9.1. TravTek system.

cient percentage of their trips. TravTek was also designed to explore the viability of using the equipped vehicles, themselves, as traffic probes or collectors of traffic data for the Traffic Management Center.

These functions were achieved by a three-segment system architecture detailed by Rillings and Lewis (1991) and Blumentritt, Balke, Seymour, and Sanchez (1995) and shown in Fig. 9.1. The TravTek system and its major subsystems, the Vehicles and Drivers, the Traffic Management Center (TMC) and the TravTek Information and Services Center (TISC) are described briefly next. They were linked by digital communications, including 800-MHz FM data radio, dedicated telephone lines, and automatically dialed cellular telephone. In addition to fulfilling their expressed ATIS purposes, all three subsystems had substantial evaluation support functions such as automatic data logging and special software to manipulate experimental variables.

TravTek Vehicles and Drivers. One-hundred specially equipped 1992 Oldsmobile Toronados were engineered and provided by General Motors. The Toronado was chosen because of its built-in color cathode ray tube (CRT) with infrared touch-screen matrix, steering-wheel mounted control keys, hands-free cellular telephone, and internal digital communications bus. The Toronado's production features served both as enablers of and constraints for the ultimate TravTek design (Fleischman et al., 1991). The vehicles were equipped with a combination of dead-reckoning, map-matching, and global-positioning system (GPS) navigation. General Motors engineers developed an in-vehicle system that integrated navigation functionality, external communications, real-time traffic data, traffic probe functionality, and data logging for evaluation support.

The drivers, of course, were an integral part of the vehicle subsystem. Although people with a special interest in TravTek, such as employees of partner organizations and members of ITS AMERICA, drove during the year of operation, most of the 4,354 TravTek drivers were visitors to Orlando who rented vehicles through Avis. The design team viewed drivers as key actors in the system, as decision makers and users of information. For example, system-levels benefits such as congestion reduction can only be achieved if drivers can safely extract and understand the information provided and are willing to use features such as route guidance. The TravTek partners also recognized the importance of the experience of the TravTek drivers and the user friendliness of the driver–system interface from both the design and evaluation standpoints. As such, the system design would be judged successful, not only by fulfilling its technical objectives, but to the extent that drivers could learn to use it easily and extract information from displays safely and on demand.

Because of the extensive system functionality and multiple display choices, TravTek drivers had various alternatives for data entry and information retrieval. Illustrations of many TravTek screens can be found in Carpenter et al. (1991) and Krage (1991). At the start of a given trip, a driver could choose to enter a destination, in which case TravTek selected a route and provided route guidance throughout the trip. Drivers were offered three methods of specifying a destination:

- Use a touch screen alphanumeric keyboard to spell a street address or street name.
- Select a destination from a stored list of previously entered destinations.
- Select an establishment from the Services and Attractions directory.

Once the driver had specified a valid destination, a choice of routing criteria was offered. The driver could request either the fastest route, a route that avoided interstates, or a route that avoided toll roads. All three options used real-time traffic data from the TMC to optimize the route for travel time.

En route, the driver had two visual guidance displays available, as well as auditory guidance instructions. The visual displays consisted of the Route Map, a moving map on which the selected route was highlighted (see Fig. 9.2), and the Guidance Map, a maneuver-by-maneuver display that depicted the geometry of the upcoming maneuver intersection (see Fig. 9.3). Drivers

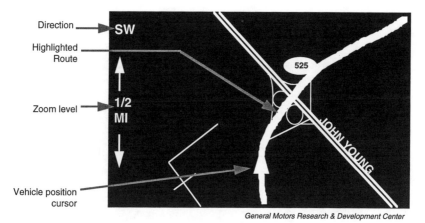

General Motors Research & Development Center

FIG. 9.2. Route map.

could toggle between the two visual displays with a SWAP MAP button on the steering wheel. Voice guidance consisted of maneuver instructions spoken by a computer-generated voice that could be switched on and off with a VOICE GUIDE button on the steering wheel (Fig. 9.4).

Drivers were not required to enter a destination for each trip. In the absence of a planned route, navigation assistance was available through the moving map, as well as the WHERE-AM-I? key on the steering wheel. WHERE-AM-I? provided a voice message specifying the vehicle's heading, the name of the street on which the vehicle was traveling, and the next cross-street ahead.

Traffic information was provided to drivers in three ways:

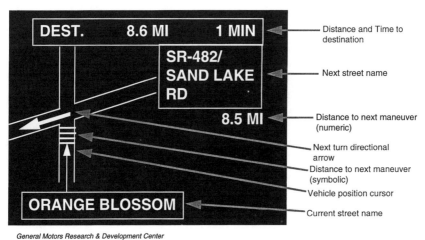

General Motors Research & Development Center

FIG. 9.3. Guidance map.

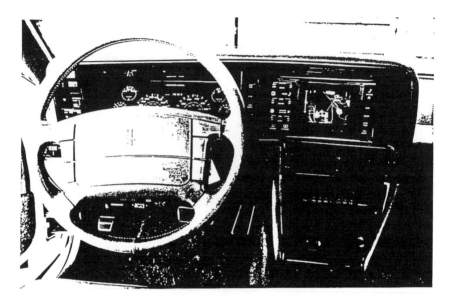

FIG. 9.4. TravTek controls and displays.

- Icons representing traffic congestion, incidents, and blocked roads were displayed on the moving map.
- The maneuver-by-maneuver display provided a yellow dot indicating the presence of traffic ahead on a route.
- Drivers could request an auditory traffic report, filtered for geographic relevance, with the TRAFFIC REPORT button on the steering wheel.

Other driver controls included touch screen keys on the moving map display that enabled the driver to zoom in and out, pan the map, toggle between the vehicle's heading or north at the top of the map, and show names of streets that were not labeled.

In addition to the steering wheel buttons, which have already been mentioned, other steering wheel controls enabled the driver to correct the position of the vehicle cursor on the moving map, accept a new route offered by the system, and repeat the most recent voice message. The volume of voice messages could be controlled by using the radio volume control while a voice message was being spoken.

Traffic Management Center (TMC). The TMC was the responsibility of the public sector partners. Its function was to collect, analyze, and fuse traffic information from several sources and broadcast data to the computers in the fleet of vehicles. A subset of the greater Orlando road network was established as the TravTek Traffic Network by the Florida Department of

Transportation and the City of Orlando. This network is used for traffic routing and consists of freeways, arterials, and collector roads.

TravTek was designed to increase trip efficiency by selecting routes that minimized individual drivers' travel times to their destinations. To that end, current link times, as well as a historic database of measured travel link times, were maintained on the traffic network segments. Sources of current link times and other data were collectively referred to as the Traffic Information Network (TIN). These sources included loop detectors, video surveillance on a portion of I-4, Metro Traffic Control, and the local police and Florida Highway Patrol. Once per minute, the TravTek vehicles automatically transmitted probe reports to the TMC. These communications included the time, vehicle position, and vehicle identification, as well as other data.

Once per minute, the TMC broadcasted information messages to the vehicles that included traffic network link-time ratios, incident and congestion messages, and other current information available to the drivers such as weather, parking lot status, and special event information. The TMC report also contained date and time synchronization.

Much of the data collection and fusion process was automated or computer assisted. In TravTek, the TMC operator had an important role in the handling of incoming traffic information, particularly incident information. TMC personnel were hired and trained by the City of Orlando.

Some ITS engineers assume that traffic management soon will be fully automated and therefore give little thought to the operator–system interface. However, as described in the ATMS chapter of this book, operators will always have a substantial role in ATMS operations (Kelly et al., this volume) primarily in cooperative or supervisory roles. Human factors practitioners can apply the considerable knowledge gained from other control room environments and ought to be included on TMC design teams. Moreover, human factors of TMC evaluation are critically important to system success.

TravTek Information and Services Center (TISC). The TISC was the responsibility of AAA and served several functions. Most apparent to the TravTek drivers was the online help desk accessible toll-free by telephone. Drivers could reach motorist assistance counselors from the cellular phones onboard the TravTek vehicles. The motorist assistance counselors were specially trained and familiar with the vehicles and system functions. In addition, a TravTek in-vehicle system simulator assisted counselors in answering questions. The driver's vehicle location was acquired by a leased line from the TMC computer while the simulator replicated the displays seen by a driver. The TISC personnel also made in-service phone calls to each TravTek rental driver to detect any problems and to underscore personalized support.

The TISC supplied the map and local information databases for the vehicles and daily event information to the TMC. AAA recruited TravTek rental drivers, served as liaison to Avis, and developed TravTek written and video training materials for drivers. All TISC interface design and training was conducted by AAA.

TravTek Driver–System Interface Design

From the earliest stages of the in-vehicle system conceptualization, TravTek was designed to be easy to learn and use, as well as to help drivers find their way efficiently. A multidisciplinary design and implementation team of hardware, software, and communications engineers included human factors practitioners and researchers. The design process, described by Fleischman et al. (1991) was iterative. From the creation of scenarios to the development of functional requirements, trade-offs, and testing, human factors and engineering psychology was involved through the application of human factors principles, problem solving, and research in support of the design process.

As is apparent from the other chapters in this book, there are a number of human factors concerns and issues that are hotly debated. It is also usually true that there are a number of competing forces in the design of any system. When possible, TravTek used empirical methods to gain a better understanding of these issues and to guide decision making. Figure 9.5 shows that several laboratory studies were conducted early in the TravTek

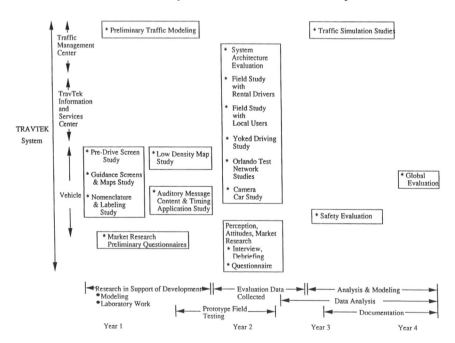

FIG. 9.5. TravTek research project stream.

research project stream. Ease of use and learning were the objects of several studies focusing on predrive screens (Dingus et al., 1991) system nomenclature, and key labeling. Concerns about the potential for distraction and the visual processing demand (Rockwell, 1972; Dingus, Antin, Hulse, & Wierwille, 1989) motivated laboratory studies of the extraction of information from route guidance and map screens under dual-task conditions. Rather than relying solely on the intuitions of the design team, TravTek used these experiments, along with over-the-road application studies of auditory messages, to guide the design by testing hypotheses and observing people of different ages and educational levels.

There were a number of constraints on the driver interface design imposed by scheduling, practical and technical limitations, and the TravTek user population. Most of the drivers were to be visitors to Orlando who would rent their vehicles from Avis. Novice users, ranging in age and education, would need to cope with an unfamiliar car in an unfamiliar place. In addition, they would face a great deal of high-tech functionality. Therefore, in many instances, the design team decided to trade off versatility and complexity for ease of use and simplicity. As there was minimal opportunity for training, TravTek needed to be as self-teaching as possible. A number of different display types were available in TravTek to give the system flexibility in meeting the needs of drivers with different cognitive and perceptual styles in a wide variety of driving situations.

Although the driver interface was designed as an integrated dynamic system, a special emphasis was placed on the displays and controls that the user would operate while driving. Several strategies were adopted for informing drivers without overloading them.

The use of visual displays and controls was restricted. A conservative approach was adopted that divided TravTek functions into predrive, drive, and zero-speed functions. Reading text and using soft key inputs on the touch screen may require long glances and search times. Therefore, the use of trip planning services, such as paging through the Service and Attractions directory, browsing maps, and entering destinations, were possible only when the car was in PARK. Only TravTek functions that supported driving and way-finding tasks were available while the cars were in motion. Driving functionality was limited to traffic information and navigation screens. Steering wheel buttons provided tactile feedback and convenience while allowing drivers to swap maps, select auditory navigation and traffic messages, and correct the vehicle position cursor on the map displays.

The issue of how restrictive a system should be continues to be important. Many drivers would like to be able to cancel and enter a new destination while driving. Some system designers want drivers and passengers to have access to all information while the car is in motion, suggesting that drivers might pull over and stop the vehicle in inappropriate or unsafe situations in order to have access to restricted functions. Many factors, such as the complexity of the system, the size and placement of visual displays, and the attention required to enter data, should be factors in such design decisions.

The information density of drive screens was reduced. For example, a simplified route maneuver-by-maneuver Guidance Map was designed as an alternative to the more traditional moving map. Examples of both visual route guidance displays are shown in Figs. 9.2 and 9.3 and described by Carpenter et al. (1991). Laboratory tests, mentioned above, indicated that the guidance display required less glance time to extract information and resulted in better performance than the map with a highlighted route. The route map was retained because some drivers prefer it and because the map information is useful in certain driving situations. The laboratory studies could not recreate the utility of the information on the displays.

The auditory mode was used to present information. Maneuver-by-maneuver route guidance instructions, orientation, and traffic information were available by voice in an attempt to decrease the need to glance away from the roadway and mirrors. The TravTek auditory interface was designed to augment the visual displays. For example, a computer-synthesized voice informed drivers of their location when they pressed the WHERE-AM-I? steering wheel button. More importantly, maneuver-by-maneuver route guidance supplemented the information on the Guidance Map. The information included the distance to and name of the next turn street as well as the direction of the turn. Details of the design can be found in Means et al. (1993). Because auditory information is transitory, a REPEAT button was located on the steering wheel.

The specification of default displays. The design team picked default displays to encourage drivers to use the recommended displays without being restricted. For example, the Guidance Map and VOICE GUIDE were designed to reduce visual attention directed at the CRT display. However, the inclusion of a more information-rich Route Map allowed more flexibility, so the design was such that when the vehicle entered the first link of a route during guidance, the simplified guidance map was automatically displayed. The driver needed to press the SWAP MAP steering wheel button to toggle to the Route Map. Voice guidance on was also the default condition; the driver needed use a steering wheel button to turn the voice off.

The TravTek Evaluation Program

Process and Concepts. Multiple partners and objectives require structures and processes to successfully exploit the potential for learning from an LSE. The TravTek team understood the desirability of a research plan that was objectives driven and cohesive, so early in the planning stages the TravTek partners organized an Evaluation Working Group (EWG) to plan for the evaluation portion of the LSE. Each partner organization was represented by at least one person who was a researcher or who had an interest in the research and evaluation aspects of TravTek. The Evaluation Working Group, along with the Technical Working Group, reported to the Partners Working Group that met once every 4 to 6 weeks to plan and implement the LSE. Eventually, this partners group was joined by the members of Science Applications International Corporation (SAIC), the research team contracted to detail the experimental designs and to implement the studies. Over the course of its first year, the group went through the process shown in Fig. 9.6. At each stage, there were dialogues with the Partners Working Group and the parent organizations. During animated discussions, engineers learned about the requirements and costs of controlled experiments while researchers

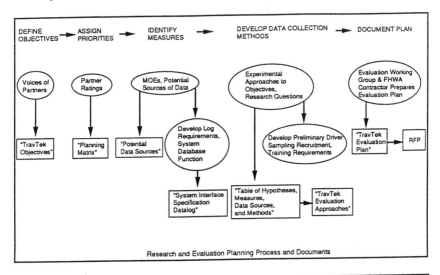

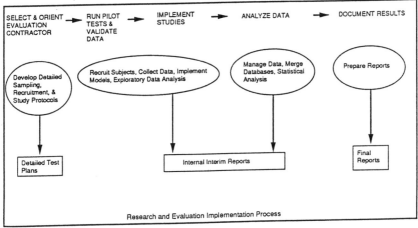

FIG. 9.6. TravTek research and evaluation process.

learned about the technical constraints and realities of fielding a complex dynamic system. In fact, grappling with how to evaluate the system was a catalyst to a deeper and shared understanding of how the TravTek system would work.

The TravTek partners committed themselves to a plan for a program of research and evaluation that was more ambitious in scope and sophistication than initially envisioned and that would require more resources. There also was more emphasis on human factors and behavioral science than usual. This was due to the presence of human factors researchers and the recognition that many of the objectives had behavioral components. Keys to the success of this process were: (a) each partner had the opportunity to state research interests and understanding of the TravTek objectives, (b) a negotiated list of objectives was distilled and agreed on by all the partners (see Fig. 9.7), and (c) there was a mix of traffic researchers, behavioral scientists, and market research professionals on the team.

An overarching objective of the TravTek research team was to apply as much scientific rigor as possible in the real-world test situation. Especially important were the commitments to collecting performance data in parallel with self-reported data, to using comparison groups and control conditions, and to using analytic models. Formative TravTek evaluation plan concepts are described by Fleischman (1991). The complexity of the TravTek system and its numerous objectives and research questions required evaluation at different levels of analysis. Therefore, a modular, multimethod programmatic approach was developed so that each objective could be tested by the use of appropriate methods and each study could be conducted by appropriately trained researchers (behavioral scientists, traffic engineers and modelers, market research experts, etc.). Each study was designed to stand on its own. Data collected in some studies were used in others. The studies were tied together by a final, integrative report. A complex operational field test was well served by obtaining converging data from divergent methods. Furthermore, our understanding about behavior, safety, and attitudes are better understood by the gathering of performance data as well as self-reported data.

TravTek, therefore, was a constellation of studies that included naturalistic and controlled over-the-road studies, questionnaires and interviews, and measures of subjective workload and individual differences, as well as more traditional studies of system architecture (see Fig. 9.5). As of this writing, all studies have been completed and final reports are in press.

TravTek Studies

General Independent Variables. The TravTek partners were interested in evaluating the effectiveness of the system by manipulating two variables: the level of information and display type. TravTek was designed to produce benefits as a result of navigation–route guidance and real-time traffic informa-

A. **TRIP/NETWORK EFFICIENCY**
  1. **Congestion Avoidance**
  2. **Time Savings**
  3. **Pollution Reduction**
  4. **Fuel Savings**
  5. **Reduced Vehicle Operating Cost**
B. **Benefits to Non-TravTek Users**
C. **Driver Performance/Behavior/Satisfaction**
  1. **Driving/Navigation Behavior**
  2. **Perception of Congestion, Time, Safety**
  3. **Usability/Learnability**
  4. **Feature Use (Route guidance, voice guidance, phone, local info. etc.**
  5. **User Friendliness/Driver Satisfaction**
D. **Safety**
E. **System and Subsystem Performance**
  1. **Hardware**
      a. **Reliability**
      b. **Compatibility**
  2. **Software**
      a. **Reliability**
      b. **Compatibility**
  3. **Data**
      a. **Accuracy (map, local information, etc. databases)**
      b. **Timeliness**
  4. **Operations/Procedures**
      a. **Data Collection (map, traffic, events, local information)**
      b. **Data Input-TMC, TISC**
      c. **Driver Recruitment, Training, Debriefing**
      d. **Helpline Management**
      e. **Vehicle Management**
      f. **Vehicle Maintenance**
F. **Image**
G. **Impact Future Transportation/Travel**
  1. **Generalizable/Transportable**
  2. **Technology Transfer**
H. **Feature Preferences**
I. **Price/Cost**
  1. **Willingness, personal vehicle**
  2. **Willingness, rental vehicle**
  3. **Infrastructure (TMC, TISC)**
J. **Local Area Impact**
  1. **Improvements Beyond TravTek Operating Period**
  2. **Routing Through Sensitive Areas (neighborhoods, hospitals, etc.)**
  3. **Local Driver Usage**
  4. **Local Jurisdiction, Policy Issues**
  5. **Macroeconomic Benefits**

FIG. 9.7. TravTek objectives and evaluation goals.

tion. To test the incremental effects of in-vehicle navigation and route guidance, as well as the value added by the real-time traffic data supplied by the
infrastructure, a level of information variable was manipulated by assigning
drivers to three groups. Drivers assigned Services cars had access to service
information only. This information was available only when the cars were in
PARK and included the Services and Attractions directory, communications
with the TISC, 911 emergency services, and maps with no vehicle position
provided. The Services (S) groups were used as control conditions against

which other groups were compared. Drivers assigned to the Navigator (N) groups had access to all autonomous navigation and route guidance information as well as services information, but no real-time information from the TMC. The Navigator Plus (N+) drivers had access to all of the TravTek information. All cars in the TravTek fleet could be set in the various configurations by the experimenters.

The second independent variable was the differential effect of the type of route guidance display on ease of navigation and several measures of attention and driving performance. The research investment in studying display type was driven by human factors issues such as visual attention, which has been discussed widely in the literature. At the time of the evaluation planning, designers were debating the relative merits of traditional map displays, less information-dense symbolic displays, and auditory information presentation. Therefore, the displays chosen for study were a moving map with color-highlighted route, a simplified maneuver-by-maneuver symbolic guidance map, and maneuver-by-maneuver voice guidance. Participants in the naturalistic field studies could choose among the displays. In controlled field experiments, the experimenters set the display type.

Person variables of interest to the human factors researchers included age, gender, geographic familiarity, educational level, computer experience, map skill, and purpose of trip. Unfortunately, because of insurance constraints, drivers under the age of 25 years could not drive TravTek cars. The exception to this was that young drivers could participate in the Camera Car Study. Additional demographic variables of interest to market researchers were region, marital status, occupation, and income. Traffic and environmental variables included peak versus off-peak traffic, traffic density, road class, and day versus night.

Data Collection Instruments and Resources. All three TravTek subsystems kept extensive data logs. The TMC, the TISC, and the TravTek vehicles automatically logged a substantial amount of data by computer. Other sources of data were vehicle maintenance logs, police and traffic reports, optional driver diaries, and debriefing. Specific studies yielded additional data described in the following discussion. These data collection sources were designed so that the vehicle, demographic, traffic, and other data could be merged to address the wide variety of questions generated by the partners and researchers.

All files included the driver ID number, vehicle number, configuration, and all data were date and time stamped. The TMC maintained a log that included traffic incident reports, event data, communications, data fusion inputs, history, winners, link histories, failures, and so forth. The TISC maintained logs that included calls to the HELP desk, in-service calls to drivers, and demographic and reservation information. The TravTek vehicles were instrumented for automatic data collection. Driver interactions with the system such as key presses were logged as well as display changes, communications, power up/down, shifter (PRNDL) position change, vehicle position, speed, GPS status, and so on. The vehicle logs contained a history of each trip including all links traveled, functions selected, and displays shown. They were periodically downloaded and processed for analysis by GM, AAA, and SAIC researchers. Details of these logs can be found in the Final Detailed Test Plan (Science Applications International Corporation, 1993a, 1993b).

Naturalistic Field Studies. The quasi-experimental field studies were the heart of the evaluation program. They most especially took advantage of the real-world setting, enabling researchers to study how people used the ATIS functions that TravTek provided and whether they behaved as ITS conceptualizers presumed. In addition, the field studies were sources of accident data for the Safety Study and reliability data for the System Architecture Evaluation.

The questionnaires addressed most TravTek objectives and concerns. Drivers rated attributes such as system reliability and benefits such as time savings, congestion avoidance, and helpfulness in finding their way around. The questionnaires also included questions about their perceptions of safety and distraction. Numerous questions were aimed at the overall user friendliness of TravTek as well as at the ease of learning, the use of the displays, and the usefulness of the information. The questionnaire data was supplemented by face-to-face interviews. A subset of rental drivers (19%) were debriefed as they returned their TravTek vehicles to Avis, and all local drivers were debriefed. In this way, immediate feedback was obtained about drivers' impressions, most- and least-preferred features, problems, and suggestions.

Because of the extensive automatic logging of the driver's actions, it was possible to assess what people actually do along with what they report about their own attitudes and perceptions concerning the system, its benefits, and deficiencies. A measure of system usefulness is whether or not people actually use it!

Rental and Local User Studies. The facets of these studies, which are of particular importance from a human factors standpoint, include the observation and analysis of the choices drivers actually made when unconstrained in the selection of TravTek functions, displays, and destinations. Driver preference, perception, and marketing data were gathered by means of questionnaires given to all participants and by more detailed interviews of a subset of participants. Because Orlando is a major destination for travelers, TravTek was originally conceived with visitors as primary users of the system by means of vehicle rentals. The addition of a local drivers field study allowed an exploration of how information might be used by people driving in their home locale and enabled observations to be made over a longer period of time.

The following results are just highlights and are more fully reported and explored by Inman et al. (1996). Various facets of the results also have been reported by Fleischman, Thelen, and Dennard (1993), Perez, Fleischman, Golembiewski, and Dennard (1993), and Peters, Mammano, Dennard, and Inman (1993).

Participants in the Rental User Study were recruited primarily through AAA clubs representing different regions of the United States and through the AAA national travel reservations "800" number. Additional drivers were recruited with the help of the Orlando Chamber of Commerce and Avis rental agents at the Orlando airport. Of the 100 TravTek cars, 75 were part of the Avis rental fleet. The analyses of this study concentrated on 2,568 renters who identified themselves as primary drivers of the TravTek vehicles between March 1992 and March 1993. Of these drivers 84% were male and 16% were female. They were distributed across age categories as follows: 19% were between 25 and 34 years, 64% were between 35 and 54 years, and 17% were 55 years or older. Drivers were assigned to Services (14%), Navigation (36%), or Navigation Plus (50%) vehicles. They received materials and training appropriate to their vehicle configurations. Informed consent was obtained from all participants. The rental drivers questionnaire return rate was 41% of Services, 65% of Navigation, and 65% of Navigation Plus drivers (Peters et al., 1993).

An attempt was made to balance vehicle configurations across variables. Clearly, this balance was not achieved, but as an average rental period was 5.3 days for Services, 6.1 days for Navigation, and 5.2 days for the Navigation Plus group, enough trips were generated to pursue most research questions.

Participants in the Local User Study were recruited by the SAIC experimenters. The participants reported that they used their vehicles an average of 121 km per day at the time of recruitment into the study. Of the 53 local drivers, 33 were male and 20 were female. They ranged in age as follows: 14 (26%) were between 25 and 34 years, 26 (49%) were between 35 and 54 years, and 13 (25%) were 55 years or older. They drove TravTek cars for an average of 8 weeks. Each participant drove both the Navigation (N) and Navigation Plus (N+) vehicle configurations. There was no Services condition because the local drivers could compare their TravTek driving experience with their non-TravTek experience in Orlando. One half of the drivers started in the Navigation configuration and half started in the Navigation Plus configuration: they were switched halfway through the study. Drivers were debriefed at several points in addition to filling out questionnaires for both N and N+ vehicle configurations.

Overall Use and Perceptions of TravTek. Both rental and local drivers used some TravTek feature for most of their trips. A comparison of the S, N, and N+ rental users data showed, not surprisingly, that as additional features were available, drivers used the system more often. The biggest jump in vehicle system use was with the addition of navigation–route guidance features.

TravTek was rated very favorably, particularly by the users in the N and N+ groups. Both renters and local drivers who had route guidance (N and N+) perceived that TravTek helped them save time in reaching their destinations and aided them in paying more attention to driving. This was not the case with drivers who did not have route guidance (S). Those who had real-time traffic information (N+) did not perceive that they avoided traffic congestion. Overall, the users rated TravTek very highly on its human factors aspects, reporting that the vehicle system and its displays and controls were easy to learn, easy to use, and useful.

Driver Use and Perceptions of TravTek Features and Displays. Rental and local drivers in both the N and the N+ conditions used route planning and guidance for approximately one half of their trips. That the local drivers used the route guidance feature this much, both without (N) and with (N+) traffic information, was a surprise. Some ATIS system designers have speculated that local drivers would use route guidance much less without traffic information. Although the frequency of route guidance usage by locals declined somewhat over time, even after extended use, locals used route guidance on more than 40% of their trips. The high level of route guidance use bears out the results of questionnaires and interviews citing route guidance features as favorites.

While on a route, drivers used the simplified maneuver-by-maneuver Guidance Map most of the time, over 80% of renters and 70% of local drivers. N+ rental drivers used the Route Map slightly more than the N drivers, possibly because of the traffic information available on the route map in that configuration. N+ drivers used the SWAP MAP steering wheel button more often to toggle between the visual displays. No difference was found, however, between N and N+ for local users. There was an indication that males and younger drivers used the Route Map more than females and older drivers.

Rental and local drivers in both the N and N+ conditions left the VOICE GUIDE on most of the time while using the route guidance feature. Renters had the VOICE GUIDE on over 90% of the time, and locals used it more than 70% of the time. Although some system designers had hypothesized that drivers would not use a synthesized voice display, the high level of use in TravTek suggests that a well-designed auditory display can be met by user acceptance. Furthermore, the VOICE GUIDE feature was rated by drivers as most helpful in keeping their attention on their driving. Some interesting results emerged from the driver debriefings: Although the quality of the voice display topped the list of drivers' least favorite features, the VOICE GUIDE feature itself was among their most favorite.

The Guidance Map use results should be considered with some caution. It should be noted that the "on" condition for the Guidance Map and VOICE GUIDE was purposely designed as a default display. The results probably reflect, in some measure, this human factors design strategy. The questionnaire data showed that drivers liked both the Guidance Map and the Route Map and rated them more equally than their level of use would indicate. Whereas renters strongly preferred both visual displays supplemented by voice guidance, locals did not. Including two visual displays and an auditory display provided flexibility that users seemed to regard favorably. At the same time, presenting the simplified maneuver-by-maneuver visual and auditory displays automatically encouraged drivers to use these as much as possible.

WHERE AM I? This steering wheel button was available for use in both N and N+ conditions. When pressed, it activated a voice message that gave drivers their vehicles' current position. Logged data indicates that this navigation function was used at least once by 92% of rental drivers and by all of the local drivers. WHERE AM I? was used infrequently on a per trip basis, but was rated favorably in terms of usefulness and timeliness.

Traffic Information. In the N+ configuration, the TravTek system continuously searched for faster routes. If a new route was detected that could significantly avert delays because of traffic, the driver was alerted that an alternative route was available. A message was presented by voice, and a banner appeared on the visual display. Drivers could accept this route by pressing the OK NEW ROUTE steering wheel button. The vehicle logs indicated that the occurrence of a faster route was rare. New routes were offered on only 21 rental user trips and on a mere 15 local user trips during the entire LSE. In all except four instances, renters accepted the new routes. Local users accepted the new route in 8 out of the 15 instances. However, the voice was off in 6 of the 7 occasions when a route was not accepted, whereas the voice was on for all of the new-route-available occurrences for rental users.

A voice traffic report was also available to drivers in the N+ condition. The TRAFFIC REPORT steering wheel button was pushed at least once by 85% of rental drivers and 94% of local drivers. The traffic report was on only 11% to 14% of the time. Nevertheless, those who used the traffic report feature rated it favorably.

Willingness to Pay for the System. The market research questions focused on the respondents' willingness to pay for a system with TravTek functions such as navigation and route guidance, traffic information, and services attractions information. Overall, the participants estimated that they would pay around $1,000 for a system like the one they used in TravTek. They indicated that they would pay the most for navigation features such as route planning and guidance and somewhat less for traffic information and local information. This is not surprising given their perceptions of the relative benefits of way-finding versus congestion avoidance.

Much additional data can be found in the Rental and Local Users Field Study report (Inman et al., 1995a). Data that is valuable for future system design, marketing, mobility, and safety assessments can be obtained from naturalistic field studies like these. The collection and analysis of such data required careful planning and coordination from all parties involved in the OFT process.

Controlled Field Experiments. Controlled field experiments have the advantage of allowing cause-and-effect inferences to be drawn. In addition, experimenters can focus on situations of special concern. It is possible to hold constant or account for factors such as trip length and weather in addition to the level of information and display variables discussed previously (Inman, Fleischman, Dingus, & Lee, 1993).

The most central claims for the TravTek system rested on hypotheses that drivers would save time, would navigate more effectively, and would perform driving tasks adequately while doing so. To determine the impact of the TravTek information features and displays, three controlled experiments were conducted. In all three experiments, paid participants drove carefully constructed origin–destination (O-D) pairs. The O-D pairs were designed to hold trip length and difficulty constant. The development of the O-D pairs was one of the most challenging tasks in these experiments. Many pairs of equal difficulty needed to be generated to allow proper counterbalancing and to avoid effects because of learning the test routes. For the most part, the three controlled field experiments used the same O-D pairs to enhance interexperiment comparability. The participants were geographically unfamiliar with the Orlando area. They attended a brief training session and drove several practice routes. All participants were given vision, hearing, and map skills tests. SAIC researchers recruited and trained the participants and served as in-vehicle observers. A measure of subjective workload was taken at several points during each drive.

Camera Car Study. This was the most microscopic study of driving behavior. The overall purpose was to investigate whether drivers are able to extract information from route guidance displays without a decrement in driving performance that might result in unsafe driving behavior. A combination of safety surrogates including eye-glance behavior, driving performance measures, and experimenter observations contributed to the assessment of the impact of the TravTek system and displays on safety risk when compared with control conditions. For this study, a taxonomy of potentially risky maneuvers such as near misses and abrupt braking was developed to assist the safety analysis.

Specific questions were as follows: Do different display types result in more or less driving task intrusion; that is, are there differences in measures of visual attention, performance, and workload as a function of display type? How is safety affected? Are there differences in navigation performance? How user-friendly is the vehicle system? Does attention and driving and navigation performance vary as a function of driver age, experience with the TravTek system, and familiarity with the local area?

The Camera Car Study tested these TravTek route guidance display conditions:

- Guidance Map
- Guidance Map with voice guidance
- Route Map
- Route Map with voice guidance

These were compared with two control conditions:

- paper maps
- paper direction lists.

Under naturalistic field conditions, drivers could switch among the TravTek options as suited each individual and situation. However, the experimental design of this study required that drivers use one display type at a time for navigating an assigned O-D pair.

Participating in this study were 30 drivers. They ranged in ages from 16 to 73 years. To test effects of novelty, learning and experience, and area familiarity on driving performance and attention, 12 Local User Field study drivers also participated in this experiment. They were tested twice in the Camera Car, first, as novices with the TravTek system and, again, 6 to 8 weeks later having gained experience by driving in the field study. Although these drivers were generally familiar with Orlando, they were unfamiliar with the specific O-D pairs in this study.

The Camera Car, itself, was an instrumented TravTek vehicle specially equipped with four video cameras to record simultaneous views of the driver's eyes, the controls and displays, the environment external to the vehicle, and the position of the vehicle relative to the lane. Steering wheel position, two-axis acceleration, and brake light status were also collected as well as auditory comments. A custom control panel allowed the experimenter who accompanied all drivers to enter data such as roadway type, traffic density, start and stop times, workload ratings, and events such as near misses. A detailed description of this vehicle and study can be found in Dingus and Hulse (1993), Dingus, Hulse, McGehee, Manakkal, and Fleischman (1994), Dingus et al. (1995), and Hulse, Dingus, McGehee, and Fleischman (1995). Following are some findings generated by the Camera Car Study.

Overall Performance Results. Maneuver-by-maneuver guidance information, whether presented by auditory instruction, visually displayed graphic, or text directions on paper, resulted in effective driving and navigation. The maneuver-by-maneuver Guidance Map with voice and the paper direction list produced the best overall performance. Using the TravTek Route Map with no supplemental voice information required the most visual attention and produced the greatest number of safety-related errors. Although voice guidance instructions improved overall performance in the use of all TravTek displays, this was especially true for drivers over the age of 65 years. The paper map control condition required high cognitive attention. It was the least usable means of trip planning and produced substantially worse navigation performance than any other condition.

Driving Task Intrusion. The highest visual attention demand was created by using the route map without supplementary voice guidance. This was evident in the number and duration of glances to the display, resulting in the disruption of eye-scanning patterns. The presence of voice guidance reduced the visual attention requirements of both the Route Map and the Guidance Map. Although using a paper map required relatively little visual attention in this study, measures of subjective workload, vehicle speed, and braking indicated that this condition did intrude on the driving task.

Safety. No accidents and very few near misses occurred during the Camera Car runs. An analysis of safety-related events and several measures of driving performance show that there were considerably more safety-related errors when drivers used the TravTek Route Map without voice rather than a paper map or direction list and all other TravTek display configurations. This was, however, mitigated by experience. Using the Guidance Map with voice guidance, a paper map or direction list resulted in the lowest number of safety-related errors. In comparison, using the Guidance Map without voice guidance or the Route Map with voice guidance produced similar numbers of safety-related errors, but slightly poorer measures of driving performance. Adding maneuver-by-maneuver voice instructions to a TravTek visual display generally improved overall safety performance.

Navigation Performance. When drivers used TravTek, they saved time planning routes to a destination. Using TravTek or a paper direction list resulted in shorter overall trip times. Drivers became lost the most while using a paper map. The TravTek displays that provided some maneuver-by-maneuver information (Guidance Map with or without voice, Route Map with voice) produced positive navigation performance results.

Usability. Overall, test participants performed extremely well. The TravTek driver interface was found to be easy to learn and use. This was found even for older drivers, some of whom had little or no computer experience. Measures of usability and ease of learning indicated that TravTek trip planning and route guidance functions were easy to learn and use, particularly when compared to the use of a paper map. Of the TravTek route guidance display options, the Route Map without voice was the most difficult to use. Recommendations for improving the driver interface are included in Dingus et al. (1995).

Age. Older (65+) drivers were especially vulnerable to the attentional demands of driving while navigating. They compensated by driving more cautiously, but made a greater proportion of safety-related errors. However, as a group, older drivers had more difficulty with the control conditions, especially the paper map. Older drivers benefited the most from using the TravTek route guidance system on measures of attention, driving, and navigation performance. This was particularly true when they used the maneuver-by-maneuver Guidance Map with voice guidance. An analysis of the naturalistic field study results indicated that when free to make choices, older drivers chose to use the Guidance Map supplemented by the VOICE GUIDE most of the time.

Experience. Experience with the TravTek system reduced the attention required to extract information from the visual displays. Fewer unsafe incidences occurred as drivers gained experience. Furthermore, experienced drivers planned and drove to destinations in less time than novices.

Area Familiarity. Visitors to the Orlando area glanced at the displays more than local drivers, but drove more cautiously. Surprisingly, visitors made fewer navigation errors than locals.

The Camera Car experiments confirmed that reducing the information density and using an auditory supplement for route guidance can reduce attentional demand. A high-tech system that applies human factors engineering concepts can be of benefit to a wider range of users than is generally thought as demonstrated by the older drivers. It is important to notice how this study dovetailed with the field studies. In the absence of field-study data, designers might be tempted to be overly restrictive, leaving out a route map option because it requires greater attention. However, the data logged in field studies showed that users made appropriate choices, using the maneuver-by-maneuver Guidance Map with the VOICE GUIDE on most of the time, particularly as novices (Inman et al., 1996). Drivers also made age-appropriate decisions.

These carefully controlled experiments conducted in a real-world driving environment confirmed prior research (Bhise, Forbes, & Farber, 1986), which found that displays or activities in moving vehicles requiring long (>2.5 s) glances can have a negative effect on driving performance and safety. Displays or activities that require more cognitive effort negatively influenced ease of use and system benefits, whereas those that required high visual attention negatively affected safety-related driving performance. The simultaneous collection of driver eye movements, driving performance measures, and safety events in an over-the-road study can allow the evaluation of trade-offs between mobility benefits and attentional demands.

Orlando Test Network Study (OTNS). The purpose of this study was to discover whether navigation is improved when people are provided with route planning and guidance information and to evaluate navigation performance as a function of information display. It was an opportunity to study a range of drivers during day and night driving. Route guidance conditions tested were as follows:

- Route Map with voice guidance
- maneuver-by-maneuver Guidance Map with voice
- voice guidance (no visual display)
- Route Map (no voice guidance)
- maneuver-by-maneuver Guidance Map (no voice guidance)
- control (no TravTek navigation)

For the control condition, drivers had to use conventional means of navigating such as a paper map or asking for verbal instructions. Drivers were assigned either to a voice-guidance or no-voice-guidance group. Within those groups, each driver navigated three O-D pairs: one with no TravTek visual display, one with the Route Map, and one with the Guidance Map. Participants were given a brief orientation as well as practice runs, during which their learning was observed.

The three O-D pairs were completed by 249 drivers. Some of the dependent measures were travel and trip planning times, mental workload, navigation errors and perceptions of utility and usability.

Results of this study were reported by Inman, Sanchez, Bernstein, and Porter (1995a). Gains in trip efficiency resulted from using TravTek route guidance. There were significant savings in trip-planning times and travel times. Participants took less than 1.5 min to enter destinations into the

TravTek system and wait for a route, whereas without TravTek, route planning averaged 5.0 min. Travel time was reduced about 5 min when participants navigated with TravTek route guidance compared with that of control conditions. No significant time savings advantage was found for one type of TravTek information display over another. Overall, drivers were equally likely to make navigation errors in all conditions. However, drivers in the control condition took longer to detect an error, and the consequences of mistakes were more severe when there was no TravTek visual display. Measures of subjective workload were higher when drivers were in the control conditions than those of any in the TravTek conditions.

Questionnaire and debriefing results provided additional information about the preferences of the participants in the study. The Guidance Map and Route Map with voice guidance were rated very highly. These users agreed with participants in the other studies, preferring visual guidance displays supplemented by voice. This is probably especially true for visitors to an area.

Performance measure and subjective ratings suggested that TravTek was easy to learn and use. Particulars of this assessment, including the difficulties encountered by users during destination entry are included in the OTNS final report by Inman et al. (1995a). All participants mastered the system in relatively few trials, although younger users found the system somewhat easier to learn than older users. Once learned, the system was usable.

Yoked Driver Study. The Yoked study was designed specifically to test the trip and network efficiency objectives by measuring the differential effects of TravTek route guidance and real-time traffic information on time savings and congestion avoidance. In addition, driving performance, preferences, and attitudes were studied (Inman, Sanchez, Porter, & Bernstein, 1995b).

To study the effects of route guidance and traffic information, drivers were assigned to drive vehicles configured with navigation information (N), navigation plus real-time traffic information (N+), or neither navigation nor real-time information (S). The S group served as a control, and drivers in that condition used conventional methods of planning their trips and driving to the assigned destinations. The N and N+ groups allowed the manipulation of the level of information variable. This experiment was implemented under peak traffic conditions to maximize the potential for observing differences resulting from the use of traffic information. For each test run, a triad (S, N, N+) of drivers simultaneously drove one of three O-D pairs. The drivers in each triad departed within 2 min of each other to preclude car-following behavior, yet be close enough in time to ensure consistency of weather and traffic conditions.

The 123 drivers completed their assigned O-D pairs as triads. An additional 90 people participated in the study. The study participants were visitors to Orlando recruited at a major tourist site. Given brief classroom orientation, they drove several practice routes.

Time Savings. Results of the Yoked Driver Study demonstrated that drivers using TravTek route guidance saved time when compared with drivers who navigated by conventional methods (S). Time savings were found for both trip planning and en route proceedings to a destination, although not for all O-D pairs. This time savings result is in agreement with the results of the Camera Car and OTNS studies. The Yoked Study was the only one of the three controlled field experiments that attempted to evaluate time savings due to real-time traffic information. No difference in measured time savings were found between the N and N+ groups.

The automatically logged data and the questionnaire data from the Renter and Local Field Studies make it clear that the drivers in the N and N+ configurations clearly perceived the time-savings and way-finding benefits measured in the controlled experiments. It is also interesting that there were no differences in perceptions between the N and N+ groups, which corresponds to the findings in the Yoked Study.

Congestion Avoidance. An analysis of N, N+ dyads revealed 50 out of 60 instances in which vehicles with traffic information planned different routes. In the absence of real-time information, routes planned by TravTek between a given origin and destination should be the same for N and N+. More significant, when N+ vehicles took different routes, they tended to travel further and take lower-class (slower) roadways, yet did not have longer travel time than the N vehicles. This suggests that the N+ vehicles must have avoided congestion that the N vehicles did not. Inman et al. (1995b) concluded that, as implemented, TravTek used real-time traffic information to successfully avoid congestion, and that avoiding congestion could result in benefits to other users of the traffic network in a deployed system. The questionnaire data from the Field Studies indicated that a congestion-avoidance benefit was not perceived by users. Perhaps this is because the congestion-avoidance benefit is relatively rare and small or because it lacks salience. Users do not see the congestion they may have avoided on another route.

Network Modeling and Safety Studies. The TravTek fleet of vehicles was small in number relative to the 1,200 square miles covered by the TravTek Traffic Network, and the impacts of real-time route planning and guidance on traffic congestion, fuel consumption, and safety were not expected to be measurable. Therefore, modeling studies were employed to project what effects a TravTek-like system might generate at greater levels of market penetration.

Traffic Simulation Studies. Data collected in the other TravTek studies were used to calibrate an INTEGRATION simulation model developed by Van Aerde and Yagar (1988). The INTEGRATION simulation model is an analysis tool with the ability to represent dynamic traffic networks and controls and route guidance systems that are generic and specific to TravTek.

The calibrated model can estimate various measures of effectiveness such as travel times and fuel and emissions statistics. Various levels of market penetration were simulated to estimate how benefits would increase or decrease if more drivers have route guidance information. The potential benefits to drivers without guidance was also calculated. The findings suggested that drivers equipped with a TravTek-like system in addition to those not equipped would benefit. It is projected that at a 50% market penetration, there would be some travel time savings for all vehicles on the traffic network. Further discussion and results can be found in Van Aerde and Rakha (1995).

Safety Evaluation. This integrative study considers several aspects of safety. Data from various TravTek studies were examined to determine the potential for safety benefits and the possibility of safety detriments. As a starting point, traditional statistical safety analyses were performed. In addition, modeling was employed to evaluate how safety observed in a 100-vehicle LSE in Orlando might change as a function of market penetration level. Methodology included the establishment of facility and traffic volume effects on base accident rates, the evaluation of incidents and accidents involving TravTek vehicles, the potential safety impacts of the TravTek in-vehicle devices, and the modeling of potential safety impacts for both TravTek vehicles and non-TravTek equipped vehicles sharing a traffic network with equipped vehicles.

In the report by Perez, Van Aerde, Rakha, and Robinson (1995), although the TravTek sample was relatively small, the vehicle crash involvement rates of TravTek vehicles were compared with non-TravTek data supplied by Avis and national databases. Also, every incident and accident that occurred during the LSE was analyzed in detail. There was no indication that the TravTek crash involvement rate was greater than what would be expected based on national averages. Furthermore, there was no indication that TravTek was the cause of any accidents.

Clearly, TravTek did not have a sample of sufficient size to evaluate safety on a solely empirical basis, so the INTEGRATION model was extended to include safety. The TravTek route selection algorithm was biased toward higher-class road facilities such as freeways, which are statistically safer. However, the addition of real-time traffic information (N+) and the presence of traffic congestion can cause vehicles to be routed onto road facilities such as arterials, which are statistically less safe. The model suggests that this is only a potential disbenefit under conditions of high traffic demand and market penetration levels below 30%. This would apply only to N+ type vehicles with real-time traffic information. Perez et al. (1995) and Inman and Peters (1996) suggested that the results of this exploratory simulation be viewed with caution. Based on experience, the authors of the Safety Study report (Perez et al., 1995) offered a discussion of the challenges and issues presented in trying to associate data and modeling to predict safety impacts.

Concerns about the potential for decrements in safe driving due to distraction and information overload at the vehicle level of analysis were addressed by the Camera Car study and discussed in an earlier section. Detailed safety-related results, safety event classifications, and the accompanying error descriptions can be found in that final report (Dingus et al., 1995). It is hoped that these measures and methods as well as the event classification tool will be used by future researchers to continue to explore the relationship between attention and safe driving.

System Architecture Evaluation. The architecture evaluation emphasized the performance of the traffic network aspects of the TravTek system. These included an analysis of the hardware, software, and databases as well as the reliability of communications and the effectiveness of system operations. The focus of the analysis was on the Traffic Management Center (TMC), the Traffic Information Network (TIN), and sources of traffic data including probe vehicles (Blumentritt et al., 1995). To maximize the potential transfer of learning to other ITS designs and projects, a functional approach was used in addition to a detailed physical description of the TravTek system.

The focus of the TMC Design and System Architecture Evaluation was on the nonhuman parts of the system. Results indicated that TravTek was very reliable, that database accuracy was good, and that the TMC data fusion algorithm, probe vehicle concept, and distributed architecture all worked. However, better traffic incident reporting is needed.

Because human monitoring and decision making were essential, an ergonomic assessment of the TMC operator interface covered information, job, and workload analyses. The TMC operator interface was deemed functional, although several sources of information were identified as causing potential delays in information flow. Unpredictable intermittent reporting intervals, reduced operator vigilance because of low workload, and noise levels that masked warning cues were among the factors that contributed to missed operator data entry. Several workstation design factors were identified as having the potential to degrade operator performance. Better operator training was recommended.

It is clear that human factors participation in the analysis of work, environment, and workstation design in control rooms can greatly improve operator–system performance. The challenge is to increase the acceptance of the concept that the operator is part of the system architecture.

Global Evaluation. Because of the number of TravTek objectives, research questions, and studies, a final integrative analysis was necessary to supply a unified summation of what had been learned. Results from all of the studies were weighed to assess whether each goal was met, and if not, why not. In the final report, Inman, and Peters (1996) wrote that on the whole, TravTek performed at or above expectations. Time-saving and congestion-avoidance benefits were demonstrated. The system was user friendly and safe. Users showed in words and behavior that they would use and pay for a system like TravTek. The modeling study projected that even modest levels of market penetration could result in network benefits to nonusers.

Inman and Peters also included discussions of many lessons learned along with recommendations. Among the lessons learned that relate to human factors and the evaluation process were the following: Designing for safety of the TravTek system paid off; early consideration of the evaluation objectives paid off; building TravTek for evaluation paid off; and test subjects did not make privacy an issue. Also among the lessons learned were these: Recruitment of test subjects was resource intensive; measuring the impact of TravTek on safety was challenging; and resources required for processing, checking, and archiving TravTek evaluation data were underestimated.

A FINAL WORD

Human factors participation in shaping and implementing ITS operational field tests should be the rule rather than the exception. Human factors practitioners and scientists can contribute considerably to the safety, usefulness, and acceptability of ITS technology. In addition, human factors objectives serve to further the state of knowledge about safety and driving behavior, and to deepen the understanding of transportation systems as person–machine systems. However, to make these contributions and take advantage of the opportunities discussed in this chapter, human factors professionals must continue to educate and persuade. To do this, behavioral scientists must increase their understanding of the technology and its constraints and work with technologists and planners in their arenas as well as in their own.

The U.S. Department of Transportation requires that each ITS operational field test include an evaluation component. As a guideline, the authors believe that the evaluation portion of a demonstration program should receive roughly one fourth to one third of the total program funding. Large-scale evaluation at this level along with painstaking, detailed test planning and implementation will help ensure the meaningfulness of the test results. In addition to the compulsory OFT evaluation component, there ought to be a human factors and safety component. We suggest that proposals for planning and implementing LSEs be required to address human engineering, safety, and societal issues as well as user acceptance and marketing issues. This requirement should not be limited to particular areas of ITS such as ATIS, which have been explored to a degree, nor can an assessment be a reiteration of potential benefit claims. Rather, all LSEs should address issues and concerns such as those discussed in this volume. Proposals should discuss how the issues will be dealt with or demonstrate why the issues are not relevant to a particular project. This would help to build a partnership with the human factors community leading to increased safety and satisfaction through better design.

ACKNOWLEDGMENTS

TravTek was made possible by the ingenuity and hard work of many people. The TravTek Partners (FHWA, Florida Department of Transportation, City of Orlando, AAA and GM) were unusual in their commitment to human factors and safety, and in their willingness to participate in an extensive, innovative research program. TravTek benefited greatly from the leadership of project manager, Jim Rillings, and the dedication of the members of the TravTek Evaluation Working Group. Thanks especially to Frank Mammano (FHWA), August Burgett (NHTSA), Al Mertig (MITRE), and Deborah Dennard (AAA) for seeing the project through to the end. The authors wish to acknowledge members of the driver interface human factors team including Frank Szczublewski, Linda Means, Mark Krage (all GM) and Janeth Carpenter (Hughes). The implementation of the studies was made possible by many engineers and technicians at GM, Hughes, and AAA who spent many hours bringing automatic data logging, the S, N, and N+ vehicle configurations, and the Camera Car into existence. The findings reported in this chapter are the result of a multiyear effort by researchers who contributed in various ways. The SAIC research team, led by Joe Peters and Vaughan Inman, had the responsibility of implementing the TravTek evaluation program under FHWA Contract No. DTFH61-91-C-00106. They, the study leaders, and a multitude of experimenters went beyond the call of duty. Finally, thank you to Melissa Hulse, Dan McGehee (University of Iowa) and Lisa Thelen (EDS).

REFERENCES

Barbaresso, J. C. (1994). Preliminary findings and lessons learned from the FAST-TRAC IVHS program. *Proceedings of the 1994 Annual ITS AMERICA Meeting* (pp. 489–497). Atlanta, GA: ITS AMERICA.

Bhise, V. D., Forbes, L. M., & Farber, E. I. (1986). *Driver behavioral data and considerations in evaluating in-vehicle controls and displays*. Paper presented at the Transportation Research Board 65th Annual Meeting, Washington, DC.

Blumentritt, C., Balke, K., Seymour, E., & Sanchez, R. (1995). *TravTek System Architecture Evaluation* (Federal Highway Administration Tech. Rep. No. FHWA-RD-94-141). Washington, DC.

Bolczak, R. (1993). *Generic IVHS Operational Test Evaluation Guidelines* (Working Paper No. 93W0000367). Federal Highway Administration, Washington, DC.

Carpenter, J. T., Fleischman, R. N., Dingus, T. A., Szczublewski, F. E., Krage, M. K., & Means, L. G. (1991). Human factors engineering the TravTek drivers interface. *Proceedings of the Vehicle Navigation and Information Systems (VNIS '91) Conference* (pp. 749–755). Dearborn, MI: SAE.

Dingus, T. A., Antin, J. F., Hulse, M. C., & Wierwille, W. W. (1989). Attentional demand requirements of an automobile moving-map navigation system. *Transportation Research, 23*(4), 301–315.

Dingus, T. A., & Hulse, M. C. (1993). *Camera car study final detailed test plan, TravTek evaluation* (FHWA Contract DTFH61-91-C-00106). Science Applications International Corporation, McLean, VA.

Dingus, T. A., Hulse, M. C., Krage, M. K., Szczublewski, F. E., & Berry, P. (1991, October). A usability automobile navigation and information system "pre-drive" functions. *Conference Records of Papers at the 2nd Vehicle Navigation and Information Systems (VNIS) Conference*, Dearborn, MI.

Dingus, T. A., Hulse, M. C., McGehee, D. V., Manakkal, R., & Fleischman, R. N. (1994). Driver performance results from the TravTek IVHS camera car study. *Proceedings of the Human Factors and Ergonomics Society 38th Annual Meeting* (pp. 1118–1121). Nashville, TN: Human Factors and Ergonomics Society.

Dingus, T., McGehee, D., Hulse, M., Jahns, S., Manakkal, N., Mollenhauer, M., & Fleischman, R. N. (1995). *TravTek Evaluation Task C3—Camera Car Study* (Federal Highway Administration Tech. Rep. No. FHWA-RD-94-076). Washington, DC.

Fleischman, R. N. (1991). Research and evaluation plans for the TravTek ITS operational field test. In *Proceedings of the Vehicle Navigation and Information Systems (VNIS '91) Conference* (pp. 827–837). Dearborn, MI: SAE.

Fleischman, R. N., Carpenter, J. T, Dingus, T. A., Szczublewski, F. E., Krage, M. K., & Means, L. G. (1991). Human factors engineering in the TravTek demonstration IVHS project: Getting information to the driver. *Proceedings of the Human Factors Society 35th Annual Meeting* (pp. 1115–1119). San Francisco, CA: Human Factors Society.

Fleischman, R. N., Thelen, L., & Dennard, D. (1993). A preliminary account of TravTek route guidance use by rental and local drivers. In *Record of Papers of the Conference on Vehicle Navigation and Information Systems (VNIS'93) Conference* (pp. 120–125). Ottawa, OT: IEEE.

Green, P., Hoekstra, E., & Williams, M. (1993). *On-the-road tests of driver interfaces: Examination of a navigation system and a car phone* (Tech. Rep. UMTRI-93-21). The University of Michigan Transportation Research Institute, Ann Arbor, MI.

Hulse, M. C., Dingus, T. A., McGehee, D. V., & Fleischman, R. N. (1995). The effects of area familiarity and navigation method on ATIS use. *Proceedings of the Human Factors and Ergonomics Society 39th Annual Meeting*. San Diego, CA: Human Factors and Ergonomics Society.

Inman, V. W., Fleischman, R. N., Dingus, T. A., & Lee, C. H. (1993). Contribution of controlled field experiments to the evaluation of TravTek. *Proceedings of the 3rd Annual ITS AMERICA Meeting*. Washington, DC: ITS AMERICA.

Inman, V. W., Fleischman, R. N., Sanchez, R. R., Porter, C. L., Thelen, L. A., & Golembiewski, G. (1996). *TravTek Evaluation Rental and Local User Study Final Report* (Federal Highway Administration Tech. Rep. No. FHWA-RD-96-028). Washington, DC.

Inman, V. W., & Peters, J. I. (1996). *TravTek Global Evaluation* (Federal Highway Administration Tech. Rep. No. FHWA-RD-96-031). Washington, DC.

Inman, V. W., Sanchez, R. R., Bernstein, L. S., & Porter, C. L. (1995a). *Orlando Test Network Study* (Federal Highway Administration Tech. Rep. No. FHWA-RD-95-162). Washington, DC.

Inman, V. W., Sanchez, R. R., Porter, C. L., & Bernstein, L. S. (1995b). *TravTek Evaluation Task C1- Yoked Driver Study* (Federal Highway Administration Tech. Rep. No. FHWA-RD-94-139). Washington, DC.

ITS AMERICA. (1993). *National Program Plan for Intelligent Vehicle Highway Systems.* Washington, DC.

JHK Associates (1993). *Pathfinder Evaluation Report.* Prepared for the California Department of Transportation (Caltrans). Pasadena, CA.

Juster, R. J., Wilson, A. P., & Wensley, J. A. (1994). An evaluation of the SmarTraveler ATIS operational test. *Proceedings of the 1994 Annual ITS AMERICA Meeting* (pp. 438–446). Atlanta, GA: ITS AMERICA.

Koziol, J. S., Wagner, D. P., & Bolczak, R. (1994). Evaluation of the safety impact of the ADVANCE system. *Proceedings of the 1994 Annual ITS AMERICA Meeting* (pp. 312–317). Atlanta, GA: ITS AMERICA.

Krage, M. K. (1991). The TravTek driver information system. *Proceedings of the Vehicle Navigation and Information systems (VNIS '91) conference* (pp. 739–748). Dearborn, MI: SAE.

Mammano, F. J., & Sumner, R. (1991). Pathfinder status and implementation experience. *Proceedings of the Vehicle Navigation and Information Systems (VNIS '91) Conference* (pp. 407–413). Dearborn, MI: SAE.

McGehee, D., Dingus, T., & Horowitz, A. (1994). An experimental field test of automotive headway maintenance/collision warning visual displays. *Proceedings of the 38th Annual Meeting Human Factors and Ergonomics Society*, Santa Monica, CA: Human Factors Society, 1099–1103.

Means. L. G., Carpenter, J. T., Szczublewski, F. E., Fleischman, R. N., Dingus, T. A., & Krage, M. K. (1993). Design of the TravTek auditory interface. *Transportation Research Record*, 1403, (pp. 1–6). Transportation Research Board: Washington, DC.

Mitchell, M. W. (1989). *Determining Effective Display Format and Content Options for In-Car Moving-Map Navigation and Information Systems.* Unpublished Master's thesis, The University of Idaho, Moscow, ID.

Perez, W. A., Fleischman, R. N., Golembiewski, G., & Dennard, D. (1993). TravTek field study results to date. *Proceedings of the 3rd Annual ITS AMERICA Meeting* (pp. 667–673). Washington, DC: ITS AMERICA.

Perez, W. A., VanAerde, M., Rakha, H., & Robinson, M. (1995). *TravTek Evaluation Safety Study* (Federal Highway Administration Tech. Rep. No. FHWA-RD-95-188). Washington, DC.

Peters, J. I. (1995). Human factors evaluations in operational field tests of intelligent transportation systems. In *1995 Compendium of Technical Papers for the 65th ITE Annual Meeting* (pp. 681–685). Denver, CO: Institute of Transportation Engineers.

Peters, J. I., Mammano, F. J., Dennard, D., & Inman, V. W. (1993). TravTek evaluation overview and recruitment statistics. In *Record of Papers of the Conference on Vehicle Navigation and Information Systems (VNIS'93) Conference* (pp. 108–113). Ottawa, OT: IEEE.

Rillings, J. H., & Lewis, J. W. (1991). TravTek. *Proceedings of the Vehicle Navigation and Information Systems (VNIS '91) Conference* (pp. 729–737). Dearborn, MI: SAE.

Rockwell, T. (1972). Skills, judgement, and information acquisition in driving. In T. W. Forbes (Ed.), *Human factors highway traffic safety research* (pp. 133–164). New York: Wiley Interscience.

Science Applications International Corporation (1993a). *Renter user study final detailed test plan, TravTek evaluation* (FHWA Contract DTFH61-91-C-00106). McLean, VA.

Science Applications International Corporation (1993b). *Local user study final detailed test plan, TravTek evaluation* (FHWA Contract DTFH61-91-C-00106). McLean, VA.

Science Applications International Corporation (1995a). *SWIFT Evaluation: Task 1. Evaluation Plan.* McLean, VA.

Science Applications International Corporation (1995b). *Genesis Pilot Phase Phase Evaluation: Overall Evaluation Test Plan.* McLean, VA.

Texas Transportation Institute (1993). *System architecture evaluation detailed test plan, TravTek evaluation* (FHWA Contract DTFH61-91-C-00106). Science Applications International Corporation, McLean, VA.

Underwood, S. E. (1994). FAST-TRAC: Evaluating an integrated intelligent vehicle-highway system. *Proceedings of the 1994 Annual ITS AMERICA Meeting* (pp. 300–311). Atlanta, GA: ITS AMERICA.

VanAerde, M., & Rakha, H. (1995). *TravTek Evaluation Modeling Study* (Federal Highway Administration Tech. Rep. No. FHWA-RD-95-090). Washington, DC.

Van Aerde M., & Yagar, S. (1988). Modeling dynamic integrated freeway/traffic signal networks: A proposed routing-based approach. *Transportation Research, 22A-6* (pp. 445–453). New York: Pergamon.

10

SURVEY METHODOLOGIES FOR DEFINING USER INFORMATION REQUIREMENTS

Linda Ng
University of Washington

Woodrow Barfield
Virginia Polytechnic Institute and State University

Jan H. Spyridakis
University of Washington

This chapter reviews several methodologies for conducting surveys to define user information requirements for Intelligent Transportation Systems (ITS). Case studies of various transportation surveys administered by University of Washington research teams are provided to elucidate the suitability and advantages of surveys for the purpose of defining these requirements.

To improve traveler information, to alleviate traffic congestion, and to increase road safety, a major national effort is underway to integrate knowledge of traveler behavior and decision making into the design of ITS. To focus on two of the nation's most critical problems—road congestion and safety—ITS encompasses several aspects of advanced road travel: traffic management, vehicle control, public transportation, traveler information, and commercial vehicle operations (Mahmassani, Baaj, & Tong, 1984). The use of survey methodology for obtaining information requirements for users of Advanced Traffic Management Systems (ATMS), Advanced Traveler Information Systems (ATIS) and Commercial Vehicle Operations (CVO) is the focus of this chapter.

ATMS involves the management of freeways and surface streets with real-time information that is adaptive and responsive to traffic flow (Mobility 2000, 1989). For example, current traffic management systems such as variable message signs (VMS) are designed to make significant improvements in traffic operations. In addition, ATIS seeks to provide in-vehicle routing and navigational information as well as information on roadside services and emergency road conditions. Finally, CVO, advanced commercial vehicle systems, are being designed to automate existing manual procedures, electronically capture data, and improve the flow of information between carriers and regulatory agencies, as well as between operators and dispatchers (IVHS America, 1992).

In the past, techniques to alleviate traffic problems have typically focused on capital-intensive strategies such as the development of new roads or light rail systems to increase the system capacity. However, a less capital-intensive alternative strategy is to design an information system based specifically on the information needs of motorists (Dudek & Jones, 1970; Dudek, Messer, & Jones, 1971a). It is hypothesized that for ITS to be effective, its design must be based on a comprehensive understanding of the traffic information needs of travelers. To this end, information about drivers' attitudes, preferences, driving strategies, and reactions to traffic information must be obtained as well as information on their cognitive and information-processing abilities (see Fig. 10.1). This premise represents one of the basic design principles for any complex system: Design the system with the end-user in mind. Deviations from this principle typically result in systems that gain little support and usage from the public. The design of information systems that will be accepted by the user depends on obtaining information from the users, which requires that information be gathered from many potential users of a system, often by surveys administered to the target population. Surveys enable researchers to analyze the effects of ITS directly from perceptions of individual motorists and their trip behaviors. Unlike highly structured experiments that typically use only a small sample size (typically fewer than 30 subjects), surveys can achieve a sample size that adequately supports qualitative modeling and forecasting (Abdel-Aty, Vaughn, Kitamura, & Jovanis, 1992).

This chapter is divided into four sections. Section one discusses the benefits of using surveys to define motorist information requirements for ATIS as well as effective approaches for administering transportation surveys. Section two reviews case studies of surveys as designed and administered by researchers at the University of Washington. Section three reviews the analysis techniques that can be conducted on large- and small-

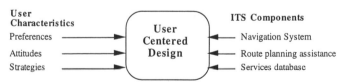

FIG. 10.1. A user-centered design strategy for the design of ITS.

scale surveys. Finally, section four summarizes the first three sections and provides recommendations for using survey data to define requirements for transportation systems.

SURVEYS FOR DEFINING USER REQUIREMENTS

For many years, transportation specialists have used survey results to guide highway design decisions. Many previous surveys have been conducted to increase our general understanding of motorist behavior (Anderson, Haney, Katz, & Peterson, 1964; Carpenter, 1979; Dudek, Messer, & Jones, 1971a; Huchingson, Whaley, & Huddleston, 1984; Shirazi, Anderson, & Stresney, 1988). In general, these surveys have focused on basic demographic factors describing motorists rather than exploring how and when motorists make driving decisions. Initiatives from ITS America (formerly IVHS America) have supported the use of surveys to identify the human factors elements of ITS. Currently, information requirements are being obtained for all driver types including noncommuting drivers, commercial vehicle operators, and commercial system operators (Ng, Barfield, & Mannering, 1995). For example, a survey conducted in 1990 on for-hire carriers, private carriers, and intercity bus fleets throughout Canada identified the highway information services preferred by commercial vehicle operators for a real-time information system (Tsai, 1991). Over 40 trucking and busing organizations participated in Tsai's studies. According to the survey results respondents felt that traffic and weather information was "very" to "somewhat" important. Furthermore, the survey found that respondents desired traffic information about traffic congestion, accidents, lane closures, bridge closures, construction updates, alternate routes, low bridges, road weight restrictions, and legal truck routes. The type of weather information desired included reports on adverse or severe weather conditions, fog conditions, and areas experiencing black ice. Tsai also studied the timeliness of weather and traffic information to determine its usefulness to the trucking industry and found that in mild to moderate weather conditions, respondents desired information hourly, twice daily, or daily, but that in severe to adverse weather conditions, they wanted information every 10 to 30 min. This survey provided important information for the design of a real-time traffic information system for commerical drivers that could not be easily determined in a laboratory setting because of the diversity and number of subjects.

Given the great diversity in the designs and methods available for use in surveys (see Table 10.1), numerous complex decisions have to be made during the survey design phase. Specifically, these include decisions about sample size and administration methods, sample generation methods, subject characteristics and questionnaire content, and finally, the statistical

TABLE 10.1
Summary of Survey Methodology

Agencies	Sample Size	Sex, Age, Income Distribution	Sampling Method	
			Generation	Administration
Stanford Res. Inst. (Anderson et al., 1964; Haney, 1964)	30 drivers	19 M, 11 F; age, income NG*	Employees from S. Palo Also	Unstructured interview
Stanford Res. Inst. (Anderson et al., 1964; Haney, 1964)	25 drivers	Sex, age, income NG*	Employees from S. Palo Alto	Structured interview
Stanford Res. Inst. (Anderson et al., 1964; Haney, 1964)	150 drivers	All male; age, income given	Residence in given block	Structured interview
Texas Trans. Inst., TX A&M (Dudek et al., 1983)	18 drivers in each survey	Sex, age, income NG*	Not given; naive about freeway system	In-car interview
Transport Studies Unit, Oxford University (Carpenter, 1979)	9 homes, 32 homes	Sex, age, income NG*	Maximized range of car use, trip purpose, household character-istics, not representative	Unstructured in-depth home interview
LA City & County, CA, Highway Patrol & DOT (Shirazi et al., 1988)	400 drivers	58% M, 42% F; Age: 18-20 = 5%; 21-40 = 67%; > 40 = 28%; Income NG	Random list of computer generated phone numbers	Structured phone interview
Trans. Res. Board, Houston, TX (Huchingson et al., 1984)	843 drivers	Sex, age, income NG*	License-plate survey at 6 sites on 3 freeways	1-page mail-in questionnaire
British Dept. of Trans. (Wooton et al., 1981)	~508	Sex, age, income NG*	Volunteers at destination sites	Not given
Lund (Sweden) Inst. of Technol., Dept. Trans. P1 & Eng. (Hansson,1975)	691 homes 830 drivers	Sex, age, income NG*	Not given	Home interview/ mail questionnaire

TABLE 10.1 (continued)

Agencies	Sample Size	Sex, Age, Income Distribution	Sampling Method	
			Generation	Administration
UCLA, Inst. of Trans. Traffic Eng. (Case et al., 1971)	304, half at each	76% M, 24% F; Mean age 37; Income NG	Volunteers at sites	On-site interview
UCLA, Inst. of Trans. Traffic Eng. (Case et al., 1971)	2971 drivers; grops of 250	48% M, 52% F; Age: < 35 = 61%, 35-49 = 24%, ≤ 50 = 16%; Income NG*	CA driver's license; recruited to represent population	Questionnaire
Texas Trans. Inst., Texas A&M (Dudek et al., 1971a, 1971b)	505 drivers	68% M, 32% F; age NG*; income given	Employees of 17 groups	On-site interview; 329 in homes, 176 in car
Univ. of Washington (Barfield et al., 1991; Spyridakis et al., 1991)	~4000 drivers	50.7% M, 48.8% F; age, income	Seattle commuters	Questionnaire
Univ. of Washington (Ng et al., 1995)	1610 drivers and dispatchers	Sex, age, income**	Randomly distributed nationwide	Questionnaire

Note. From Spyridakis et al. (1991). Adapted from *Transportation Research, 25A(1),* Surveying commuter behavior: Designing motorist information systems, pp. 17-30, © 1991, with kind permission from Elsevier Science Ltd, The Boulevard, Langford Lane, Kidlington OX5 1GB, UK.
*NG = Not given but researcher claims sample represents population.
**Distribution of sex, age, and income differ among the user groups. See Ng et al. (1995) for details.

techniques used to analyze the data. Each of these topics is discussed below. Other information regarding survey preparation was previously reviewed in chapter 8 of this volume.

Sample Size and Administration Methods

The interdependency of sample size, budget, and data-analysis techniques influences a survey's design, and this interdependency cannot be ignored. Surveys can be administered in person, by phone, by newsletter, by mail and even by computer networks. Cost and sample size relate to survey administration method: A larger sample size typically requires a less personal administration method if costs are to be controlled.

The surveys summarized in Table 10.1 exemplify a large range of sample sizes and administration methods used in previous studies. Because in-person interviews are costly and time consuming, their use often leads to the use of a smaller survey sample (i.e., fewer than 200 subjects). Stanford's 1963 and 1964 personal interviews used sample sizes of 25, 30, and 150 drivers (Anderson et al., 1964; Haney, 1964); three on-road interviews by the Texas Transportation Institute each used 18 drivers (Dudek, Huchingson, & Brackett, 1983), and an Oxford University study conducted two sets of in-person surveys in 9 and 32 homes, respectively (Carpenter, 1979).

When larger sample sizes are needed, surveys often are conducted by mail. For example, sample sizes in the midrange, from 304 to 843 drivers, have included both personal interviews (at destination sites, at home, or by telephone) and mailed questionnaires (Case, Hulbert, & Beers, 1971; Dudek, Friebele, & Loutzenheiser, 1971b; Dudek, Messer, & Jones, 1971a; Hansson, 1975; Huchingson et al., 1984; Shirazi, Anderson, & Stresney, 1988; Wooton, Ness, & Burton, 1981). Sample sizes for mailed questionnaires ranged from 505 drivers in a 1971 Houston and Dallas survey to 843 drivers in a 1984 Houston survey (Dudek et al., 1971b; Dudek et al., 1971a; Huchingson et al., 1984). The 1984 Houston survey was limited to a one-page questionnaire, perhaps because costs for data analysis, printing, and postage all increase with sample size or because response rates are higher with a shorter survey. In contrast to the general trend that larger samples are surveyed with mailed questionnaires, Case et al. (1971) conducted in-person interviews with a large sample size of 2,971 by interviewing large groups of drivers (250 drivers per group).

Sample size may also influence the conclusions that can be drawn by the researcher and also the generalizability of the results. If the sample size is too small, the utility of results diminishes, and the desired precision (as measured by the variance of the sample mean) may not be achieved (Cochran, 1977). Many standard formulas exist for determining sample size (Cochran, 1977). A simple equation for defining the sample size, n, for a simple random sample of an infinite population can be shown as

$$n = \frac{s^2}{V},\tag{1}$$

where s^2 is the population variance, and V is the acceptable margin of error squared. A larger population variance will therefore constitute a larger sample size, and a smaller acceptable margin of error will also constitute a larger sample size. For a finite population, the value, n, is corrected by multiplying n by the finite population correction (fpc) factor $(1+\frac{n}{N})$. Therefore, the sample size for a finite population, n^*, will be

$$n^*_{w/fpc} = \frac{n}{1 + \dfrac{n}{N}}, \tag{2}$$

where N is the population size. The chosen value of n must always be reviewed with the cost, labor, time and material required to obtain the proposed sample size (Cochran, 1977) as well as any anticipated nonresponse rate.

Clearly, the choice of sample size involves trade-offs. A small sample size allows for extensive questioning and personal administration methods, but a larger sample size, while restricting the number of questions and the administration methods, increases the likelihood of greater external validity.

Sampling Methods

After sample size and administration methods have been determined, a method for generating the sample must be carefully designed. There are several techniques for selecting a sample. A simple random sample of a given population is perhaps the most popular method. It involves selecting drivers by using some random generation method (e.g., a dice, random number table, or picking numbers out of a hat). This process gives an equal chance of selection to any person in the population not already drawn.

Stratified sampling involves dividing the population into distinct subpopulations called *stratas*. This method is useful when subpopulations may be the domains of the study or when subpopulations require different sampling procedures. For example, a group of motorists may be subdivided into two groups—private motorists and commercial drivers—because there are relevant differences between the two groups. There also exist different sampling methods. Obtaining a sample of drivers from an automobile club will not provide a representative sample of the commercial driving population nor would a sample of drivers from a truck inspection point provide a representative sample of private motorists.

Cluster sampling is the method of dividing one population into groups or clusters. For example, the driving population in America can be subdivided into 50 clusters or groups that represent drivers by state (i.e., drivers from Maine, New York, Texas, etc.). A sample of the states is then chosen for sampling. This method is less expensive than simple random sampling, but unfortunately, it is not as precise as simple random sampling. If, however, a nationwide sample of drivers were required, this method would reduce the cost of sampling immensely.

There are various other sampling methods such as systematic sampling, area sampling, and rotation sampling, each of which can be reviewed in standard survey sampling textbooks (Cochran, 1977; Kish, 1965; Levy & Lemeshow, 1991). In addition, the sampling methods discussed here can be combined or used with other methods. However, if the survey's sampling frame is to accurately reflect the driver population in question, the sample must be generated in a manner that creates a sample representative of that population.

Subject Characteristics
and Questionnaire Content

The characteristics of the subjects comprising a particular survey may affect the validity of the survey findings. Researchers may be quite clear about what information they wish to gather, but they also must be careful to identify variables that initially may appear to be irrelevant yet ultimately may provide the basis for interpreting other data. Specific subject characteristics that may be regarded as demographic or subject descriptor information may actually become critical variables for drawing inferences as well as for comparing results across studies. For example, Bonsall and Joint (1991) found that younger people were more likely to reject advice. They determined that in addition to targeting groups who are likely to find advice credible, drivers who go to many unfamiliar destinations, such as commercial drivers and drivers of for-hire cars, should also be targeted. Studies have shown that variations in trips have been attributed to household size, car ownership, and number of adults in a household (Said, Young, & Ibrahim, 1991). Gender and residence location also have been found to contribute to differences in trip behavior (Prevedouros & Schofer, 1991).

In any survey, subject classification variables such as age, gender, and income distribution may correlate with other variables under examination. Such variables may help the researcher to interpret data patterns as well as allow other researchers to compare their findings. The survey literature on motorist behavior is inconsistent in its collection of data on subject characteristics (see Table 10.1). Although difficult, it is crucial to identify all critical subject variables (e.g., gender, income, age, and education). For example, a surveyor seeking information on drivers' responses to road sign messages may need to include questions assessing subjects' visual acuity so that the relationship between vision and the drivers' responses to road signs can be determined.

Statistical Analysis

After the data have been collected, investigators can use several statistical tools to analyze their information. Descriptive statistics (e.g, means, variances) will organize and summarize the surveyed population data, whereas inferential statistics will go beyond the sample to draw conclusions about the population. Econometric models also can be used to predict the likelihood of trip behavior given various characteristics. The studies cited in Table 10.1 provide some inferential statistics, but very few mathematical modeling techniques were used. Vaughn, Abdel-Aty, Kitamura, Jovanis, and Yang (1992) showed how inferential statistics have been used to provide insight on the impact of route-choice behavior under ATIS. An interactive route-choice simulation experiment was conducted that had the following significant analysis of variance (ANOVA) findings relating to socioeconomic variables:

- As information accuracy increases, acceptance of ATIS advice also increases.
- Males are more willing than females to accept advice.
- Males make decisions faster than females.
- Males are less likely to purchase an information system when the system accuracy is below 90%.
- Compared to less experienced drivers, experienced drivers are not as willing to accept advice.
- Experienced drivers make decisions faster than inexperienced drivers.
- Experienced drivers are more likely to purchase an information system.

In summary, to achieve the goal of attaining user information requirements for ITS, it is recommended that two principles must be considered: (a) More powerful statistical techniques often are required rather than basic descriptive statistics, and (b) innovative applications of inferential and descriptive techniques as well as mathematical modeling can enhance the external validity of the research design and increase the generalizability of the findings.

CASE STUDIES OF SURVEY METHODOLOGIES

This section discusses experiences and recommendations from researchers at the University of Washington on the methodology of designing and distributing surveys over a period of several years and under the guidance of separate project initiatives. These surveys are presented as case studies to focus on the trade-offs between various procedures and methodologies. The surveys discussed in this section encompass various sample sizes and addresses a multitude of driving groups including commuters, commercial vehicle drivers, recreational travelers, and commercial dispatchers. The first survey focuses on defining commuter information requirements using commuters from Seattle, Washington, as a representative sample. This survey evolved into several smaller sampled surveys to specifically address requirements for a real-time traffic information system. The last set of surveys focuses on the distribution of three nationwide surveys to several driver groups to define user requirements for advanced traveler information systems (ATIS).

Case Study 1: Inital Survey—Defining Commuter Information Requirements

ATIS should affect four aspects of commuter behavior: (a) departure time, (b) transportation mode (buses, trains, car pools, etc.), (c) pretrip route choice, and (d) on-road route modification. However, to impact this commuting behavior positively, an understanding of the decision-making processes and the commuting behavior of motorists must be achieved. In 1988, as part of the Freeway Arterial Management Effort (FAME) in Seattle, Washington, information was gathered concerning motorist activities and behaviors, particularly as they related to the design and delivery of motorist information. In the past, an understanding of motorist behavior was often obtained through household-based trip surveys or general driver population surveys. The approach taken in the FAME project, however, targeted commuters who used a specific freeway corridor and consisted of a large sample mail-in survey distributed on-road. One goal of this large on-road survey was to identify commuting characteristics of Seattle motorists and their use of and preferences for motorist information. This approach proved cost efficient for the purpose, resulted in a high response rate, and gathered complex and varied data.

The motorist information questionnaire was designed by a multidisciplinary research team of individuals with experience in human factors, technical communications, psychology, and statistics. The study focused on one particular subset of the Seattle commuter population: motorists who commuted to work from north of Seattle to a downtown location during the morning rush hour (5:45 a.m. to 8:45 a.m.). The selected population involved approximately 25,000 drivers of which 9,652 were given surveys. The survey was distributed to commuters as they exited a major freeway (I-5) and stopped at the first intersection. Respondents completed and returned (by mail) the survey within two weeks. The survey was designed to collect data on 62 variables covering four main topics: (a) basic demographic information, (b) route choice information, (c) driver traffic information needs, and (d) information for driver classification (see Fig. 10.2). The return rate for the survey was 40%.

Survey Respondents

The first steps in the project's design were to identify the survey population, the sampling frame, and the actual sample size.

A. Your Commute

1. In an average week, how many days do you drive I-5 to or from work anywhere between Lynnwood and downtown Seattle?

☐ 7   ☐ 6   ☐ 5   ☐ 4   ☐ 3   ☐ 2   ☐ 1   ☐ 0

(If zero, please skip to Section C, next page)

2. Please tell us where you usually enter and exit I-5 when you commute.

Southbound --   Enter I-5: _____

                Exit I-5: _____

Northbound --   Enter I-5: _____

                Exit I-5: _____

3. Estimate you driving...

Distance between home and work, excluding detours and errands: _____miles

Time from home to work, excluding detours and errands: _____ minutes

Time from work to home, excluding detours and errands: _____ minutes

4. How much flexibility is there in the time when you...

Leave home for work	☐ A lot	☐ Some	☐ Very little
Leave work for home	☐ A lot	☐ Some	☐ Very little

5. How much stress do you experience during your usual commute to and from work?

☐ A lot   ☐ Some   ☐ Very little

6. During your commute, how much importance do you place in...

Saving commute time	☐ A lot	☐ Some	☐ Very little
Reducing commute distance	☐ A lot	☐ Some	☐ Very little
Increasing commute safety	☐ A lot	☐ Some	☐ Very little
Increasing commute enjoyment	☐ A lot	☐ Some	☐ Very little

7. How many people (including yourself) usually are in the car when you commute?

☐ 5 or more   ☐ 4   ☐ 3   ☐ 2   ☐ 1

B. Your Route Choice

1. How familiar are you with north/south routes that can be used as alternatives to I-5?

☐ Very   ☐ Somewhat   ☐ Not at all

2. How often do you modify or change the route you travel from...

Home to work	☐ Frequently	☐ Sometimes	☐ Rarely
Work to home	☐ Frequently	☐ Sometimes	☐ Rarely

3. How often do the following factors affect your choice of commuting routes?

	Frequently	Sometimes	Rarely
Traffic reports and messages	☐	☐	☐
Actual traffic congestion	☐	☐	☐
Time of day	☐	☐	☐
Weather conditions	☐	☐	☐
Time pressures	☐	☐	☐

4. Where are you most likely to choose your commuting route?

☐ At home or work   ☐ On city streets   ☐ Near entrance ramps   ☐ On I-5

5. When you are commuting, what length of delay on I-5 would cause you to divert to...

An alternate route that you know   _____minutes

An alternate route that you do not know   _____minutes

C. Traffic Information

1. From which media have you ever received traffic information? (Check all that apply in each column)

	Column A Before Driving	Column B While Driving
TV	☐	- - -
Electronic message sign over I-5	- - -	☐
Advisory radio indicated by flashing lights on highway sign	- - -	☐
Commercial radio station	☐	☐
Phone	☐	☐
CB Radio	☐	☐
None	☐	☐

FIG. 10.2. *(Continued)*

2. From whch medium would you prefer to receive traffic information? (Check one only in each column)

	Column A Before Driving	Column B While Driving
TV	☐	---
Electronic message sign over I-5	---	☐
Advisory radio indicated by flashing lights on highway sign	---	☐
Commercial radio station	☐	☐
Phone	☐	☐
CB Radio	☐	☐
None	☐	☐

3. How much help do you get from traffic information delivered by...

	A lot	Some	Very little	Never used
TV	☐	☐	☐	☐
Electronic message sign over I-5	☐	☐	☐	☐
Advisory radio indicated by flashing lights on highway sign	☐	☐	☐	☐
Commercial radio station	☐	☐	☐	☐
Telephone highway construction hot line	☐	☐	☐	☐

4. When you are in I-5, how often does traffic information cause you to divert to an alternate route?
   ☐ Frequently   ☐ Sometimes   ☐ Rarely   ☐ Never receive information

5. Before you drive, how often does traffic information influence...

	Frequently	Sometimes	Rarely	Never receive
The time you leave	☐	☐	☐	☐
Your means of transportation (e.g. car, bus)	☐	☐	☐	☐
Your route choice	☐	☐	☐	☐

6. At what point do you prefer to receive traffic information? (Check one only)
   ☐ Before driving   ☐ On city streets   ☐ Near entrance ramps   ☐ On I-5

7. If continual up-to-the-minute traffic information were available in the following ways, would you use them?

Traffic information delivered via phone hot line	☐ Yes	☐ No
Radio station dedicated to traffic information	☐ Yes	☐ No
Traffic information delivered via computer	☐ Yes	☐ No
Cable TV station dedicated to traffic information	☐ Yes	☐ No

8. Which of these services would you like to see developed first? (Check one only)
   ☐ Traffic information delivered via phone hot line
   ☐ Radio station dedicated to traffic information
   ☐ Traffic information delivered via computer
   ☐ Cable TV station dedicated to traffic information

9. Which of the following are available to you? (Check all those items that are usually in working order.)

Radio:	☐ Home	☐ Office	☐ Car
Phone:	☐ Home	☐ Office	☐ Car
TV:	☐ Home	☐ Office	
TV cable hook-up:	☐ Home	☐ Office	
Computer:	☐ Home	☐ Office	

D. For Classification Purposes

1. What is your home Zip Code? _____ your work Zip Code? _____
2. Are you: ☐ Male   ☐ Female
3. What is your age?   ☐ Under 31   ☐ 31-40   ☐ 41-50   ☐ 51-64   ☐ 65 and over
4. What is your annual income, before taxes, for you entire household?

	Total Household		Total Household
No income	☐	40,000-49,999	☐
Under $10,000	☐	50,000-59,999	☐
10,000-19,999	☐	60,000-74,999	☐
20,000-29,000	☐	75,000-100,000	☐
30,000-39,000	☐	Over 100,000	☐

5. Would you be willing to take part in a follow-up interview about your use of traffic information? If so, please fill out the following. A more detailed discussion of your commute would help us improve your travel on Seattle freeways. All information will be kept confidential. [Spaces were provided for name, occupation, address, and work and home phone]

FIG. 10.2. Questions from motorist information survey (1988).

Survey Population and Sampling Frame. The goal in selecting the surveyed population was to identify a large number of Washington State freeway commuters who experience normal traffic congestion and who have access to various forms of traffic information. In response to this goal, the Washington State Department of Transportation (WSDOT) suggested two Seattle corridors: north/south Interstate 5 (I-5) or east/west State Route 520 (SR520). It was assumed that each freeway corridor would be unique in certain ways. These corridors were initially specified for three reasons: (a) these Seattle corridors have extremely large traffic volumes that would allow the obtaining of a sizable sample; (b) motorists who use these corridors during commuter rush hours frequently experience traffic congestion; and (c) these corridors contained various message-delivery systems. The northern I-5 corridor was finally selected because it contained the largest variety of motorist information delivery media, including two on-road traffic message delivery systems: Variable Message Signs (VMS) and Highway Advisory Radio (HAR). The population was then narrowed to include all motorists who use I-5 to commute from the North into the greater Seattle downtown area.

From the population described in the preceding discussion, a sample frame was selected that would provide a sufficient number of motorists to represent the population. To be included in the sample frame, motorists had to commute south on some portion of the northern I-5 corridor to downtown Seattle during peak morning rush hours (5:45 a.m. to 8:45 p.m.), commute on this corridor at least once a week, and be the driver of a commuting vehicle. According to Washington State DOT exit ramp data, approximately 25,000 drivers per day met these criteria.

Sample Size. A minimum sample size of 500 would have supplied findings with sufficiently small standard errors to accurately represent the driving population for the selected highway corridor, provided that the selected sample truly represented the population. However, there was to be a follow-up in-depth survey that required about 100 participants randomly selected from the initial survey. Thus, it was necessary to start with a much larger initial sample size than the 500 minimum first suggested. Assuming that 25% of the total initial survey respondents were likely to volunteer for the in-depth survey, a minimum of about 2,000 respondents to the initial survey would be needed to assure the success of the follow-up survey. Furthermore, with a minimal expected response rate of 20% for the mail-in survey, the desired motorist sample grew to almost 10,000.

Survey Administration

The surveys were administered in Seattle, Washington, a city with the sixth worst traffic congestion problem in the nation, which is representative of large cities with a large number of commuters to a central business district. Moreover, for the 3.5 million licensed drivers in Washington (FHWA, 1991), there is only one major freeway that passes north and south through downtown Seattle (I-5) and two major freeways that pass east and west (I-90 and SR520). The following describes the various distribution methodologies evaluated for administering the surveys and the specific chosen distribution procedure and survey collection.

Distribution Method. Five distribution approaches were evaluated, and were weighed against the following criteria: time frame for accessing motorists, potential response rate or bias, accuracy of contacting the desired sample frame, and time differential between time of commute and receipt of survey. The goal was to identify a method that would allow for accessing motorists easily and without much delay, obtaining a high response rate, obtaining a representative sample, and reducing the time between motorists' commute and receipt of the survey. The five distribution methods that were evaluated are described in the following discussion.

1. *Geographical Selection by Business Location.* This method required questionnaire distribution and data collection at specific building sites in downtown Seattle. Access to potential participants in these buildings would have to be handled through the city's building transportation coordinators. This method required potentially cumbersome coordination for access to participants, and targeting of the northern I-5 corridor commuters would be complex. Furthermore, a potential sample bias could occur because in Seattle only buildings with large employee bases have transportation coordinators.

2. *License Plate Survey Method.* This method required recording license plate numbers of motorists on the northern I-5 corridor during the morning rush hour and then acquiring motorist names, addresses, and phone numbers from the Washington State Department of Motor Vehicle Licensing. This method would be very labor intensive, could provide inaccurate sampling because people often move or list incorrect addresses, would require a large time differential between commuting time and survey receipt, and would be impersonal.

3. *Geographic Selection by Home Address or Telephone.* This method required door-to-door sampling, a selected mailing, and/or telephone solicitation. It would be very labor intensive given the desired sample size, pose some population targeting problems, and require a large time differential between commuting time and survey receipt.

4. *General Distribution.* This method required placing the survey in a large circulation Sunday newspaper. A newspaper–newsletter survey usually produces a very low response rate because of the impersonal distribution method.

5. *On-Road Solicitation Method.* This method involved acquiring potential respondents by accessing them at freeway off-ramps. Although somewhat labor intensive, this method was selected because it offered an in-person distribution method during commuting time, a timely distribution method, and accurate targeting of the correct sample frame.

After identifying specific details of these methods, the five-member design team individually assigned ratings to each method based on the stated criteria. These averaged ratings appear in Table 10.2.

The method of on-road solicitation was chosen as the survey distribution method. As Table 10.2 shows, this method best met the selection criteria, receiving the highest rating of 20 out of the 25 possible points.

Distribution and Collection. Surveys were distributed at eight north–south off-ramps in downtown Seattle on two consecutive weekday mornings (5:45 a.m. to 8:45 a.m.) in September 1988. Survey distribution times were based on peak-hour data from Washington State Department of Transportation (WSDOT) ramp figures. A statement on the survey requested that motorists return it within two weeks. In addition, radio media coverage of the survey distribution was used to generate publicity encouraging motorists to fill out and return their questionnaires. The surveys, when folded and stapled or taped, displayed the WSDOT address and a postal permit number. The daily rate of return was initially very high but decreased by the end of September. By November 4, 1988, a total of 3,893 surveys had been returned; any that arrived after this date were not processed.

TABLE 10.2
Rating of Distribution Methods

	Evaluation Criteria					
Method	Time Frame	Sampe Accuracy	Response Proximity	Method Complexity	In-Person Method	Total Rating
Business location	2	3	3	3	5	16
License Plate ID	1	4	1	2	2	10
Address or telephone number	3	2	1	1	3	10
Newspaper	5	1	1	5	1	13
Off-ramps	4	4	5	3	4	20

Note. 1 = low, 5 = high.
From Spyridakis et al. (1991), *Transportation Research, 25A*(1), Surveying commuter behavior: Designing motorist information systems, pp. 17-30, © 1991, with kind permission from Elsevier Science Ltd, The Boulevard, Langford Lane, Kidlington OX5 1GB, UK.

Questionnaire Design

The questionnaire needed to address specific information needs of this study in four main categories: (a) characteristics of the commute itself, (b) motorist choices and behavior, (c) delivery of traffic information, and (d) descriptive data for driver classification. To develop the questions, an extensive review of other motorist surveys, an analysis of human factors issues relevant to the design of traffic messages and motorist decision making, and suggestions from WSDOT personnel were used. After discussions with WSDOT representatives, a pretest of 25 drivers, and elimination of questions (some of which became part of an in-depth follow-up survey), the final questionnaire was completed. The questionnaire included a name, address, and telephone number section to be completed if the respondent would be available for a follow-up in-person interview.

Case Study 2: Follow-Up Surveys—Defining Commuter Information Requirements

After the initial analysis of the large-sample motorist survey, several smaller surveys were conducted to obtain more specific information on how respondents would use real-time traffic information. This section briefly describes two of the surveys designed to identify motorist needs for information systems based on knowledge of the behavior of identifiable subgroups of commuters (defined through analysis in the initial survey).

Graphics Questionnaire

A survey was developed in an effort to determine if the commuter groups or clusters identified from the initial survey preferred different forms of traffic information and to test the effectiveness of several traffic information screen prototypes. The questionnaire presented a series of sample screens and provided mostly check-the-box questions related to the effect of the screens on the commuter choice of driving options. The screens are briefly discussed in the following section and in detail in Gray (1990).

Prototypes of Traffic Information Screens. The questionnaire, based on a typical commuting scenario and pertaining to a graphics-based system, was administered to 97 commuters. Commuters were asked to imagine viewing sample screens as they might appear on their televisions before they left home for work in downtown Seattle in their private vehicles. The motorists were asked questions about how the prototypic screens affected their choice of driving options. Five different samples were used that simulated television screens (Gray, 1990). Development of these screens was based on a literature review of existing screen designs, results of the initial survey, and human factors guidelines for screen design.

The five screens displayed the following information and pictures:

Screen 1: A color-coded map representing traffic speeds, along with weather information and text to report traffic conditions.

Screen 2: A photograph of very light traffic volumes with numerical time estimates indicating the travel times between key locations.

Screen 3: A photograph of heavy traffic volumes, and a bar chart of parking information.

Screen 4: A photograph of heavy traffic volumes, and a color-coded map of traffic speeds.

Screen 5: A photograph of heavy traffic volumes along with travel time estimates.

Therefore, the screen designs presented five different forms of graphic traffic information (i.e., bar graphs, maps, text messages, photographs of actual traffic conditions, and travel time estimates). These five general forms of traffic information were combined to produce the graphic screens.

Questionnaire Content. The questionnaire contained two main sections: an "individual screen" section and a general "all-screen" section. The "individual screen" section dealt with each screen in sequence, for every one posing the same questions, which addressed the subject's commuting and decision- making behavior. Commuters about to leave home were able to modify their behavior in three ways (other than staying home): They could delay departure, choose an alternate route, or change mode of transportation. After the subjects' response to these three options, they were also asked to indicate which form of information within each screen influenced the choice (e.g., map, color coding, bar chart, etc.). The "all-screens" section included general questions pertaining to the design of traffic information screens.

Information Requirement. It was important for the design of ATIS to obtain rankings of the screens according to their value in planning a commute, so subjects were asked to rank-order the five screens across several variables related to the delivery of traffic information. Next, and perhaps most critical, was the question pertaining to the specific form of information. Subjects were asked to rank-order all forms of traffic information as displayed in the prototype screens to determine how helpful they were in selecting a driving option (e.g., barchart, text, numerical time estimate, map). A significant difference in answers to this question across motorist groups would allow screens to be designed and tailored specifically for the present grouping of motorists.

The commuters' preference for map orientation was also determined. It was felt that individuals commuting north–south might prefer the vertical orientations, whereas east–west commuters might prefer the horizontal.

Therefore, the traffic map was presented in each orientation in the screens section of the questionnaire, and subjects' preference of orientation was asked in the section of general questions.

When providing physical reference points for a commuter, it was deemed important to provide landmarks most familiar to the commuters. The lowest level of geographic familiarity is termed landmark knowledge (Thorndyke, 1980), which consists of orienting oneself by visual landmarks. Therefore, subjects were asked to identify the landmarks they typically used for travel-time estimates. Also, the subjects were asked if they consider the Express Lane an alternate route.

In the initial survey, 73.4% of the subjects stated they would not use a continual up-to-the-minute traffic information service delivered via cable TV. Subjects were reasked this question after they had seen actual examples of traffic information presented on sample screens. This question was asked for two reasons: (a) because the subjects' acceptance of the media by which traffic information is conveyed is critical, and (b) because subjects have had limited exposure to traffic information delivered via TV. In addition to knowing what to display on a screen, the time of delivery is important. To answer the question of broadcast timing, subjects were asked what time they would use the TV-based information. Finally, to identify potential subjects for the screen usability tests, subjects were asked if they were willing to participate in other follow-up tests.

The response scales used in the questionnaire were based on response scales used in previous, similar studies and are consistent with survey design methodology. In the "all screen" section, subjects were asked about their delay time, route choice, and mode choice. In addition to simple yes–no questions, a scale pertaining to time was provided for the question on time delay. The scale included 5, 10, 15, and greater than 15 min. In the "all-screen" section the subject was asked to rank-order the five forms of information. The scale was mutually exclusive, from 1 to 5, with 1 designating the most helpful. The remainder of the questions in this section were answered by marking yes or no.

Pilot Test. A pilot test was conducted to identify deficiencies in the questionnaire and to estimate the time required for its completion. Based on the pilot test and the need to limit total interview time to less than 1 hour, a time limit of 15 minutes was placed on the administration of the graphics questionnaire. Because of this constraint, the questionnaire was limited to four different screens plus a control screen and a section of general questions.

Administration Method. The questionnaire was administered by researchers at the University of Washington. Each subject was provided with a questionnaire and binder containing the simulated screens of traffic information and instructed to begin. Any clarifications the subject needed during the questionnaire were answered by the administrator. The graphics question-naire was administered in sequence with an in-depth interview focusing on traffic route choice. The first 17 subjects took the questionnaire before the interview, and the last 80 subjects took the questionnaire after the interview.

Telephone Survey

In another subsequent study, a total of 100 Seattle commuters were surveyed by telephone (25 from each of the four defined motorist groups as discussed in Barfield, Haselkorn, Spyridakis, and Conquest, 1991) as se-lected aspects of motorist information needs were examined. The survey consisted of questions relating to motorists' preferences for screen-based traffic information. The survey's scenario involved delivery of information at home via television. Subjects were asked to respond as if experiencing normal traffic conditions. The responses to the questions were selected from a list of options, ranked, or open ended.

Subjects were contacted by interviewers Monday through Sunday, usu-ally in the evening between 6 a.m. and 9 p.m. These times were based on previous experience with scheduling subjects for in-depth surveys. Subjects were reminded of their participation in the initial on-road survey, and the graphics survey if applicable, and were then asked if they would be willing to particpate in a 10-minute telephone survey. If they responded positively, the survey continued; if they responded negatively, they were asked if they would be willing to particpate at a more convenient time. If so, arrangements were made.

Surveys were coded and entered into a datafile on the University of Wash-ington IBM mainframe computer and analyzed using SAS (Statistical Analysis Software). Randomly selected, 5% of the records were checked for errors. The following list summarizes the results of the telephone survey:

1. Commuters tended to expect a TV-based system to be more accurate than any other current information source.
2. The commuters expressed a strong interest in knowing the timeliness of the displayed information.
3. Traffic information presented via radio is currently perceived as not being as accurate, up-to-date, or geographically specific as it could be.
4. Motorists who desired the reason for traffic congestion or type of incident want this information to help them decide the severity of an incident so they can generate their own time estimates.
5. Motorists tended to be very familiar with typical commuting conditions and alternate routes and therefore desire information that lets them make their own decision about which driving action to take. They also

want to know about unusually severe weather, freeway conditions, and so on, rather than about typical traffic rush-hour occurrences.

6. When traffic conditions were very heavy, motorists tended to want the range of displayed traffic speeds to be shifted downward. The higher speeds 45+ mph are unrealistic and unnecessary during this time period.

To summarize, the findings from case studies one and two are used to provide specifications for the design of a real-time graphics-based traffic information system derived from a comprehensive understanding of commuters' traffic information needs. An existing system has been developed (termed Traffic Reporter) that takes data collected from freeway sensors and presents that data in formats similar to those discussed in this section (Haselkorn, Barfield, Spyridakis, & Conquest, 1990). This system is currently integrated into the Seattle traffic information system and can be viewed online at various kiosks in the Puget Sound area and through the Internet.

Case Study 3: Surveys Defining User Requirements for Advanced Traveler Information Systems and Commercial Vehicle Operations

The University of Washington and Battelle Seattle Research Center, with a contract from the FHWA and USDOT, designed three surveys administered nationwide to define user requirements for ATIS and CVO. Whereas the previously discussed surveys were conducted primarily to gather user information requirements for an out-of-car traffic system, this case study focused on the user requirements for an in-car traffic system. In designing these surveys, a review was conducted on the design of other motorist surveys and survey reports to determine the applicability of relevant data for this effort (Conquest, Spyridakis, Haselkorn, & Barfield, 1993; Dudek, Messer, & Jones, 1971a; King, 1986; Mannering, Koehne, & Kim, 1995). The remainder of this section defines the principle steps required to design these surveys with the objective of defining user requirements for an in-car traveler information system.

Population

The targeted population for the surveys were licensed drivers in America. It was further determined that there were three major categories of drivers for this project: (a) private motorists, (b) commercial drivers, and (c) commercial system operators.

Private Motorists. Private motorists include people who commute to and from work, the business traveler, and those who travel for enjoyment in both urban and rural settings and under normal, congested, poor-weather, and emergency conditions.

Commuters. Commuters with regularly scheduled working hours account for the majority of morning and afternoon rush-hour traffic. Commuters are more interested in reducing travel time than they are in trip enjoyment (Barfield, Conquest, Spyridakis, & Haselkorn, 1989). However, most of the recurrent congestion is caused by this group of drivers. When weather conditions worsen during rush hours, accidents tend to occur. ATIS can help alleviate the congestion for commuters by providing alternate route information.

Nonworking Trips. According to Highway Statistics (FHWA, 1991), over 900,000 vehicle miles were traveled in the United States for nonworking purposes in 1990. Work-related trips constituted nearly 500,000 miles. These data reveal that about 65% of driving is nonwork related. Recreational vehicle drivers are also concerned about road limitations. Because this mode of traveling usually is not done on a recurrent basis, there are possibilities for ATIS to enhance trip enjoyment, allow for searches for roadside services, and improve navigational and routing requirements.

Commercial Drivers. Commercial vehicle operators, such as truck drivers and bus drivers, are targeted audiences for the Commercial Vehicle Operations (CVO) segment of ITS. Trucks, specifically, are a viable means of transporting goods or materials from various areas. From 1970 to 1985, trucks have moved more tons of freight than railroads at a national level (Moon, 1986). The number and size of trucks are increasing to such a degree that the FHWA has become very concerned with highway safety and the life of pavement structures. Changes in pavement conditions affect highway users by increasing or decreasing vehicle repair cost, speed, and fuel economy. Furthermore, with highway maintenance, other drivers suffer time delays during pavement reconstruction, rehabilitation, and maintenance (TRB, 1990).

Commercial System Operators. These individuals assist commercial drivers to arrive at certain destinations in a timely and reliable manner. They may also assist in routing and navigating commercial drivers. These drivers include dispatchers for buses, delivery vans, taxis, and emergency vehicles. Dispatchers or commercial operators are also targeted users of CVO. These groups are currently researching new computer-aided dispatch systems that will provide mapping information to users for enabling safe and efficient arrival times. For dispatchers of emergency vehicles, the accuracy of such information is crucial because they need to ensure that vehicles not only are located in the best manner possible but also whether shifts are spaced correctly and demand has been accurately forecasted. Therefore, information systems will be of great benefit to them for providing optimum route information.

Older Drivers. In all driver categories, older drivers are a major focus. In the United States, there are over 41 million people ages 55 and over who are licensed drivers (FHWA, 1991). This represents approximately 25% of the entire licensed population, and the numbers are increasing as the U.S. population ages. A navigation system can assist drivers in unfamiliar areas, thereby providing a means of congestion avoidance (Kitamura, Jovanis, & Owens, 1991). Thus, the elderly population is a large targeted audience for ATIS owing to the nature of their vehicle trips.

Older drivers face several difficulties in freeway usage (Malfetti & Winter, 1987). These difficulties include driving next to large trucks, traffic speed variance, rudeness of "young" drivers, and tasks such as merging into freeway traffic. Although such complaints are heard from all age categories, the stress felt by older drivers because of these driving maneuvers appears to be more severe. Older drivers have complained about the failure of freeway and nonfreeway signs to provide adequate advance warning, and have expressed discontent that signs typically display inappropriate or confusing information (Lerner & Ratte, 1991). In a real-time system, elderly drivers may not be able to react in a safe time frame if messages are continually changing. This was confirmed in a study conducted by Ranney and Simmons (1992) who observed that older drivers may have more difficulty than younger drivers locating targets in a visual search while driving.

Survey Content

Given the diversity of the three driver categories, three surveys were designed to address the unique driver differences. However, for the sake of comparing results among the three surveys, certain questions were purposely made similar. Questions such as, "Will a driver use an in-vehicle traffic system?" and "How much would a driver pay for an in-vehicle system?" were asked in all surveys so data from the three driving subpopulations could be compared.

The surveys consisted of closed-ended questions whenever possible to eliminate interpretive responses. Furthermore, there were options to add other information if respondents wanted to include additional concerns. The surveys were entitled Private Vehicle Survey, Commercial Vehicle Operator Survey, and Dispatcher Survey.

An advanced traveler information system will be able to provide a user with a variety of travel and traffic information. Therefore, the surveys needed to address all the necessary components of the system.

Commuting and Work Trips. This section had questions about the information that drivers use while commuting to help them arrive at their final destinations in a manner perceived as most efficient. This includes the decision to take alternate routes and utilize media information for their decisions (see Fig. 10.3).

20.　　　What is the most likely reason for changes in your commuting route? (**Check only one**)

☐ Road conditions　　　　　　　　　　　　　☐ Traffic congestion
☐ Road restrictions　　　　　　　　　　　　☐ Weather information
☐ Other (please specify) _____

FIG. 10.3. Question from Commuting section of Private Vehicle Survey. This question was asked to determine why most drivers need to change their commuting route.

For commercial vehicle operators, a definite commuting time is not identified; rather, each driver has routes designated by either a dispatcher or the customer. Therefore, questions related to planning a trip were asked to assess what information is considered crucial (see Fig. 10.4).

Trip Planning and Recreation. This section asked questions on what drivers require and prefer when traveling on long trips and on other non-work related trips such as visiting a friend, going to the park, and so forth. Also addressed were the type of routing and navigational information considered useful and what information is absolutely necessary. (see Fig. 10.5). The responses from this section were used to determine how specific information must be for nonwork-related trips.

Driving Stress. Undue stress can cause drivers to react in an undesirable manner that can lead to accidents. Therefore, information on what types of situations are very stressful to drivers is needed (see Fig. 10.6). For example,

8.　How important are the following factors for planning your work travel route?
　　(**Check one box for each item**)

	Very important		Moderate importance			Not important	
Height limits	☐1	☐2	☐3	☐4	☐5	☐6	☐7
Size limits	☐1	☐2	☐3	☐4	☐5	☐6	☐7
Speed limits	☐1	☐2	☐3	☐4	☐5	☐6	☐7
Weight limits	☐1	☐2	☐3	☐4	☐5	☐6	☐7
Railroad crossing	☐1	☐2	☐3	☐4	☐5	☐6	☐7
Road construction	☐1	☐2	☐3	☐4	☐5	☐6	☐7
Shoulder types	☐1	☐2	☐3	☐4	☐5	☐6	☐7
Shoulder widths	☐1	☐2	☐3	☐4	☐5	☐6	☐7
Steep grades	☐1	☐2	☐3	☐4	☐5	☐6	☐7
Traffic volume	☐1	☐2	☐3	☐4	☐5	☐6	☐7.
Turning restriction	☐1	☐2	☐3	☐4	☐5	☐6	☐7
Weather condition	☐1	☐2	☐3	☐4	☐5	☐6	☐7
Other (please specify) _____	☐1	☐2	☐3	☐4	☐5	☐6	☐7

FIG. 10.4. Question from A Typical Working Trip section of Commercial Vehicle Operator Survey. This question was asked to access the importance of road limitarion information.

26. For planning a trip how important is the following?
(**Check one box for each item**)

	Very important		Moderate importance			Not important
Address of destination	☐1	☐2	☐3	☐4	☐5	☐6	☐7
Knowledge of alternative routes	☐1	☐2	☐3	☐4	☐5	☐6	☐7
Anticipated time of arrival	☐1	☐2	☐3	☐4	☐5	☐6	☐7
Freeway access	☐1	☐2	☐3	☐4	☐5	☐6	☐7
Trip mileage	☐1	☐2	☐3	☐4	☐5	☐6	☐7
Stop-off points	☐1	☐2	☐3	☐4	☐5	☐6	☐7
Turn-off points	☐1	☐2	☐3	☐4	☐5	☐6	☐7
Others (please specify) _____	☐1	☐2	☐3	☐4	☐5	☐6	☐7

FIG. 10.5. Question from "Route Planning for Recreations and Traveling" section of Private Vehicle Survey. This question was asked to determine how important was various information for arriving at a nonwork destination.

if the majority of elderly drivers are too stressed when making left turns in nonregulated intersections, then the system should not give an alternate route with too many left turns. If the majority of drivers get very stressed driving on icy roads, then information on icy roads and ground temperature should be provided to caution drivers about slippery roads.

Use of In-Vehicle Systems. For this system to be successful, the number of drivers in each subpopulation willing to purchase this system needed to be determined. Therefore, in each survey, questions specifically related to using an in-vehicle traffic information system were included (see Fig. 10.7). If respondents were willing to use this system, questions were then asked concerning when and how they wanted to see the information displayed and what would be deciding factors for the user in purchasing this system.

23. Rate the level of stress experienced from the following <u>major highway/freeway</u> driving activities?
(**Check one box for each item**)

	Very stressful		Somewhat stressful			No stress
Merging into heavy traffic	☐1	☐2	☐3	☐4	☐5	☐6	☐7
Driving behind a vehicle that is constantly braking	☐1	☐2	☐3	☐4	☐5	☐6	☐7
Driving behind a vehicle that is moving slower than the speed limit	☐1	☐2	☐3	☐4	☐5	☐6	☐7
Driving next to trucks	☐1	☐2	☐3	☐4	☐5	☐6	☐7
Night driving	☐1	☐2	☐3	☐4	☐5	☐6	☐7
Driving on icy roads	☐1	☐2	☐3	☐4	☐5	☐6	☐7
Driving in heavy rain, sleet or snow	☐1	☐2	☐3	☐4	☐5	☐6	☐7
Moving across lanes to an exit	☐1	☐2	☐3	☐4	☐5	☐6	☐7

FIG 10.6. Question from Driving Stress section of Private Vehicle Survey.

38.   Let's say you were given an in-vehicle traffic information system (e.g., a computer screen in your
        vehicle), that had the capability to show you current traffic conditions. The system can also
        provide roadside motorist services, such as nearest rest stop and next gas station as well as provide
        you information about oncoming road conditions. Would you use it?

                                                                        ☐ Yes   ☐ No

39   How much would you pay for this in-vehicle traffic information system? $ _____

        FIG. 10.7. Question from Use of an In-Vehicle Traffic Information System
        section found on all three surveys.

Background Information.   Basic background information was also needed
to categorize the various driving groups. Information on gender, income, age,
and occupations was thus requested. As described previously in this chapter,
studies have shown that age, gender, household size, car ownership, and
number of adults in a household are major contributors to differences in trip
behavior (Bonsall & Joint, 1991; Prevedouros & Schofer, 1991; Said et al., 1991).

For commercial vehicle operators, inquiries about other important demographic information included the number of miles driven per day, the type
of vehicle driven, and the type of material or cargo carried. This information
is important for understanding how important the system will be for the
various drivers. For commercial dispatchers, the only unique question was
"How many years have you been a dispatcher?" to determine if experience
is important for reacting appropriately to an information system.

Pilot Test

A pilot study of the surveys was conducted by Ng, Barfield, and Mannering
(1995) in the Puget Sound region of Washington State. Reviews also were
coordinated among the University of Washington, Commercial Vehicle Safety
Alliance (CVSA), and Battelle Seattle Research Center. The organizations
that participated in the pilot study are shown in Table 10.3.

After feedback was received from these organizations, the survey was
revised. Questions considered ambiguous to individuals who read the survey
were rewritten, and other information determined to be necessary was
included in the final survey. For example, an initial review revealed that
some private motorists do not have a working commute (e.g., retired people,
people who work out of their home, and people who do not have a job).
Thus, the question "Do you have an occupation outside your home?" was
added at the beginning of the sections on commuting. Other revisions corrected ambiguous wording ("What is meant by an arterial road?"), grammar
problems, and the inclusion of relevant options and questions that were not
considered in the original write-up.

Other organizations also provided user requirements without reviewing
the survey, such as Greyhound of Vancouver and the Salem Sheriff's Depart-

TABLE 10.3
Organizations Participating in Pilot Survey

Organization/Business Participating	No. of Reviewers
Private Vehicle Survey	
American Automobile Association of Washington	5
(employees and members of AAA)	
Department of Licensing employees)	5
Students at University of Washington	4
Dispatcher Survey	
Snohomish county Police staff and Auxiliary Service Center	1
(SNO-PAC-Pacific Northwest)	
American Automobile Association (dispatchers)	1
METRO	1
Greyhound	
Commercial Vehicle Operator Survey	
Continental Van Lines	5
United Van Lines	2

Note. From Ng et al. (1995). Reprinted from *Trans-portation Research, 3C*(2), A survey-based methodology to determine information requirements for advanced traveler information systems, pp. 113-127, © 1995, with kind permission from Elsevier Science Ltd, The Boulevard, Langford Lane, Kidlington OX5 1Gb, UK

ment in Salem, Oregon. The information from all organizations was used to design the initial questions in the survey.

Sampling Units

For a survey to be representative of a population, the population must be divided into parts or sampling units (Cochran, 1977). For each of the three surveys, the sampling units were these:

- Individuals with drivers' licenses for private vehicles
- Individuals with a commercial drivers' licenses
- Individuals paid to dispatch vehicles (commercial system operators)

Sampling Frame

Because the population of licensed drivers included three driver categories, three separate sampling frames were determined. Private motorists were further subdivided into two subpopulations.

To identify where most private motorists of all age groups were located, sampling frames included the following:

- People who go to an automobile club for maps, travelers checks, auto insurance, and other information (e.g., AAA, Cross Country)
- People who obtain drivers' licenses or license renewals at a Department of Licensing facility
- People pumping gas at gas stations (e.g., Chevron, BP, Hess, Petro)

To concentrate on private motorist's ages 55 years and over, sampling frames included the following:

- Licensed drivers at a retirement home
- Licensed drivers attending the 55 Alive Defensive Driving class given by the American Association of Retired Persons (AARP)

For commercial drivers, sampling frames included the following:

- Truck drivers at weigh stations and inspection points
- Drivers at a truck stop
- Members of bus and trucking organizations (e.g., American Bus Association and American Trucking Association)

For commercial operators or dispatchers, sampling frames included the following:

- Members of a dispatching organization (American Public-Safety Communications Officers)
- Taxi and limousine service organizations

There was also a particular interest in male and female responses because studies have shown that gender differences have an influence on the usability and acceptability of traveler information systems (Barfield et al., 1991; Prevedouros & Schofer, 1991; Spyridakis et al., 1991; Vaughn, Abdel-Aty, Kitamura, Jovanis, & Yang, 1992). Defining separate sampling frames for male and female drivers can be difficult because there are no known automobile organizations that are exclusively male or female. However, motorist surveys will typically provide a representative sample from each gender when distributed randomly to the motorist population. Therefore, an initial stratification on males and females was not done (i.e., a separate sample size was not defined for males and females). Rather, poststratification will be done given that the approximate percentage of females and males in the driving population are known. Poststratification is the technique by which a sample can be classified into separate samples after the data have been collected, given that the population percentages are known (Cochran, 1977). In this case, the population percentage of male and female drivers can be obtained from several sources (FHWA, 1991; MVMA, 1992).

Over a 2-month period, letters, faxes, and phone calls were coordinated with organizations and businesses. Whenever necessary, a surveyor went out to the coordinated site. Coordination with each organization and business depended on the time and cost involved. The organizations that agreed to participate in the distribution of the surveys are listed in Table 10.4.

Approximately 9,000 surveys were distributed by an initial person-to-person contact. In addition, 10,000 Dispatcher Surveys were enclosed in a newsletter distributed nationwide through the American Public Safety Communications Officer (APCO). Because the members of APCO are located nationwide and not readily conglomerated into a classroom setting, such as the defensive driving class for AARP or into a business location such as the Department of Licensing or AAA, a newsletter was deemed the most feasible method for reaching an important group of emergency dispatchers.

Survey Administration

Distribution of the surveys was divided into four regions as defined by the Commercial Vehicle Safety Alliance (CVSA). Three states from each region were initially chosen for survey distribution. There was an additional survey distribution in the Washington State area because a face-to-face contact with various organizations was established by researchers at the University of Washington and Battelle Seattle Research Center. Because the surveys were distributed nationwide to individuals who were to be anonymous, it was not feasible to have a complete person-to-person survey interview. Therefore, a postage-paid marking with a return address was imprinted on the back of each survey for easy mailing by responders.

The surveys were between 7 to 11 pages long and professionally printed with blue ink on white paper for easy reading. A registration card was enclosed with each survey to be filled out by respondents if they were willing to participate in other research associated with this project. A serial number also was placed on all surveys distributed face to face to establish the number of respondents from each surveyed area. The dispatcher surveys that were part of the APCO newsletter did not have a serial number, but the appearance of these surveys was slightly different. They were printed on different paper stock with black instead of blue lettering. Therefore, they were easily distinguished when returned.

In distributing the surveys, to ensure that people understood the importance of the information to be collected, face-to-face contact was initiated when possible. In some cases, such as the Department of Licensing in

TABLE 10.4
Organizations Participating in the Distribution of the Survey

Organization/Business Participating in Survey Distribution	Number of Surveys Distributed
Private Driver Survey	
American Association of Retired Persons (AARP)	
• Washington, Oregon, and Idaho	300
• California	350
• Texas	300
American Automobile Association (AAA) of Washington	500
Commuters employed in private businesses	
• Washington	150
• California	350
• New York	150
Department of Licensing (DOL)	
• Washington	400
• Arizona	1,250
• Kentucky	1,250
• Illinois	1,250
TOTAL	6,250
Dispatcher Survey	
Snohomish County Police Staff and Auxiliary Service Center (SNO-PAK - Pacific Northwest)	400
American Public-Safety Communications Officer (distributed nationwide via newsletter)	10,000
Greyhound (Washington)	2
Continental Van Lines	2
Rescue Rooter	5
Associate members of Commercial vehicle Safety Alliance and various inspection points	150
TOTAL	10, 559
Commercial Vehicle Operator Survey	
Greyhound (Washington)	15
Continental Van Lines (Washington)	10
Associate members of Commercial Vehicle Safety Alliance and various inspection points	2,400
TOTAL	2,425

Note. From Ng et al. (1995). Reprinted from *Transportation Research, 3C*(2), A survey-based methodology to determine information requirements for advanced traveler information systems, pp. 113-127, © 1995, with kind permission from Elsevier Science Ltd, The Boulevard, Langford Lane, Kidlington OX5 1GB, UK.

Washington and the AAA in Washington State, a surveyor was available to answer as many questions as necessary. The surveyor was also able to collect the surveys if they were completed that day. Involvement by CVSA (Commercial Vehicle Safety Alliance) helped to facilitate the distribution of the surveys on a nationwide basis to the Department of Licensing facilities in three other states and at various inspection points.

Survey Data Collection

Surveys were sent to a central location for data processing. From these surveys, there were approximately 1,800 respondents. Each survey had over 200 data entries. Therefore, developing a key code that was easily used by the data processor was very important. This key code had to be transferable to ASCII format to be used with various statistical packages for data analysis and to be checked by a separate person.

DATA ANALYSIS

This section reviews several approaches to analyzing transportation data. Data analysis is an art form developed from experience. Each analyst can approach the same dataset with different statistical and mathematical perspectives. However, when the data interpretation is done accurately, it will be observed that each analyst has supportive findings. This section discusses the following data analysis topics: descriptive statistics (e.g., means, histograms, and standard deviations), inferential statistics (ANOVAs and cluster analyses), and mathematical modeling (e.g., poisson models and logit models). There are numerous statistical packages commercially available to assist in evaluating data (e.g., SAS, SPSS, SYSTAT and EXCEL). The use of statistical techniques are demonstrated in examples and analyses conducted from the surveys administered in the case studies when possible.

Descriptive and Inferential Statistics

Descriptive statistics provide summary data such as percentages, overall means, and standard deviations. This information is usually represented graphically or in tabular form. From this information, an analyst may be able to provide inferential statistics that bring about insights and draw upon conclusions.

In the Seattle commuter survey, frequencies were calculated for all variables of the total sample and for the sample grouped by gender. Gender differences were then assessed with t tests for interval data and Mann-Whitney U tests for ordinal data. Next, Pearson correlations were conducted on interval scaled values and Spearman correlations on all ordinal and a few nominal scaled variables if the coding met the assumptions of the Spearman routine. An analysis of variance was then conducted on relevant variables. Finally, cluster analyses and chi squares within clusters were conducted to identify possible motorist groups and the significant differences between the groups' responses on specific questions.

The sampled commuter group consisted of 50.7% male and 48.8% female respondents, with the majority (61.1%) under 41 years of age. Furthermore, the majority of the surveyed motorists reported living in households with combined earnings above $50,000 per year. The average one-way distance traveled on the I-5 freeway corridor was 14 kilometers, which exposed the commuters to several forms of traffic media. In general, the morning commute took about 31 min.

ANOVA. Analyses of variance models are used for studying the relationship between a dependent variable and one or more independent variables for experimental and observational data. In the 1993 University of Washington study, there were several independent variables (e.g., the utilization of ATIS, route choice behavior, and mode of preferred transportation) that were believed to be affected by several dependent variables, such as the stress experienced from various driving modes and the importance placed on several driving needs. The results of this data provided valuable insights for significant differences in driving behaviors and for the design of an ATIS.

Cluster Analysis. Cluster analysis is a statistical technique that uncovers an underlying structure in a dataset by grouping cases or subjects into similar groups according to a specified distance metric, such as Euclidean distance. The objective of cluster analysis is to group a large number of cases into a smaller number of relatively homogeneous groups based upon similar responses on specifically chosen variables. Ideally, the data points within each cluster should be relatively close to the cluster center, and the cluster should be easily describable, thereby allowing interpretation and communication of cluster results (Conquest et al., 1993).

For the 1988 motorist survey data, a cluster analysis was performed based on the willingness of commuters to adjust their commuting behavior in relation to motorist information. In performing the cluster analysis, there was a focus on a particular group of variables that characterized the influence of traffic information with respect to time, route choice, and mode of transportation. The cluster analysis separated the 3,893 cases into four major commuter groups:

1. *Route changers* (RC), those willing to change routes on or before entering I-5 (20.6%)

2. *Nonchangers* (NC), those unwilling to change time, route, and mode (23.4%)

3. *Route and time changers* (RTC) (40.1%)

4. *Pretrip changers* (PC), those willing to make time, mode, or route changes before leaving home (15.9%)

The four clusters proved to be stable both over other noncluster variables and through subsequent in-person interviews of random group members. These groups were further investigated to uncover similar characteristics with respect to variables from the survey related to motorists' responses to traffic information.

Table 10.5 reveals the significance of using multivariate techniques such as cluster analysis to identify commuter subtypes. The combined data for all surveyed commuters revealed an extreme diversity in commuters' willingness to alter departure time based on traffic information received at

TABLE 10.5
Influence of Traffic Information on Three Commute Decisions Before Driving

Commuter Subgroup	Never Receive	Rarely	Sometimes	Frequently
The time you leave:				
Nonchangers	42.9%	55.8%	1.2%	0.1%
Route changers	7.1%	92.9%	0.0%	0.0%
Route and time changers	0.0%	0.3%	79.2%	20.5%
Pretrip changers	0.0%	0.5%	68.5%	30.9%
All	12.5%	37.4%	37.2%	12.9%
Your means of transportation:				
Nonchangers	62.1%	37.0%	0.9%	0.0%
Route changers	30.6%	66.4%	3.0%	0.0%
Route and time changers	34.5%	64.0%	1.3%	0.1%
Pretrip changers	2.3%	60.1%	26.6%	11.0%
Your route choice:				
Nonchangers	39.5%	60.4%	0.1%	0.0%
Route changers	0.0%	9.9%	79.3%	10.9%
Route and time changers	0.5%	29.4%	55.0%	15.1%
Pretrip changers	0.2%	6.6%	67.3%	25.9%

Note. From Barfield et al. (1991). Reprinted from *Transportation Research, 25A*(2/3), Integrating commuter information needs in the design of a motorist information system, survey-based methodology to determine information requirements for advanced traveler information systems, pp. 71-78, © 1991, with kind permission from Elsevier Science Ltd, The Boulevard, Langford Lane, Kidlington OX5 1GB, UK.

home. However, based on the awareness of commuter types, we focused on a carefully defined group of commuters and study factors that influenced their departure time. Then, assuming that information could be delivered to address these specific factors, we could be confident of a high degree of success in influencing when certain commuters begin their commute.

This picture of the two groups most willing to alter departure times suggests how to speak to them and what to tell them. They are under pressure to complete a complex commute on a rigid schedule, and these needs must be addressed. Most important, they need to know commute time information. Ideally, they need to be told what time they would arrive if they left immediately under current conditions and followed their primary route. They also need to know time estimates for alternate routes and alternate modes of transportation. These groups are flexible before leaving the house, and they will change departure times, routes, and modes if necessary to arrive on time.

Mathematical Modeling. Survey data also can be analyzed using mathematical models or econometric models that predict traveler decision making. For example, the dynamic traffic equilibrium model is used to investigate traveler decision making that follows the structural form developed by Mannering, Abu-Eisheh, and Arnadottir (1990). One of the model's main purposes is to forecast traveler response to congestion, given congestion information. The form of this model accounts for all traveler responses to congestion information: the ability to change departure times, vary driving speeds, change routes, and cancel trips. This is a vast improvement over most urban traffic models that consider only changes made en route.

Mannering, Kim, Barfield, and Ng (1993) conducted an analysis of travelers' estimates of the traffic delay length required for them to change their route to familiar and unfamiliar alternate routes using the 1988 Seattle commuter survey. The impact of traffic information was evaluated by studying the frequency with which travelers reported that pretrip information influenced their trip departure time, mode of transportation, and route choice.

In previous work on route changing, Mannering (1989) used a Poisson regression to predict the frequency of commuters' route changes per month and found that both highway networks (e.g., the availability of alternate routes, the level of traffic congestion) and commuters' socioeconomic characteristics played an important role in the frequency of route changes.

For the Ng et al. (1995) surveys, a series of econometric models can also be estimated for the work trip. The model structure will account for congestion avoidance, choices of route, speed, and departure time. The route-choice model assumes that travelers select the route that provides the highest level of utility and that will be of the standard multinomial logit form.

Estimation Procedures.   Survey data also can provide estimations for the target population. The following estimations can be calculated on transportation survey data:

- The overall population of people who will use real-time information as an in-vehicle traveler information system
- The population of people willing to choose alternate routes when they drive
- The population of people who use media information when choosing alternate routes
- The population of people who are highly stressed in various situations
- The population of people who use specific roadside information

In the surveys conducted in 1993, for each estimate, the population distribution by gender, age, income, private drivers, commercial drivers, and dispatchers also were needed. Because the population was divided into distinct populations, separate stratum statistics for private drivers, commercial drivers, and dispatchers were also calculated. Each stratum was then subdivided into various socioeconomic levels. This reduced the variance of the sample estimate and provided the required subpopulation estimates for the study.

An initial analysis of the survey data may also provide some direct correlations between variables. If the correlation coefficient, r, is close to one, a ratio estimation may be performed in addition to the stratified sampling estimate.

SUMMARY AND CONCLUSIONS

The general issue of how to design motorist information to impact a target audience is closely tied to the type of motorist information available and our understanding of the behavior and decision-making characteristics of the target commuters. Therefore, it is very important to understand the needs of the motorists by receiving feedback from them. Surveys provide a very effective means of collecting information that is pertinent to the design of a motorist information system. Conclusions drawn then become the basis for converting traffic data into motorist information. Furthermore, surveys help to limit the scope of future in-laboratory and on-road experiments and help to identify specific focuses for future studies.

In conducting a survey, sources of bias are usually apparent. It is important to make an attempt to recognize these sources of bias and then try and avoid them. For example, in the Ng et al. (1995) surveys, a high number of nonresponders was expected owing to the size of the surveys (7 to 11 pages).

The Barfield et al. (1991) survey, consisting of only 4 pages, provided a 40% response rate. Therefore, a response rate of anything higher for a longer survey is not anticipated. However, these surveys address a system that probably will not be purchased by everyone when it is first implemented, but only by those individuals who see a need for it. Therefore, the survey will be biased toward individuals who see a need for this type of system. This is an acceptable bias for understanding the concerns of those individuals who feel such systems are needed to help reduce congestion and relieve stress resulting from congestion.

Another source of bias is the population who completes the survey. Because the survey is written in English, it will be completed only by drivers who speak English. However, a percentage of the driving population does not speak English as a first language, and are thus not literate in English. Unless there are sufficient funds to provide interpreters or conduct in-person interviews, this bias cannot be averted.

Other potential sources of error include the number of respondents who actually complete the entire survey. In large surveys, portions will be left blank because of time constraints or weariness of answering a "test." Therefore, questions that absolutely need to be answered should be placed near the beginning of the survey.

To eliminate as many nonresponses as possible, a pilot test on a small sample of the target audience always should be conducted. This will help to eliminate any ambiguous questions and wording that may be too difficult. Furthermore, a face-to-face contact should be initiated when possible during distribution of the survey. This will enable the person distributing the survey to provide as much information as needed. If the survey will be distributed by people who also are not familiar with the project, a standard introductory speech can be scripted. After the survey is completed, no extra costs for stamps or envelopes should be necessary on the part of the participant, because there should be a postage-paid marking or self-addressed envelope. In both the 1988 and 1993 mail in survey, postage-paid markings were printed on the backs of the surveys.

There are various tools that can be used to provide descriptive and inferential statistics. Each tool has its own unique benefits and provides various insights for the analyst. The mathematical modeling method also should lead to some interesting findings. In terms of the frequency of commuters undertaking route changes, results confirm the importance of both traffic network and commuter socioeconomic characteristics.

In summary, the use of surveys to obtain user information requirements for components of ITS is a good procedure when a large representation of the target population is to be obtained easily and less expensively than with structured laboratory experiments.

REFERENCES

Abdel-Aty, M. A., Vaughn, K. M., Kitamura R., & Jovanis, P. (1992). *Impact of ATIS on driver's travel decisions: A literature review* (Res. Rep. No. UCD-ITS-RR-92-7). UC Davis, Institute of Transportation Studies, Davis, CA.

Anderson, J. L., Haney, D. G., Katz, R. C., & Peterson, G. D. (1964). *The value of time for passenger cars: Further theory and small-scale behavioral studies*. Stanford Research Institute. Prepared for Bureau of Public Roads, U.S. Dept. of Commerce, Menlo Park, CA.

Barfield, W., Conquest, L., Spyridakis, J., & Haselkorn, M. (1989). Information requirements for real-time motorist information systems. In D. H. M. Reekie, E. R. Case, & J. Tsai (Eds.), *1989 Vehicle, Navigation and Information Systems*. New York: Institute of Electrical and Electronics Engineers.

Barfield, W., Haselkorn, M., Spyridakis, J., & Conquest, L. (1991). Integrating commuter information needs in the design of a motorist information system. *Transportation Research, 25A*(2/3), 71–78.

Bonsall, P. W., & Joint, M. (1991). Driver compliance with route guidance advice: The evidence and its implications. *1991 Vehicle Navigation and Information Systems Conference* (Vol. 1, No. 912733, pp. 47–59). Warrendale, PA: Society of Automotive Engineers.

Carpenter, S. (1979). *Driver's Route Choice Project Pilot Study*. Oxford, England: Transport Studies Unit, Oxford University.

Case, H. W., Hulbert, S. F., & Beers, J. (1971). *Research development of changeable messages for freeway traffic control*. Los Angeles: Institute of Transportation and Traffic Engineering, School of Engineering and Applied Science, UCLA.

Cochran, W. (1977). *Sampling techniques* (3rd ed.). New York: Wiley.

Conquest, L., Spyridakis, J., Haselkorn, M., & Barfield, W. (1993). The effect of motorist information on commuter behavior: Classification of drivers into commuter groups. *Transportation Research, 1C*(2), 1–19.

Dudek, C. L., Friebele, J. D., & Loutzenheiser, R. C. (1971b). Evaluation of commercial radio for real-time driver communications on urban freeways. *Communications and emergency services: 6 report. Highway Research Record 358*. Washington, DC: Highway Research Board.

Dudek, C. L., Huchingson, R. D., & Brackett, R. Q. (1983). Studies of highway advisory radio messages for route diversion. *Highway information systems, visibility, and pedestrian safety, Transportation Research Record 904*. National Research Council, Washington DC: Transportation Research Board.

Dudek, C. L., & Jones, H. B. (1970). *Real-time information needs for urban freeway commuters* (Res. Rep. No. 139-4, Study 2-8-69-139). College Station, TX: Texas A&M University, Texas Transportation Institute.

Dudek, C. L., Messer, C. J., & Jones, H. B. (1971a). Study of design considerations for real-time freeway information systems. *Operational improvements for freeways: 4 reports. Highway Research Record 363*. Washington, DC: Highway Research Board.

Federal Highway Administration. (1991). *Highway statistics*. Washington, DC: U.S. Department of Transportation.

Gray, B. G. (1990). *Analysis of Washington State Traffic System Management Center: Information system and screen design*. Unpublished master's thesis, University of Washington, Seattle, WA.

Haney, D. G. (Ed.). (1964). *The value of time for passenger cars: A theoretical analysis and description of preliminary experiments*. Stanford Research Institute. Prepared for Bureau of Public Roads, U.S. Dept. of Commerce, Menlo Park, CA.

Hansson, A. (1975). *Studies in driver behavior with applications in traffic design and planning, Two examples*. Lund Institute of Technology, Department of Traffic Planning, University of Lund, Bulletin 9, Lund, Sweden.

Haselkorn, M., Barfield W., Spyridakis, J., & Conquest, L. (1990). *Improving Motorist Information Systems, Final Report.* Washington State Department of Transportation.

Huchingson, R. D., Whaley, J. R., & Huddleston, N. D. (1984). Delay messages and delay tolerance at Houston work zones. *Transportation Research Record 957.* Transportation Research Board, National Research Council, Washington, DC.

IVHS America. (1992). *Strategic plan for intelligent vehicle-highway systems in the United States.* Washington DC: Author.

King, G. (1986). Driver attitudes concerning aspects of highway navigation. *Transportation Research Record, 1092,* 11–21.

Kish, L. (1965). *Survey sampling.* New York: Wiley.

Kitamura R., Jovanis, P., & Owens, G. (1991). *Driver decision making with route guidance information: Background conceptual issues and empirical results* (Res. Rep. No. UCD-ITS-RR-91-8). University of California Davis, Institute of Transportation Studies, Davis, CA.

Lerner, N. D., & Ratte, D. J. (1991). Problems in freeway use as seen by older drivers. *Transportation Research Record, 1325,* 3–5.

Levy S., & Lemeshow, S. (1991). *Sampling of populations: Methods and applications.* New York: Wiley.

Mahmassani, H. S., Baaj, M. H., & Tong, C. C. (1984). Characterization and evolution of spatial density patterns in urban areas. *Transportation, 15,* 233–256.

Malfetti, J., & Winter, D. (1987). *Safe and unsafe performance of older drivers: A descriptive study.* Washington, DC: AAA Foundation for Traffic Safety.

Mannering, F. (1989). Poisson analysis of commuter flexibility in changing routes and departure times. *Transportation Research, 23B*(1), 53–60.

Mannering, F., Abu-Eisheh, S., & Arnadottir, A. (1990). Dynamic traffic equilibrium with discrete/continuous econometric models. *Transportation Science, 24*(2), 105–116.

Mannering, F., Kim, S., Barfield, W., & Ng, L. (1994). Statistical analysis of commuters' route, mode and departure time flexibility. *Transportation Research, 2C*(1), 35–47.

Mannering, F., Koehne. J., & Kim, S. (1995) Statistical assessment of public opinions toward conversion of general-puprose lanes to high-occupancy vehicle lanes. *Transportation Research, 1485,* 168–176

Mobility 2000. (1989). *Proceedings of a workshop on intellgent vehicle/highway systems.* San Antonio, TX: Author.

Moon, S. A. (1986). Keeping up with big trucks: Experiences in Washington State. *Transportation Research, 1052,* 17–22.

MVMA. (1992). *Motor vehicle facts and figures, '92.* Detroit, MI: Motor Vehicle Manufacturers Association.

Ng, L., Barfield, W., & Mannering, F. (1995). A survey-based methodology to determine information requirements for advanced traveler information systems. *Transportation Research, 3C*(2), 113–127.

Prevedouros, P. D., & Schofer, J. L. (1991). Trip characteristics and travel patterns of suburban residents. *Transportation Research Record, 1328,* 49–57.

Ranney, T. A., & Simmons, L.A. (1992). The effects of age and target location uncertainty on decision making in a simulated driving task. *Proceedings of the Human Factors Society 36th Annual Meeting* (pp. 166–170). Santa Monica, CA: Human Factors Society.

Said, G. M., Young, D. H., & Ibrahim, H. K. (1991). Trip generation procedure for areas with structurally different socioeconomic groups. *Transportation Research Record, 1328,* 1–9.

Shirazi, E., Anderson, S., & Stresney, J. (1988). *Commuters' attitudes toward traffic information systems and route diversion.* Los Angeles: Commuter Transportation Services, Inc.

Spyridakis, J., Barfield, W., Conquest, L., Haselkorn, M., & Isakson, C. (1991). Surveying commuter behavior: Designing motorist information systems. *Transportation Research, 25A*(1), 17–30.

Tsai, J. (1991). Highway environment information system interests and features survey. *1991 Vehicle Navigation & Information Systems Conference* (Vol. 1, No. 912743, pp. 113–122). Warrendale, PA: Society of Automotive Engineers.

Thorndyke, P. W. (1980). *Performance models for spatial and locational cognition* (Tech. Rep. No. R-2676-ONR). Washington, DC: Rand Corporation.

Transportation Research Board (1988). *Special report 218: Transportation in an aging society: Improving mobility & safety of older persons* (Vol. 1). Washington, DC: Transportation Research Board, National Research Council.

Transportation Research Board (1990). *Special report 225: Truck weight limits, issues & options.* Washington, DC: Transportation Research Board, National Research Council.

Vaughn, K., Abdel-Aty, M., Kitamura, R., Jovanis, P., & Yang, H. (1992). *Experimental analysis and modeling of sequential route choice behavior under ATIS in a simplistic traffic network* (Res. Rep. No. UCD-ITS-RR-92-16). Davis, CA: University of California Davis, Institute of Transportation Studies.

Wickens, C. D. (1984). *Engineering psychology and human performance.* Columbus, OH: Charles E. Merrill.

Wooton, H. J., Ness, M. P., & Burton, R. S. (1981). Improved direction signs and the benefits for road users. *Traffic Engineering and Control, 22*(5), 264–268.

11

DETERMINING USER REQUIREMENTS FOR INTELLIGENT TRANSPORTATION SYSTEMS DESIGN

Linda Ng
University of Washington

Woodrow Barfield
Virginia Polytechnic Institute and State University

An exciting aspect of Intelligent Transportation Systems (ITS) is that it will allow real-time traffic information (e.g., level of congestion, speed limits) and other valuable sources of information (e.g., roadside services) to be delivered to motorists in their vehicles. Because there is a tremendous range of potential information for delivery to motorists, an important goal for researchers in ITS is to determine what information users would like to receive in their vehicles. This will be an especially difficult task for transportation specialists because of the wide variety of users who will access information using ITS. For example, these users will include private drivers who commute to work, commercial drivers who transport goods and people, and dispatchers who help drivers arrive at various locations in a timely and efficient manner.

Based on results obtained primarily from surveys of potential users of ITS, this chapter provides some of the basic information requirements for ITS that are beginning to emerge from the literature. These requirements are based on the information needs of different user groups representing the driving population, on how ITS information should be presented to these different groups, and on what groups would benefit the most from different types of information. In this chapter, specific focus is placed on the Advanced Traveler Information Systems (ATIS) segment of ITS and the ATIS portion of Commercial Vehicle Operations (CVO). We begin this chapter by providing a list of questions that will need to be answered in designing ITS based on user requirements. A review of the relevant research relating to each topic is also presented to provide the reader with additional sources of information on each topic. The three sections that follow provide basic ITS information requirements pertinent for all potential user groups of ITS, for revealed subgroups, and for users based on trip and socioeconomic characteristics.

RESEARCH ON USER REQUIREMENTS

Many researchers have expressed the importance of defining exactly what information users would want in traveler information systems (Lunenfeld, 1989; Mast, 1991; Transportation Research Board, 1991; Wierwille, 1993). Definitions of user requirements must include human factors considerations that influence system safety and performance through the demands imposed on the human operators of ITS technology (including drivers and operators of traffic control systems). Furthermore, to design ITS, standard human factors guidelines, many of which are based on knowledge gained from designing human–computer interfaces, will have to be considered. A review of this knowledge base is presented in chapters 12 and 13. However, the focus of this chapter is on the type of information that users of ITS would like to receive in their vehicles.

For the design of ITS, there are several areas for which user requirements need to be specified (e.g., determining route choices and roadside services and defining information requirements specifically for commercial vehicle drivers and commercial system operators). The following sections list pertinent questions for ITS design and the subsequent work which has been done. Some of these topics are explored further in various case studies presented later in this chapter.

What Information Do Users Want?

Tsai (1991) conducted a survey in Canada to uncover highway information services requirements for commercial vehicle drivers. Thirty-six trucking companies and seven intercity bus companies participated in the study. The analysis of the survey revealed that traffic and weather information was considered highly important. The desired traffic information included reports of traffic congestion, accidents, lane closures, bridge closures, construction updates, alternate routes, low bridges, road weight restrictions, and legal truck routes. Adverse or severe weather conditions, fog conditions, and areas experiencing black ice represented other desired weather information. In more recent studies, Ng, Barfield, and Mannering (1995) revealed that there were both similarities and differences among private drivers, commercial drivers, and dispatchers in terms of their information needs.

Specifically, detailed traffic information in a traveler information system was most desired by private and commercial drivers. Dispatchers, in contrast, desired personal communication features to convey important messages to the drivers.

In What Format Should the Information Be Presented?

This question has generated a great deal of research to determine the effectiveness of different formats in presenting information. The following is a brief list of some research topic areas in this field that are relevant for ITS design (see chapters 12 and 13 for additional information):

- Static maps versus maps with highlighted alternate routes and traffic information (Allen et al., 1991)
- Moving maps versus. paper maps (Antin, Dingus, Hulse, & Wierwille, 1990)
- Auditory and displayed symbols versus map-based navigational information (Parkes, Ashby, & Fairclough, 1991)
- Head-up displays versus console-mounted displays (Green & Williams, 1992).

Where Do Travelers Need and Desire Information?

Studies have shown that some travelers desire information before their trip begins (at home or at work), whereas others would rather observe the traffic conditions (on the freeway) prior to making a route change (Barfield, Haselkorn, Spyridakis, & Conquest, 1991). Understanding where motorists would like to receive traffic information is an important goal for several reasons. First, this information will allow the design of an integrated information system consisting of ITS, information from variable message signs (VMS), and information highway advisory radio (HAR). Second, where people want to receive traffic information (at home or work, at a mall, or on the road), will influence what information will be delivered. This chapter explores this topic further as a case study in a later section.

When Do Travelers Need to Access Traffic Information?

Commuters will most likely require traffic information in the early morning and late afternoon, whereas commercial drivers and dispatchers will need information of certain types at any time (Barfield et al., 1991). For example, in Tsai's study (1991), survey respondents indicated that for weather and traffic information to be useful for the trucking industry, it needed to be transmitted at specific time intervals. Responses ranged from needing information every hour to every three hours for moderate weather conditions and every 10 to 30 min in severe or adverse weather conditions. Of the 25 firms who valued weather and traffic information, the study also revealed that 11 were willing to pay for it.

Are There Different User Groups in Terms of Their Motorist Information Needs?

Will separate guidelines need to be derived for ITS based on information about different subgroups in the driving population? That is, are there different classes of users in terms of their traffic information needs? If so, how should these groups be categorized and what are the characteristics and information requirements of these subgroups? In one study, Chang, Lin, and Lindely (1992) analyzed the commuting behavior of motorists in Dallas, Texas. Their findings suggest that by using multivariate cluster analysis, commuters could be classified into six distinct groups:

1. Commuters with the highest frequency of stops in "work to home" and "home to work" commutes
2. Commuters with a high ratio of stops on their "home-to-work" compared to their "work-to-home" commute (2.29 vs. 0.18)
3. Commuters representing the majority of drivers with very few home to work stops
4. Commuters with the longest mean travel time and distance
5. Commuters with a high ratio of stops on their "work-to-home" compared to their "home-to-work" commute (2.36 vs. 0.29)
6. Commuters with a high number of midday trips

Our assumption is that each of these commuter groups may have different information needs based on the type of homebound and workbound commute characteristics.

How Do Socioeconomic Characteristics Influence Information Needs?

It is important to determine whether drivers with different socioeconomic characteristics such as gender, age, and income react differently to the delivery of traffic information. If so, this knowledge will enable designers to tailor in-vehicle systems for specific needs. For example, several studies considering the age of the driver have shown the high probability of accidents associated with left-hand turns with or without a left turn signal (Datta 1991; Hancock, Caird, Shekjar, & Vercruyssen, 1991). The number of accidents because of left turns is especially high for drivers 65 years of age and older

(Malfetti & Winter, 1987). Because ITS is being designed for the 21st century, the group of drivers 65 years of age and older is expected to increase to about 50 million by the time the system is implemented (Bishu, Foster, & McCoy, 1991). Therefore, these systems need to account for the fact that left-hand turns are difficult maneuvers for many older drivers who would choose a longer route just to avoid the confrontation.

How Does Stress Influence Driving Behavior?

Studies have shown that stressed drivers have more accidents than drivers who are more relaxed (Evans, Palsane, & Carrere, 1987; McMurray, 1970). The level of stress also has been shown to vary inversely with age and driving experience (Gulian, Glendon, Matthews, Davies, & Debney, 1990). Although no driver expects to become a fatality, studies have shown that certain drivers are potential risk seekers (Mannering & Grodsky, 1995). Moreover, Smiley (1989) noted that inexperienced drivers find it more difficult to drive in high-density traffic than experienced drivers.

The stress associated with the long hours of driving by commercial drivers has been studied by several researchers. Kaneko and Jovanis (1992) found that consecutive hours of driving are strongly associated with accident risks, and Raggatt's (1990) findings showed that high job demands (e.g., long hours) can cause stressful situations (e.g., speeding to make up for lost time) that may lead to maladaptive behavior (e.g., use of stimulants).

In the next sections, after discussing the identification of information requirements for the ATIS–CVO portion of ITS, we explore the requirements for various subgroups revealed through statistical analyses.

USER REQUIREMENTS FOR ATIS–CVO

Surveys designed by Ng et al. (1995) were used to define user information requirements across potential users for the design of ATIS–CVO. Three nationwide surveys targeted these ATIS user groups: private drivers, commercial drivers, and dispatchers. In this section, we explore some of the ITS information requirements revealed by these surveys.

Specific details on the content and distribution of the surveys can be found in chapter 10 by Ng, Barfield, and Spyridakis in this volume. Each survey addressed many facets of driving: commuting, trip planning, routing, navigation, roadside services, and safety requirements. Summary statistics for private and commercial drivers and dispatchers are provided in Table 11.1. This study included a small number of female respondents for commercial drivers, but other studies have shown that most commercial drivers are males, a fact reflected in other studies on commercial drivers that have

TABLE 11.1
Summary Statistics From the Three Surveys

	Private	Commercial	Dispatcher
Number of respondents	938	325	348
Percentage of males/females who responded	60% / 40%	99% /1%	63.8% /34.3%
Average amount willing to pay to ATIS	$275.00	$400.00	$2033.40
Age:			
under 25	10.1%	1.3%	3.9%
25 - 44	48.8%	51.9%	66.4%
45 - 64	29.9%	46.2%	29.4%
65 and over	11.1%	0.6%	0.3%
Income:			
under $ 30,000	26.6%	24.5%	not
$30,000 - $59,000	46.3%	58.8%	reported
$60,000 - $74,999	12.4%	9.4%	
over $75,000	14.7%	7.3%	

focused only on the male population (Evans et al., 1987; Raggatt, 1990). The commercial and dispatcher survey also included a small percentage of people 65 years of age and older. This small percentage is to be expected because most people retire from such occupations when they reach 65 years of age. For the private survey, 11.1% of the respondents were in the "65 and over" age group and 20.5% were in the "55 and over" group. Highway Statistics (Federal Highway Administration, 1991) indicate that 13.6% of all licensed drivers in America are 65 and over and 24.6% are 55 and over. Thus, the study included a fairly representative sample of older drivers.

Alternate Route Information

One type of information that may be delivered by ATIS–CVO is alternate route information. In our surveys over 80% of the respondents in each group (commercial and private drivers and dispatchers) reported that accidents were a reason for choosing an alternate route (see Table 11.2). Road construction was the second biggest factor in this choice. Interestingly, less than 20% in each surveyed group perceived the "estimated time on the main route" to be a reason for choosing an alternate route, thus indicating that the time it takes to travel on a primary route is not necessary information unless there are changes in this time. This finding is quite conceivable because most drivers have a good approximation of the time it takes to

TABLE 11.2
Some Reasons Why an Alternate Route Is Chosen
(Respondents checked all that applied)

	Surveyed Group		
Reasons for Choosing an Alternate Route	Private	Commercial	Dispatcher
Accidents	86.4%	86.2%	82.8%
Traffic volume	71.2%	59.4%	55.5%
Time of day	43.4%	58.5%	37.1%
Road construction	79.5%	76.9%	82.8%
Weather condition	42.0%	54.5%	55.2%
Estimated time on main route	16.4%	14.8%	13.2%
Time gain by rerouting	49.8%	46.8%	35.6%
Proximity to existing route	31.2%	21.5%	20.7%

travel on their usual route. Reports showed that drivers attribute great value to the time gained if an alternate route is to be chosen. In summary, ITS should consider delivering information about accidents, road construction, and time gained by an alternative route to convince drivers that an alternative route is necessary.

If alternative route information is desired, how do users currently access this information? One human factors approach to systems design is determining what information users currently access and how they access it, then using this knowledge in designing the next system. As shown in Table 11.3, private drivers currently use commercial radio information and their observations of actual traffic conditions more than any other source for determining if an alternate route should be used, whereas commercial drivers and dispatchers value communication with other drivers. Cellular phones were rarely used by drivers. According to this preliminary data concerning information requirements for ATIS, private drivers also may desire auditory information and live video of actual traffic (possibly some distance ahead of the driver).

For private drivers, a component of the survey included questions on what length of delay caused them to divert to an alternate route. The questions specifically asked what length of delay would cause a driver to divert to either a known or unknown alternate route. The respondents were asked to choose only one time interval in each category. These choices included less than 5 min, 5 to 15 min, 15 to 30 min, 30 to 60 min, and more than an hour. As shown in Fig. 11.1, private drivers would divert to a known alternate route rather than an unknown alternate route for short delays (15

TABLE 11.3
Responses to the Question: What Resources Do You Use to Determine if an Alternate Route Should Be Used?

Resource for Alternate Route Information	Surveyed Group		
	Private	Commercial	Dispatcher
Cellular phone	2.0%	6.8%	21.3%
Commercial radio	57.0%	57.2%	23.3%
Commercial TV	14.4%	8.6%	16.1%
Communication with other drivers	11.0%	74.2%	54.9%
Weather radio station	12.8%	29.5%	23.3%
Observation of traffic conditions	54.6%	57.2%	N/R
CB radio	4.2%	69.5%	N/R
Company radio system	N/R	N/R	38.5%

Note. Respondents checked all that applied.
N/R: information was not reported.

min or less). For longer delays, more private drivers were willing to venture to an unknown alternate route. This indicates that if delays are too long, drivers will opt for any route other than the route they currently are using.

Utilization of Traveler Information Systems

The data from the three surveys showed that approximately 80% of the respondents in each user group would use an in-vehicle system if it had the capability to provide traffic and travel information (Table 11.4). The respond-

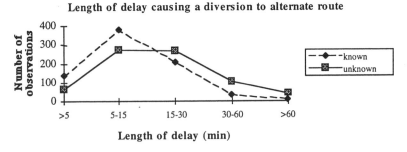

FIG. 11.1. Length of delay that would cause private drivers to divert to either a known or unknown route.

TABLE 11.4

Number and Percentage of Respondents Who Would or Would Not Use an ITS

Surveyed Group	Would Use	Would Not Use	Total
Private	784 (87.7%)	110 (12.3%)	894 (100.0%)
Commercial	266 (84.4%)	49 (15.6%)	315 (100.0%)
Dispatcher	250 (76.7%)	76 (23.3%)	326 (100.0%)

Note. From Ng et al. (1995). Reprinted from *Transportation Research, 3C*(2), A survey-based methodology to determine information requirement for advanced traveler information systems, pp. 113-127, © 1995, with kind permission from Elsevier Science Ltd, The Boulevard, Langford Lane, Kidlington OX5 1GB, UK.

ents, who would use an in-vehicle system, were then asked additional questions that are discussed later in this section.

For ATIS design it is significant that a high percentage of the respondents were willing to use an in-vehicle system. If this were not the case, considerable effort would have to be expended in educating the public about the usefulness of such a system. The positive findings in our surveys were expected, though, because the type of people willing to respond were those most likely to find in-vehicle systems of interest.

The 20% that indicated that they would not use ATIS is a plausible finding as well. In fact, studies have shown that for the benefit of the entire driving population, it is better to have a percentage of the population not change their driving behavior (Al-Deek & Kanafani, 1991; Garrison & Mannering, 1990). If all drivers divert to the same alternate route, the congestion problem will just be transferred to a new location.

The results of the surveys showed that accuracy of information (mean = 1.48 on a rating scale of 1 to 7) was the most important factor for drivers, with cost (mean = 1.51) and how recently the information is displayed (mean = 1.58) being next in importance. This finding indicates that the accuracy of in-vehicle system data is very important for the success of such a system. For example, if the data were accurate only 50% of the time, drivers would find it difficult to give credibility to the system and would be reluctant to use the provided information. Previous work on the accuracy of real-time traffic information supports this conclusion (Dudek, Messer, & Jones, 1971; Trayford & Crowle, 1989). Cost of the system was also an important consideration, implying that users do not want to pay excessive amounts.

Dispatchers responded differently than drivers for several questions including amount they were willing to pay and the value of ATIS services. These differences can be attributed to differences between the task of driving and the task of dispatching. Thus, there are perceived differences in how ATIS can benefit these two tasks. In general, dispatchers work for companies or organizations in which the system is recognized as an integrated unit to be utilized throughout the organization. Dispatchers were willing to pay the most for ATIS (mean = $2,033.40) whereas commercial and private drivers did not want to pay, on the average, more than $500.00 for this system. Studies by Tsai (1990) and Marans and Yoakum (1991) revealed that commercial drivers also were not willing to pay an excessive amount for road and traffic information. In Tsai's study, the only price reported was a total of between $10.00 and $20.00 per month for all fleet highway information needs. Marans and Yoakum (1991) found that 50% of the respondents were willing to pay 50 cents a day (approximately $10.00 a month) for information on tie-ups and alternate route information, whereas only 5% were willing to pay $2.00 per day for such information.

According to our survey data, providing traffic information was considered the most important ATIS service for private and commercial drivers, with the least important service being roadside service information. Dispatchers indicated that they valued personal communication services as more important than traffic information. That is, a reliable two-way communication system is a most important requirement if dispatchers are to perform their job effectively. This two-way communication link may encompass the ability to make and receive calls regarding emergency services and changes in destinations, as well as the ability to relay traffic information. Our analysis showed that there were still significant differences among the ITS services, but the most important factor for dispatchers was personal communication (mean = 2.23).

Driving Stress

The surveys revealed that icy roads and bad weather created very stressful driving situations on and off the freeway for both private and commercial drivers. Icy roads generate many problems for drivers including inability to steer properly. Also, bad weather conditions can impair the visual field of the driver causing added stress. Interestingly, for commercial drivers, an activity with as much stress as driving on icy roads or in bad weather was driving behind a vehicle that was constantly braking. Stopping is not an easy task for drivers of large, heavy vehicles, and the impact of a rear-end collision caused by large vehicles can be severe. The most stressful situation for dispatchers occurred when the number of calls or requirements exceeded the number of drivers available, creating an environment where quick decisions need to be made. The second most stressful activity was dispatching when weather conditions were severe. Dispatching when weather conditions are severe also can be very stressful because numerous emergency situations can develop. In severe weather conditions, roads may be blocked or impossible for large vehicles. Thus routes that were originally determined as alternates may no longer be available. The least stressful activity was dispatching during rush hour or heavy traffic. Our data revealed that there were no significant differences between gender, age groups, or experience levels.

USER REQUIREMENTS BASED ON IDENTIFIED SUBGROUPS OF THE DRIVING POPULATION

As stated earlier, it is important to determine if there are different subgroups of the driving population with varied information requirements. If so, how should these different requirements be accounted for in the design of ATIS–CVO? The next sections address these two issues.

Generally, there are three major groups of people who use traffic information systems: private motorists, commercial drivers, and commercial system operators, or dispatchers. There are also potential subgroups within each of these major categories. For example, in terms of information requirements we postulate that private motorists make up several subgroups including commuters, business travelers, recreational travelers, and nonwork-related travelers. Commuters may want to know the quickest route to their destination, whereas recreational travelers may want to know the most scenic route. Determining whether subpopulations exist within larger groups is a problem that typically can be solved by appropriate statistical techniques. For example, cluster analytic techniques can be useful in finding embedded groupings among drivers and, in fact, have been used already by researchers in the transportation field to purposely seek subgroups hypothesized in sampled populations. In chapter 10 of this volume, Ng, Barfield, and Spyridakis provide a review of this multivariate statistical tool, and further reading on this topic is also available from Hair, Anderson, Tatham, and Black (1992) and Kranowski (1988). The following case studies provide descriptions of commuter subgroups, private driver subgroups, and commercial driver subgroups, along with the type of motorist information that would benefit each group.

Case Study I: Defining Commuter Subgroups

Barfield et al. (1991) conducted an extensive study in Seattle, Washington, to better understand commuting behavior and decision-making practices for the purpose of designing and delivering motorist information. This work focused on the data collected from surveys aimed at carefully building a theoretical and design base for a complete motorist information system. The survey established a baseline on current usage of traffic information systems used in Seattle and provided a list of requirements for future commuters' needs. Chapter 10 in this volume describes the survey content, including the questions that were asked. The following section focuses on the commuter subgroups found in the surveyed population, their driving patterns and behaviors, and their user needs for a real-time information system. It should be noted that there was a 40% response rate (n = 3,893) to the survey (delivered to motorists on the road), and that the sampled commuter group consisted of 50.7% male and 48.8% female responders, with the majority (61.1%) under 41 years of age.

The Commuter Groups. A cluster analysis was performed on the surveyed data based on the willingness of commuters to adjust their traffic behavior to motorist information (Conquest, Spyridakis, Haselkorn, & Barfield, 1993). The cluster analysis focused on a particular group of variables that characterize the influence of traffic information in terms of time, route choice, and mode of transportation. These questions are reproduced in Fig. 11.2.

The analysis revealed four distinct subgroups with respect to traffic information:

Group 1: Route changers (RC): those willing to change routes on or before entering the freeway (20.6%)

Group 2: Nonchangers (NC): those unwilling to change time, route, and mode (23.4%)

Group 3: Route and time changers (RTC): those willing to change routes and departure times (40.1%)

Group 4: Pretrip changers (PC): those willing to make time, mode, or route changes before leaving home, but unwilling to change en route (15.9%)

Other subgroups of the clusters (see Table 11.5) were revealed by an analysis of seven commuting factors: departure flexibility on "home to work," departure flexibility on "work to home," amount of stress, importance of commuting time, commuting distance, commuting safety, and commuting enjoyment. Conquest et al. (1993) found that all four groups gave similar responses for their flexibility in departure time. Overall, all groups had more

5. Before you drive, how often does traffic information influence...

	Frequently	Sometimes	Rarely	Never receive
The time you leave	☐	☐	☐	☐
Your means of transportation (e.g., car, bus)	☐	☐	☐	☐
Your route choice	☐	☐	☐	☐

FIG. 11.2. Questions used for cluster analysis.

TABLE 11.5
Commuter Responses to Seven Commuting Factors (Percentage of Each Group)

Factors		Frequently	Sometimes	Rarely
Flexibility in departure time: Leave home for work				
	RC	14.7	47.2	38.1
	NC	13.5	49.7	36.7
	RTC	12.2	51.0	36.8
	PC	12.4	48.2	39.4
Flexibility in departure time: Leave work for home				
	RC	29.3	48.7	22.0
	NC	30.6	49.2	20.2
	RTC	29.5	50.1	20.3
	PC	27.8	52.1	20.0
Amount of stress experienced during commute				
	RC	12.3	58.7	29.0
	NC	12.7	53.2	34.1
	RTC	16.9	60.9	22.3
	PC	16.7	60.3	23.0
Importance of saving commuting time				
	RC	67.8	28.8	03.3
	RTC	69.6	26.6	03.8
	PC	66.6	30.5	02.9
	NC	60.3	33.5	06.3
Importance of reducing commuting distance				
	PC	22.4	40.5	37.1
	RTC	18.0	40.5	41.5
	RC	15.9	40.2	43.8
	NC	14.0	34.7	51.2
Importance of increasing commute safety				
	RTC	57.4	34.7	08.0
	PC	59.1	35.5	05.4
	RC	49.3	38.2	12.4
	NC	49.0	37.5	13.5
Importance of increasing commuting enjoyment				
	RTC	41.5	42.3	16.2
	PC	41.4	44.4	14.5
	RC	33.9	42.9	23.2
	NC	29.8	44.7	25.5

Note. RC = route changer; PC = pretrip changers; NC = nonchanger; and RTC = route and time changer. Adapted from Conquest et al. (1993). Reprinted from *Transportation Research*, 1C(2), The effects of motorist information on commuter behavior: Classification of drivers into commuter groups, pp. 183-201, © 1993, with kind permission from Elsevier Science Ltd, The Boulevard, Langford Lane, Kidlington OX5 1GB, UK.

flexibility in time leaving work than in time leaving home. This indicates that the ATIS can provide more options for alternate routes on the home commute. The findings revealed that commuters classified as changers (willing to alter mode, route, or time) cared more about saving commuting time than nonchangers. Pretrip changers placed the most emphasis on reducing commuting distance.

Knowledge of Alternate Routes. The study showed that those commuters classified as route changers were most familiar with alternate routes to the freeway (see Table 11.6). This is quite plausible because a commuter who changes routes often in response to traffic information needs to be familiar with available alternate routes. Route changers also were more likely to seek out information regarding traffic conditions on their primary route. Pretrip changers sought out information regarding traffic conditions more frequently prior to departure than did nonchangers, who indicated more frequently than members of the other clusters that they did not know of any alternate routes to their primary routes. This finding provides several implications for the design of ITS. Specifically, commuters more likely to seek out information will benefit greatly from alternate route information. Commuters who are satisfied with their primary route will be reluctant to use alternate route information.

Route Choice. The route changers had the highest response for frequently modifying or changing the route from home to work and the route from work to home (see Table 11.7). All groups modified the homebound route more frequently than the workbound route. As might be expected, the nonchangers were the least likely to modify either route.

TABLE 11.6
Familiarity With North-South Alternate Routes to I-5

Group	Very Familiar	Somewhat Familiar	Not at All Familiar
RC	73.0	24.2	02.7
NC	59.1	25.4	05.4
RTC	60.8	34.5	04.7
PC	57.7	35.7	06.6

Note. RC = route changer; PC = pretrip changers; NC = nonchanger; and RTC = route and time changer. Adapted from Conquest et al. (1993). Reprinted from *Transportation Research, 1C*(2), The effects of motorist information on commuter behavior: Classification of drivers into commuter groups, pp. 183-201, © 1993, with kind permission from Elsevier Science Ltd, The Boulevard, Langford Lane, Kidlington OX5 1GB, UK.

TABLE 11.7
Frequency of Modifying Route and Influence of Factors on Route Choice

Factors		Frequently	Sometimes	Rarely
Frequency of modifying route from home to work				
	RC	07.7	38.0	54.5
	RTC	0.70	31.5	61.7
	PC	05.4	36.6	58.0
	NC	03.4	20.1	76.5
Frequency of modifying route from work to home				
	RC	17.4	52.3	30.3
	RTC	15.0	44.2	40.8
	PC	14.6	47.6	37.8
	NC	10.8	31.6	57.6
Influence of traffic reports and messages on route choice				
	RC	31.8	54.6	13.6
	RTC	34.9	48.8	16.4
	PC	32.4	52.2	15.4
	NC	09.2	39.7	51.1
Influence of traffic congestion on route choice				
	RC	36.0	52.5	11.5
	RTC	30.1	50.6	19.3
	PC	27.8	54.8	17.4
	NC	18.1	45.0	36.9
Influence of time of day on route choice				
	RC	23.2	41.3	35.4
	RTC	26.5	39.3	34.2
	PC	30.6	37.7	31.7
	NC	14.3	31.9	53.8
Influence of weather conditions on route choice				
	PC	13.2	33.8	53.0
	RTC	09.1	34.1	56.8
	RC	16.5	28.4	65.1
	NC	03.3	17.6	79.1
Influence of time pressures on route choice				
	PC	17.1	41.3	41.5
	RTC	15.8	40.5	43.7
	RC	09.7	34.2	56.1
	NC	06.0	25.7	68.3

Note. RC = route changer; PC = pretrip changers; NC = nonchanger; and RTC = route and time changer. Adapted from Conquest et al. (1993). Reprinted from *Transportation Research, 1C*(2), The effects of motorist information on commuter behavior: Classification of drivers into commuter groups, pp. 183-201, © 1993, with kind permission from Elsevier Science Ltd, The Boulevard, Langford Lane, Kidlington OX5 1GB, UK.

Table 11.7 also shows how route choice is influenced by several factors: traffic reports and messages, actual congestion, time of day, weather conditions, and time pressures. Appropriately, the route choice of nonchangers was rarely influenced by traffic reports and messages or by the time of day. In general, to the items on the questionnaire, nonchangers consistently responded the highest on "rarely" and lowest on "frequently" for all factors. All four groups were less likely to respond to time pressures or weather conditions. The responses of commuters who would change departure times (i.e., route and time changers, and pretrip changers) were much the same; similar percentages of each group would make route changes based on traffic reports and messages, traffic congestion, time of day, and, unlike the other two groups, on time pressures as well. In a follow-up study to the effort just described, Wenger, Spyridakis, Haselkorn, Barfield, and Conquest (1989) reported that nonchangers used more landmarks (as opposed to street names) in their descriptions of their primary routes and their first alternate route. This finding indicates that nonchangers would benefit from in-vehicle messages that include information regarding landmarks on the road.

Use of Traffic Information. When asked about their use and preferences for media delivery of traffic information, the route changers, route and time changers, and pretrip changers (all the changers) tended to exhibit similar characteristics (see Table 11.8). The changers reported a high percentage of commercial radio use before and during driving. Electronic message signs and highway advisory radios also were used quite frequently (46% to 57%) by the changers. Commuters who were less likely to modify a given behavior (nonchangers) also were less likely to receive information relevant to that behavior.

Help From Current Traffic Information Delivered by Various Sources. Commuters reported that traffic information delivered from commercial radio provided the most help with percentages ranging from 95% (pretrip changer) to 75% (nonchanger; see Table 11.9). This exemplifies the effectiveness of clear, detailed information, only available from commercial radio, at the present time. An in-vehicle information system will also need to provide the same detailed level of traffic information.

Preferred Location for Receiving Traffic Information and Choosing Commuting Route. The clusters also differed in preferences for where and when they wanted to receive traffic information (see Table 11.10). As expected, the pretrip changer (PC) group had the highest response of preferring to receive information before driving. Pretrip changers were also the most likely of all the groups to choose their commuting route at home or work and the least likely to make a change en route. The route and time

TABLE 11.8
Media From Which Information Is Received
(Percentage Responding Yes)

Group	TV Before Driving	Electronic Message Sign While Driving	Highway Advisory Radio	Phone
RTC	34.9	56.1	47.8	10.5
PC	35.3	54.9	46.2	08.5
RC	29.8	57.1	46.9	05.2
NC	16.6	46.3	35.2	04.0

	Commercial Radio Only Before Driving	Commercial Radio Only While Driving	Commercial Radio Before and While Driving
RTC	06.1	18.7	73.7
PC	07.3	17.3	74.3
RC	05.9	25.1	67.7
NC	04.2	49.6	40.6

Note. RC = route changer; PC = pretrip changers; NC = nonchanger; and RTC = route and time changer. Adapted from Conquest et al. (1993). Reprinted from *Transportation Research, 1C*(2), The effects of motorist information on commuter behavior: Classification of drivers into commuter groups, pp. 183-201, © 1993, with kind permission from Elsevier Science Ltd, The Boulevard, Langford Lane, Kidlington OX5 1GB, UK.

TABLE 11.9
Traffic Information Help From Various Sources
(Percentage Finding Source Somewhat Helpful to Very Helpful)

Group	TV	VMS	HAR	Commercial Radio	Highway Construction Phone Hot Line
PC	27.7	39.6	38.8	94.9	05.6
RTC	23.4	38.7	34.1	94.3	02.8
RC	13.9	34.6	26.6	94.3	01.6
NC	06.7	27.6	18.5	75.0	01.3

Note. RC = route changer; PC = pretrip changers; NC = nonchanger; and RTC = route and time changer. Adapted from Conquest et al. (1993). Reprinted from *Transportation Research, 1C*(2), The effects of motorist information on commuter behavior: Classification of drivers into commuter groups, pp. 183-201, © 1993, with kind permission from Elsevier Science Ltd, The Boulevard, Langford Lane, Kidlington OX5 1GB, UK.

TABLE 11.10
Preferred Location for Receiving Traffic Information and Choosing Commuting Route

Group	Receipt of Traffic Information Preferred			
	Before Driving	*On City Streets*	*Near Entrance Ramps*	*On I-5*
PC	77.0%	11.7%	10.2%	1.1%
RTC	64.0%	19.4%	14.1%	2.6%
RC	47.3%	32.4%	18.1%	2.1%
NC	34.1%	30.8%	25.4%	9.7%

	Where Commuting Route Is Chosen			
	Before Driving	*On City Streets*	*Near Entrance Ramps*	*On I-5*
PC	47.0%	19.8%	22.1%	11.2%
RTC	35.2%	23.1%	25.9%	15.8%
NC	34.7%	20.6%	25.9%	19.3%
RC	22.4%	30.6%	30.6%	16.3%

Note. RC = route changer; PC = pretrip changers; NC = nonchanger; and RTC = route and time changer. Adapted from Conquest et al. (1993). Reprinted from *Transportation Research, 1C*(2), The effects of motorist information on commuter behavior: Classification of drivers into commuter groups, pp. 183-201, © 1993, with kind permission from Elsevier Science Ltd, The Boulevard, Langford Lane, Kidlington OX5 1GB, UK.

changer group had the second highest percentage of those preferring to receive information before driving: A lower percentage actually chose the route at home or at work, whereas half chose the route on city streets or near entrance ramps. The majority of route changers were most likely to choose their commuting route on city streets or near entrance ramps. Almost half, however, preferred to receive traffic information at home or at work. About 10% of the drivers in the nonchanger group preferred to receive traffic information on Seattle's major freeway (I-5) (as opposed to a much smaller percentage from the other groups).

Case Study 2: Defining Private Driver and Commercial Driver Subgroups

The preceding case focused primarily on one aspect of the private driving population, that is, commuters. This case study not only focuses on private drivers but on information requirements for commercial drivers as well.

11.   During your commute, how important is ...
       (**Check one box for each item**)

	Very important			Moderate importance			Not important

Saving commute time?   ☐ 1 ☐ 2 ☐ 3 ☐ 4 ☐ 5 ☐ 6 ☐ 7
Reducing commute distance?   ☐ 1 ☐ 2 ☐ 3 ☐ 4 ☐ 5 ☐ 6 ☐ 7
Increasing commute safety?   ☐ 1 ☐ 2 ☐ 3 ☐ 4 ☐ 5 ☐ 6 ☐ 7
Increasing commute enjoyment? ☐ 1 ☐ 2 ☐ 3 ☐ 4 ☐ 5 ☐ 6 ☐ 7

FIG. 11.3. Questions for cluster analysis from private driver survey.

Using responses from nationally distributed surveys, Ng et al. (1995) found several subgroups within the private and commercial driver populations. Cluster analysis was again used in the discovery of these groups.

Four trip factors used as the basis for the cluster analysis included the importance to drivers of decreasing trip time, decreasing trip distance, increasing trip safety, and increasing trip enjoyment. Survey respondents rated each factor on a scale of 1 (*very important*) to 7 (*not important*). These trip factors, common to both private and commercial drivers were considered because of their significant implications for the design of traveler information systems as discussed by other researchers (Barfield et al., 1991; Conquest et al., 1993). The questions used for the analysis are reproduced in Figs. 11.3 and 11.4.

The response mean of each trip factor for private and commercial drivers is shown in Table 11.11. One question of interest was whether there were differences in trip behavior as a function of whether one was a private or commercial driver. If so, then statistical procedures to determine if different subgroups existed in each separate group of drivers would need to be done. This question was answered by doing a two-sampled t-test (Devore, 1987). The results of this procedure (see Table 11.12) revealed that private drivers did in fact respond differently than commercial drivers to the questions concerning these trip factors. Thus, the cluster analysis on these variables, done to determine if different subgroups existed, were conducted separately for private and commercial drivers.

Major characteristics of each cluster group were determined by examining the number of responses in two importance categories. To do this, responses to the survey questions using the 7-point Likert scale were clas-

2.   During your work trip, how important is ...
      (**Check one box for each item**)

	Very important			Moderate importance			Not important

Saving trip time?   ☐ 1 ☐ 2 ☐ 3 ☐ 4 ☐ 5 ☐ 6 ☐ 7
Reducing trip distance?   ☐ 1 ☐ 2 ☐ 3 ☐ 4 ☐ 5 ☐ 6 ☐ 7
Increasing trip safety?   ☐ 1 ☐ 2 ☐ 3 ☐ 4 ☐ 5 ☐ 6 ☐ 7
Increasing trip enjoyment? ☐ 1 ☐ 2 ☐ 3 ☐ 4 ☐ 5 ☐ 6 ☐ 7

FIG. 11.4. Questions for cluster analysis from commercial driver survey.

TABLE 11.11
Mean Response (and Standard Deviation) for Commercial and Private Drivers as a
Function of Trip Factor

Driver Group	Trip Factor			
	Reducing Time	Decreasing Distance	Increasing Safety	Increasing Enjoyment
Private	2.67 (1.98)	3.47 (2.11)	2.37 (1.78)	3.28 (2.05)
Commercial	2.03 (1.41)	2.64 (1.79)	1.34 (0.91)	2.89 (1.94)

Note. Rating scale for mean response: 1, very important to 4, moderate importance to 7, not important.

sified as follows: Responses recorded as 1, 2, or 3 were placed into a category labeled "high importance," and responses recorded as 4 through 7 were placed in another category labeled "low importance." Chi-squared goodness-of-fit tests were conducted to determine if significant differences existed between the frequency of responses for high importance and low importance among the trip factors in each cluster group. Cluster groups were then named based on the number of observations in the two importance categories and the results of chi squared analyses. Ng, Barfield, and Mannering (1996) present these findings in detail. The groupings for the private and commercial drivers are labeled as follows:

Private Drivers

Group 1: *None of the factors are important* ($n = 55$, 7.3%). These private drivers placed very little to no importance on any trip factors

TABLE 11.12
Results of Two-Sampled t Test for Trip Differences Between Private and Commercial
Drivers

Driver Group	Trip Factor			
	Reducing Time	Decreasing Distance	Increasing Safety	Increasing Enjoyment
Result of t test	5.20	6.12	9.83	2.86
Two-tailed p value	0.0001	0.0001	0.0001	0.004

Group 2: *All factors important (n = 298, 39.8%)*. This cluster represented the largest group of drivers, who reported all trip factors as having high importance for a successful commute trip.

Group 3: *Safety and enjoyment are important (n = 145, 19.4%)*. These private drivers felt that increasing safety (mean = 1.46) and enjoyment (mean = 2.90) were important trip factors to achieve. Decreasing trip time and trip distance was considered to be of low importance to private drivers falling in this cluster.

Group 4: *Time and distance are important (n = 79, 10.5%)*. Reducing trip time (mean = 1.41) and decreasing trip distance (mean = 2.13) were very important to these drivers, and trip enjoyment was of little priority for this cluster group as indicated by the small proportion of respondents who rated enjoyment as having high importance (3 out of 79 respondents).

Group 5: *Time is important (n = 172, 23.0%)*. Decreasing trip time (mean = 2.61) for a successful commute was the highest priority for these drivers.

Commercial Drivers

Group 1: *All factors important (n = 133, 42.1%)*. These commercial drivers felt that all trip factors were important goals to achieve.

Group 2: *Safety, time and distance are important (n = 98, 31.0%)*. A large number of respondents in this category (97 out of 98 respondents) reported low importance for seeking trip enjoyment. For them, all other factors were of greater value.

Group 3: *Safety and enjoyment are important (n = 67, 21.2%)*. Increasing trip safety was very important to these drivers (mean = 1.15) with all drivers reporting high importance. Increasing trip enjoyment (mean = 1.94) was also of considerable high importance.

Group 4: *Indifferent to all factors (n = 18, 5.7%)*. The commercial drivers in this cluster reported only moderate importance for all trip factors.

Figures 11.5 and 11.6 show how the drivers in each cluster responded to the survey questions, given the two separate categories of importance: high importance and low importance. Our results revealed that common cluster groups were found for the two sets of drivers. For both commercial and private drivers there were cluster groups identified as "all factors important" and "safety and enjoyment are important." There was also a small percentage of private drivers (7%) that identified none of the trip goals as important (the "none of the factors are important" group). However, this group was not revealed in the commercial driver group.

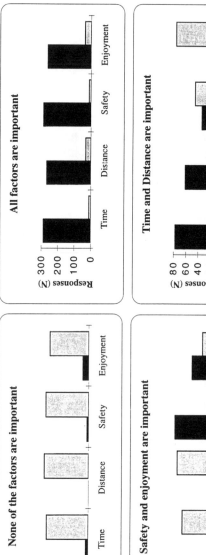

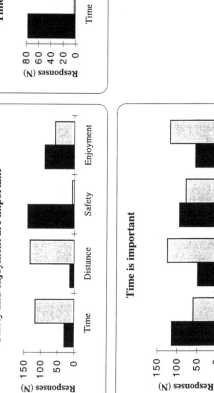

FIG. 11.5. Comparison of responses from "high importance" (ratings of 1, 2, and 3) to "low importance" (ratings 4 to 7) for private driver cluster groups.

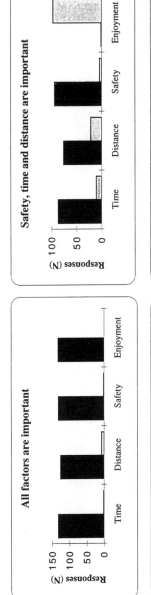
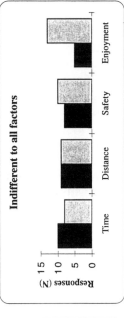
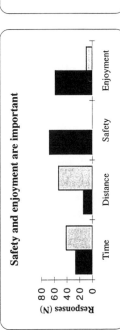

FIG. 11.6. Comparison of responses from "high importance" (ratings of 1, 2, and 3) to "low importance" (ratings 4 to 7) for commercial driver cluster groups.

The analysis revealed that there were distinct differences between cluster groups, suggesting that different information requirements will need to be considered in the design of ATIS. When the driving purpose is work related, there is common information that a majority of drivers are seeking. Questions such as "Will this route decrease the time to work?" or "Will the risk of an accident increase if I use this route?" may be answered with information supplied by ATIS, given the appropriate driving population. Furthermore, if the desired driving information cannot be attained readily, or if too much information is given, the level of experienced stress may increase. ATIS can help alleviate some of the stress and achieve the driver goal, whether the purpose is to reduce work trip time or to search for a safer or more enjoyable route. In all cases, the knowledge of alternate route information is essential for reducing driver stress and for indirectly decreasing the number of accidents. As with other systems and technologies (e.g., cellular phones, fax machines, personal computers), the success of ATIS will depend on the information needs of drivers and the subgroups that they comprise. These clusters show the importance of addressing the different needs of each driver group. The findings reveal that trip behavior, ATIS use, driver stress, and socioeconomic characteristics have an impact on the types of trip factors that drivers consider important.

USER REQUIREMENTS BASED ON TRIP
AND SOCIOECONOMIC CHARACTERISTICS

Another important consideration in the design of ITS is the information requirements for people based on their trip and socioeconomic characteristics. Trip characteristics include such items as trip chaining (whether or not you make additional trip stops besides going home or to work), flexible or fixed working hours, and the use of alternate routes. Socioeconomic characteristics include gender, income, and age. For example, in the study by Barfield et al. (1991), males predominate in the route changer (RC) and nonchanger (NC) groups, whereas females predominate in the route and time changer (RTC) and pretrip changer (PC) groups. Across groups, very few respondents (1% or less) were in the 65+ years age bracket. As for income, an analysis of variance by ranks again confirmed that the RC and NC groups (older and more males) display the higher incomes, the RTC and PC groups (younger and more females) the lower incomes. The PC group, the only group that would change mode of transportation and that has the highest percentage of female respondents, also displays the lowest income (Conquest et al., 1993).

These findings imply that females, even more than males, because of greater time demands, need an information system that provides time estimates of traffic delays and commuting routes. In a more general sense, these finding indicate that commuters differ even by gender with regard to aspects of their commute and use of traffic information, thus further supporting the notion that commuters cannot be treated as a single homogeneous group in terms of their traffic information needs. Previous studies have shown that the travel patterns of females and males are significantly different (DeJoy, 1992). Specifically, females appear to be shouldering a greater burden of household and family responsibilities in addition to making a living. These findings are also revealed in the ng et al. (1995) study showing that female private drivers were significantly more stressed than males in driving on and off the freeways. Mannering, Kim, ng, and Barfield (1995) also found that male private drivers and individuals who changed their residence frequently were less likely to view traffic information as important.

Private drivers in the age group "65 and over" also reported that they felt significantly more stressed during freeway driving than drivers under 55 years of age. This finding corresponds with other findings in which stress felt by older drivers appeared to be more severe than stress felt by drivers ranging in age from young to middle age (Lerner & Ratte, 1991; Malfetti & Winter, 1987).

Wenger et al. (1989) further analyzed the behavior and decisions of commuters before departure and while driving. Approximately 96 people from the previous Barfield et al. (1991) study participated in this second study. Some results of this second study are presented in the following section in an examination of differences in trip behavior and the implications for ITS use. Behavior patterns were investigated prior to departure, en route, and after commuting.

Behavior Prior to Departure

The majority of commuters (72.92%) received traffic information of some kind during the period prior to their departure and reported that they first received traffic information quite soon after awakening. Half of the commuters reported that they first received traffic information pertaining to their primary route almost immediately after awakening. An additional 10% of commuters (for a total of 61.43%) reported that they received their first traffic information more than one hour prior to departure. Although commuters reported that they were aware of traffic information, they reported that this information had little impact on their decisions prior to departure. This implies that most commuters will not make any departure changes based on information received too early in the morning because traffic conditions may change right before departure.

The majority of commuters reported that they rarely decided to use an alternate route (65.71%), that they rarely decided to use an alternate mode of transportation (90.00%), and that they rarely decided to change their time of departure (64.29%) based on information they received prior to departure. In an average month, commuters reported that they decided to change their route twice (mean = 2.333, SD = 2.666) prior to departure. Although these results might indicate that a large number of commuters were unresponsive to traffic information delivered before departure, the results also indicate that an important proportion of commuters could be influenced by predeparture traffic information. Indeed, if traffic information could influence one third of Seattle commuters to change their departure time or route choice or one tenth to change their transportation mode, significant improvements in peak-time traffic conditions could result.

On the whole it appears that commuters were somewhat more receptive to traffic information delivered prior to departure, reporting that the period prior to departure is not a very stressful period. They also reported that they have a relatively small number of tasks to accomplish then. The low rate of modification to route, mode, and time of departure may indicate that although commuters did receive traffic information at that time, they did not find it to be credible. This inference is supported by comments to this effect received from commuters on the initial survey (Spyridakis, Haselkorn, Barfield, Conquest, & Isakson, 1991). The low rate of route modification also may result from a temporal delay between receipt of the information and decision, because the majority of commuters reported that they received their first traffic information more than one hour prior to departure. For purposes of designing an information system, these results reinforce the conclusion that demonstrating system credibility is an extremely significant issue. Furthermore, these results indicate that commuters may have the time to use an interactive traffic information system, one that would demand some active engagement.

Behavior En Route

Commuters indicated a high degree of knowledge regarding their primary commuting route and their first and second alternate routes. Indications of route knowledge were obtained from counts of landmarks and street names used when commuters described their commuting routes. Results support the intuitive prediction that commuters have a more detailed knowledge of their primary route than of either their first or second alternate routes. Commuters, when asked to trace their commuting routes, used five times as many street names as they did landmarks in their description. Table 11.13 presents the means and standard deviations for number of street names and landmarks used by commuters to describe their primary route and their first and second alternate routes.

The vast majority of commuters (95.83%) reported that they knew of an alternate to the route they normally use, one they would use if a large portion of their normal route were inaccessible for some reason. On the

TABLE 11.13
Number of Street Names and Landmarks Used in Descriptions of Commuting Routes

		Primary Route	*First Alternte*	*Second Alternte*
Street names	*M*	8.45	5.02	4.26
	SD	6.23	3.70	4.01
Landmarks	*M*	1.67	1.03	0.79
	SD	1.89	1.48	0.90

Note. M = mean; *SD* = standard deviation. From Wenger et al. (1990). Motorist behavior and the design of motorist information systems. In *Transportation Research Record 1281*, Transportation Research Board, National Research Council, Washington, DC. Reproduced with permission.

average, commuters reported knowing two or three alternate routes (M = 2.880, SD = 1.568). However, only 75% of those interviewed reported that they actually use one of those alternate routes.

The majority of commuters (58.95%) reported that they experienced low to moderate levels of stress on their primary route. However, a sizable percentage of commuters (33.68%) did report that they experienced relatively high amounts of stress on their normal commute. If they decided to use an alternate route, 80.88% of commuters reported that the level of stress experienced changed, with 77.78% reporting that the level of stress experienced increased. Thus use of an alternate route represented an increase rather than a decrease in stress because commuters were less familiar with that route.

Although commuters reported that their choice of route was relatively stable, they also reported that they make between one and two adjustments to their normal route each day (M = 1.552, SD = 1.897). The adjustments in route primarily occurred in response to observed traffic congestion and reports of traffic congestion received in the car (i.e., traffic radio reports). The decision to use an alternate route was based first on traffic information received in the car (33.28% for the first alternate route, 35.09% for the second alternate route) and second on observed traffic conditions (23.53% for the first alternate route, 21.05% for the second alternate route). Interestingly, approximately one fourth of the commuters who used alternate routes reported that they sought out information about the use of an alternate route while at home, more than 30 minutes before departing (26.87% for the first alternate route, 24.56% for the second alternate route).

Commuters reported that they received little feedback regarding their choice to use an alternate route, and what feedback they did receive was relatively delayed. Nearly one third of the commuters indicated that they had no way of telling if their choice to use an alternate route was correct or incorrect (27.94% for the first alternate route, 31.58% for the second alternate route). The majority of commuters indicated that if they did receive any kind of information confirming or refuting their choice to use an alternate route, they received it more than five minutes after making the choice (69.57% for the first alternate route, 48.72% for the second alternate route). Only a small percentage of commuters (2.94%) indicated that they got this information from radio traffic reports.

The patterns observed for all commuters indicate that commuters have a high degree of knowledge concerning their primary and alternate routes, that a majority use alternate routes, and that nearly half of the alternate routes make use of some portion of I-5 (the primary route used by commuters into downtown Seattle). It appears that commuters are not overly burdened with tasks other than simply commuting to the workplace and that approximately one third of all commuters experience high levels of stress, with the perceived level of stress increasing if an alternate route is used. Commuters appear to make a small number of adjustments to their primary route, based mainly on their observations of traffic conditions. However, commuters appear to decide to use an alternate route based on traffic reports received either at home or in the car. Finally, commuters receive little feedback regarding their choice to use an alternate route and, when received, this feedback usually is delayed.

The implications for design of an information system for commuters en route are similar to those for such a system designed for commuters before departure. Commuters do rely on traffic reports, but they require that information to be current, specific, and verified as reliable through some type of feedback. Increasing the timeliness of information might well increase the probability of choosing an alternate route (as opposed to merely making minor adjustments to the primary route). Incorporating feedback mechanism into on-road systems for delivering information should increase their effectiveness in encouraging alternate route selection.

Behavior After Commute

Commuters reported that they are late for work because of traffic conditions approximately four times in an average month ($M = 4.097$, $SD = 4.099$). Commuters were asked to rate their flexibility in arrival times, and their responses were distributed evenly across a 5-point scale. The majority of commuters (82.29%) indicated that the penalties for arriving late for work were relatively minor.

Gender Differences

Only a small number of gender differences were uncovered in the analyses of the responses to the in-depth interview. Females indicated that they tended to have less flexibility in both the time of departure and the time of arrival. Females also rated the period prior to departure as more hectic than did males.

IMPLICATIONS FOR THE DESIGN
OF INFORMATION SYSTEMS

The findings presented in this chapter bring to light several design considerations for ATIS–CVO information systems. The majority of the surveyed respondents from the ng et al. (1995) study indicated a willingness to use an in-vehicle system if it were given to them (80%). The study showed that private and commercial drivers perceived traffic information to be the most important feature for ATIS–CVO, whereas dispatchers perceived personal communication to be the most important feature. Interestingly, all users viewed roadside services as being least important. For design requirements, this indicates that less detail should be provided on roadside services, whereas greater emphasis should be placed on the other features. Accuracy and cost of the system were very important to all user groups, a finding supported by other researchers (Dudek et al., 1971; Trayford & Crowle, 1989). In terms of cost, the entire system should cost the user no more than $500.00, the average cost commercial and private drivers were willing to pay ATIS–CVO. Other cost methods identified in the survey included the use of a monthly charge for airtime. If this is the case, studies have indicated that the cost per user should not exceed $10.00 to $20.00 per month (Marans & Yoakum, 1991; Tsai, 1991).

The majority of private drivers (over 70%) identified accidents, the volume of traffic on the road, and road construction as reasons for choosing an alternate route. A high percentage of commercial drivers and dispatchers also based their choice on accidents and road construction. Thus, appropriate traffic information will include the following:

- Average speed on primary and alternate routes
- Type of accident on primary route
- number of lanes obstructed by accidents and road construction on primary route

Other information that should be displayed in relation to alternate routes includes the length of a delay as well as a map of the proposed alternate routes. Our studies indicate that the majority of private drivers would divert to a known alternate route if the delay were 5 to 15 min, and would more likely venture to an unknown route if delays were longer. This interesting finding leads to the belief that the system could propose routes to unfamiliar areas only when delays are longer than 15 min. If the delay is longer, such a recommendation would help divert drivers to a variety of routes instead of a main alternate route, which may become congested if all drivers were rerouted to it.

The analysis of the commuter subgroup case study showed that there are different types of commuters, some more willing than others to change route, mode, or time of departure. Although it is beneficial if some drivers are not persuaded to alter their transportation routes, it appears desirable if a good majority is convinced to change routes when necessary. To do this, information should be provided to drivers in the format most appealing to them. For example, those commuters classified as nonchangers were more likely to use landmarks than street names in describing their commuting routes. Because the majority of the available traffic information sources rely heavily on the use of street names in the description of routes, members of this cluster (having a lower knowledge of the street names on their commuting routes) would be less likely to find the information usable or to act on that information. Thus, an information system targeting members of the nC cluster might provide more graphic information, including real-time displays of traffic situations such as live videos of conditions on the freeway. The information system might also need to provide greater levels of information regarding alternate routes, perhaps even offering an option that would increase commuters' familiarity with the available routes in the fashion of a tutorial.

We also observe that commuters are likely to benefit from at least two different types of information systems: one using predeparture and one using en route information. These two systems should be integrated to provide feedback and confirmation of reliability (postcommute delivery might be effective as well), and the transmitted information needs to be more current and specific so it will be used.

In the studies, changes in route were usually a result of observing actual congestion rather then information provided by current media. Furthermore, commuters rarely changed time of departure or mode of transport because of current information received at home prior to departure. Commuters, in general, were quite familiar with alternate routes, but they were rarely used. This may be partly because the use of alternate routes increases driver stress. The findings indicate that alternate route information should be displayed only when it is significantly faster or better than the primary route.

According to another finding, route modification usually occurs when actual on-road congestion is observed, rather than by information provided by current media. This suggests that the display of live pictures be restricted to conditions of heavy congestion with the intent of convincing commuters to use suggested alternate routes. Although live pictures of traffic are currently available to some commuters prior to departure, the decision of what pictures to display is not based on this thinking. Finally, commuters were resistant to changing their times of departure or modes of transport based on current traffic information received prior to departure.

CONCLUSIONS AND RECOMMENDATIONS

By consistently providing appropriately designed and delivered up-to-the-minute traffic information, one can improve short-term driver response to incidents and congestion and produce a long-term change in driver behavior that will increase the efficiency with which existing transportation facilities are used (Haselkorn, Barfield, Spyridakis, & Conquest, 1990). However, significant improvements in the level of congestion through delivery of traffic information can be achieved only if the mechanism for delivering that information is developed as an integrated system that is responsive to users' needs and perspectives. A great deal of information relating to the user subgroups was identified that reveals the importance of displaying accurate information to each group.

The results presented in this chapter support the central premise: Drivers cannot be considered as a homogeneous audience for motorist information. The method employed in this study has reinforced the idea that user groups have different behaviors and that identifying these differences will help to determine each group's use of and response to motorist information.

Targeting information for those motorists most likely to be affected does not mean that the same group will be targeted for all types of motorist information in all types of driving situations. Despite this focused approach, a single successful motorist information system can meet the needs of a wide range of motorists in varying conditions and stages of travel. This does mean, however, that a single integrated motorist information system must consist of carefully designed information modules targeted to address particular driving decisions of carefully studied and defined subgroups of receptive drivers.

As stated earlier, there are several subpopulations of motorists, each with unique user requirements that will need to be integrated for ITS. The three populations of potential users of ATIS–CVO have distinctly different preferences and requirements for information. Commercial drivers need to be aware of many more freeway restrictions and limitations. Dispatchers are very concerned about personal communication and the ability to convey sufficient information for assisting drivers in arriving at their destinations. There are many additional studies currently underway at the University of Washington to further define user requirements for other subpopulations. As more information about driver groups is obtained, more and better ways to design motorist information to meet the drivers' needs will be discovered.

REFERENCES

Al-Deek, H., & Kanafani, A. (1991). Incident management with advanced traveler information systems. *1991 Vehicle navigation & Information Systems Conference* (Vol 1., no. 912798, pp. 563–575). Warrendale, PA : Society of Automotive Engineers.

Allen, R. W., Stein A. C., Rosenthal T. J., Ziedman D., Torres, J. F., & Halati, A. (1991). Laboratory assessment of driver route diversion in response to in-vehicle navigation and motorist information system. *Transportation Research Record, 1306,* 82–91.

Antin, J., Dingus, T., Hulse, M., & Wierwille, W. (1990). An evaluation of the effectiveness and efficiency of an automobile moving-map navigational display. *International Journal of Man-Machine Studies, 33*(5), 581–594.

Barfield, W., Haselkorn, M., Spyridakis, J., & Conquest, L. (1991). Integrating commuter information needs in the design of a motorist information system. *Transportation Research, 25A*(2/3), 71–78.

Bishu, R. R., Foster, B., & McCoy, P. (1991). Driving habits of the elderly—a survey. *Proceedings of the Human Factors Society 35th Annual Meeting* (pp. 1134–1138). Santa Monica, CA: Human Factors Society.

Chang, G., Lin, T., & Lindely, J. (1992). Understanding suburban commuting characteristics: An empirical study in suburban Dallas. *Transportation Planning and Technology, 16,* 167–193.

Conquest, L., Spyridakis, J., Haselkorn, M., & Barfield, W. (1993). The effects of motorist information on commuter behavior: Classification of drivers into commuter groups. *Transportation Research, 1C*(2), 183–201.

Datta, T. K. (1991). Head-on, left turn accidents at intersections with newly installed traffic signals. *Transportation Research Record, 1318,* 58–63.

DeJoy, D. M. (1992). An examination of gender differences in traffic accidents risk perception. *Accident Analysis and Prevention, 24*(3), 237–246.

Devore, J. (1987). *Probability and statistics for engineering and the sciences* (2nd. ed.). Monterey, CA: Brooks/Cole.

Dudek, C., Messer, C. J., & Jones, H. B. (1971). Study of design considerations for real-time freeway information systems. *Highway Research Board, no. 363: Operational Improvements for Freeways.* Washington, DC: Highway Research Board, national Research Council.

Evans G., Palsane M., & Carrere S. (1987). Type A behavior and occupational stress: A cross-cultural study of blue-collar workers. *Journal of Personality and Social Psychology, 52*(5), 1002–1007.

Federal Highway Administration. (1991). *Highway statistics 1991.* Washington, DC: U.S. Department of Transportation.

Garrison, D., & Mannering, F. (1990). Assessing the traffic impacts of freeway incidents and driver information. *ITE Journal, 60*(8), 19–23.

Green P., & Williams, M. (1992). Perspective in orientation/navigation displays: A human factors test. In L. Olaussen & E. nelli (Eds.), *1992 Vehicle navigation and Information Systems Conference* (pp. 221–226). New York: Institute of Electrical and Electronics Engineers.

Gulian, E., Glendon, A., Matthews, G., Davies, D., & Debney, L. (1990). The stress of driving: A diary study. *Work & Stress, 4*(1), 7–16.

Hair, J., Anderson, R., Tatham, R., & Black, W. (1992). *Multivariate data analysis with readings* (3rd. ed.). new York: Macmillan.

Hancock, P. A., Caird, J. K., Shekjar, S., & Vercruyssen, M. (1991). Factors influencing driver's left turn decisions. *Proceedings of the Human Factors Society 35th Annual Meeting* (pp. 1139–1143). Santa Monica, CA: Human Factors Society.

Haselkorn, M., Barfield W., Spyridakis, J., & Conquest, L. (1990). *Improving Motorist Information Systems, Final Report.* Washington State Department of Transportation.

Kaneko, T., & Jovanis, P. (1992). Multiday driving patterns and motor carrier accident risk: A disaggregate analysis. *Accident Analysis and Prevention, 24*(5), 437–456.

Kranowski, W. (1988). *Principles of multivariate analysis.* Oxford: Clarendon Press.

Lerner, n., & Ratte, D. (1991). Problems in freeway use as seen by older drivers. *Transportation Research Record, 1325,* 3–7.

Lunenfeld, H. (1989). Human factor considerations of motorist navigation and information systems. In D. H. M. Reekie, E. R. Case, & J. Tsai (Eds.), *1989 Vehicle navigation and Information Systems Conference* (pp. 35–42). New York: Institute of Electrical and Electronics Engineers.

Malfetti, J., & Winter, D. (1987). *Safe and unsafe performance of older drivers: A descriptive study.* Washington, DC: AAA Foundation for Traffic Safety.

Mannering, F., & Grodsky, L. (1995). Statistical analysis of motorcyclists' self-assessed risk. *Accident Analysis and Prevention, 27*(1), 21–31.

Mannering, F., Kim, S., ng, L., & Barfield, W. (1995). Travelers' preferences for in-vehicle information systems: An exploratory analysis. *Transportation Research, 3C*(6), 339–351.

Marans, R. W., & Yoakum, C. (1991). Assessing the acceptability of IVHS: Some preliminary results. In *1991 Vehicle navigation and Information System Conference* (no. 912811, pp. 657–668). Warrendale, PA: Society of Automotive Engineers.

Mast, T. (1991). Human factors in intelligent vehicle-highway system: A look to the future. *Proceedings of the Human Factors Society 35th Annual Meeting—1991, Vol. 2* (pp. 1125–1129). Santa Monica, CA: Human Factors Society.

McMurray, L. (1970). Emotional stress and driving performance: The effect of divorce. *Behavioral Research in Highway Safety, 1,* 100–114.

Ng, L., Barfield, W., & Mannering, F. (1995). A survey-based methodology to determine information requirements for advanced traveler information systems. *Transportation Research, 3C*(2), 113–127.

Ng, L., Barfield, W., & Mannering, F. (1996). *Analysis of private drivers' commuting and commercial drivers' work-related travel behavior.* Manuscript submitted for publication.

Parkes A. M., Ashby, M. C., & Fairclough, S. H. (1991). The effects of different in-vehicle route information displays on driver behavior. In L. Olaussen & E. nelli (Eds.), *1991 Vehicle navigation and Information Systems Conference* (no. 912734, pp. 61–69). Warrendale, PA: Society of Automotive Engineers.

Raggatt, P. (1990). Driving hours and stress at work: The long distance coach driver. *Proceedings of the 15th Australian Road Research Board Conference* (Part 7; pp. 235–252). Melbourne, Australia: Australian Road Research Board.

Smiley, A. (1989). Mental workload and information management. In D. H. M. Reekie, E. R. Case, & J. Tsai (Eds.), *1989 Vehicle navigation and Information Systems Conference* (pp. 435–438). New York: Institute of Electrical and Electronics Engineers.

Spyridakis J., Barfield, W., Conquest, L., Haselkorn, M., & Isakson, C. (1991). Surveying commuter behavior: Designing motorist information systems. *Transportation Research, 24A*(1), 17–30.

Transportation Research Board (1991). *Special Report 232: Advanced Vehicle and Highway Technologies.* Washington DC: national Research Council.

Trayford, R. S., & Crowle, T. B. (1989). The ADVISE traffic information display system. In D. H. M. Reekie, E. R. Case, & J. Tsai (Eds.), *1989 Vehicle navigation and Information Systems Conference* (pp. 105–112). new York: Institute of Electrical and Electronics Engineers.

Tsai, J. (1991). Highway environment information system interests and features survey. In L. Olaussen & E. nelli (Eds.), *1991 Vehicle navigation and Information Systems Conference Vol. 1* (no. 912743, pp. 113–122). Warrendale, PA: Society of Automotive Engineers.

Wenger, M. J., Spyridakis, J., Haselkorn, M. P. , Barfield, W., & Conquest, L. (1989). Motorist behavior and the design of motorist information systems. *Transportation Research Record, 1281,* 159–167.

Wierwille, W. W. (1993). Demands on driver resources associated with introducing advanced technology into the vehicle. *Transportation Research, 1C*(2), 133–142.

12

Human–System Interface Issues in the Design and Use of Advanced Traveler Information Systems

Thomas A. Dingus
Virginia Polytechnic Institute and State University

Melissa C. Hulse
Performance and Safety Sciences, Inc.

Woodrow Barfield
Virginia Polytechnic Institute and State University

This chapter discusses human–system interface issues related to the design and use of Advanced Traveler Information Systems (ATIS). An ATIS is a subset of the more general transportation system being planned, known as the Intelligent Transportation System (ITS; see foreword and chapter 1 of this volume). The purpose of an ATIS is to regulate the flow of vehicles along roads and highways by using emerging sensor, computer, communication, and control technologies. Designing ATIS systems that humans can use safely and efficiently is a difficult undertaking. The difficulty is derived from a number of causes: (a) the inherent complexity of systems being planned or produced; (b) the widely ranging knowledge, skills, and abilities of the driving population; (c) the limited opportunities for training and instruction with regard to ATIS technology; and (d) the requirement that system use under all circumstances minimally interfere with the primary task of driving. To meet these challenges, a number of human factors issues must be successfully addressed.

SENSORY MODALITY ALLOCATION

Correctly choosing the appropriate sensory modality (auditory, visual, or tactile) to use for the delivery of in-vehicle information is an important task to consider when designing an ATIS. Sensory modality allocation (i.e., whether to present information to the visual, auditory, or haptic modalities) can greatly affect both the safety and usability of an ATIS. For example, excessive amounts of visual information delivered through an ATIS can overload the modality (visual) that already provides roughly 90% of a motorist's driving information (Rockwell, 1972). Considering auditory information, if in-vehicle auditory information is not designed according to human-factors guidelines, the use of this information can lead to a system that is unusable, frustrating, annoying, and even dangerous. In making design decisions concerning the allocation of information to the various sensory modalities, designers must carefully consider the user, the system, and the driving environment, as well as the specific capabilities of the particular sensory modality. To help designers choose the most efficient sensory modality for the presentation of information, Deatherage (1972) presents the following lists of human factors guidelines for auditory and visual information. Lists such as the following should serve only as high-level guidelines; specific applications will require close scrutiny of the following information. Furthermore, the ATIS examples are given only to provide examples of potential uses of the auditory and visual modalities to convey information. Without extensive research for determining when to use auditory or visual information for in-vehicle displays, it is not recommended that the following information be used to design actual displays. Thus these are only general guidleines. Therefore, in general, when designing information systems, use auditory or visual presentation for the information requirements presented in Table 12.1.

In-Vehicle Auditory and Visual Information

This section summarizes work done on the topic of auditory and visual displays for providing in-vehicle information. Under many circumstances, the auditory presentation of navigation information could be superior to the visual presentation of the same information. Sorkin (1987) pointed out that the omnidirectional nature of auditory displays makes them most desirable for alert and warning messages. Furthermore, a major advantage of auditory presentation in the driving environment is that it allows information-processing resources to be allocated efficiently across modalities because an additional channel for processing information is used. Therefore, allocating supplemental tasks to the auditory modality (particularly in situations of high visual attention demand) is likely to make the composite task of driving easier and safer.

TABLE 12.1
Guidelines for the Use of Auditory and Visual Information in ATIS

	Type of Information	ATIS Example
Use auditory		
information if	the message is simple.	turn right
	the message is short.	yield
	the message will not be referred to later.	stop ahead
	the message deals with events in time.	routing decisions
	the message calls for immediate action.	collision avoidance
	the visual system of the person is overburdened.	approaching a busy intersection
	the receiving location is too bright or if dark adaptation is necessary.	day or night driving
	the person's job requires continual movement from place to place.	commercial/emergency driver
Use visual		
presentation if	the message is complex.	navigational instructions
	the message is long.	list of exits or services ahead
	the message will be referred to later.	location on a map display
	the message deals with location in space.	particular destinations
	the message does not call for immediate action.	rest stop 10 miles ahead
	the auditory system of the person is overburdened.	ambulance driver
	the receiving location is too noisy.	bus driver/truck deliveries
	the person's job allows him or her to remain in one position.	dispatching

Significant evidence supports the use of auditory displays in vehicles. Many researchers (Dingus & Hulse, 1993; Dingus et al., 1995; Means et al., 1992) have found that giving turn-by-turn directions via the auditory channel leads to quicker travel times, fewer wrong turns, and lower workload. Furthermore, when route information is delivered with auditory displays, the driver can give more attention to the primary task of driving than when route information is displayed on a visual map (Labiale, 1990; McKnight & McKnight, 1992; Parkes, Ashby, & Fairclough, 1991; Streeter, Vitello, & Wonsiewicz, 1985). The idea that auditory information can be used effectively as an in-vehicle display is supported by additional evidence concerning the visual sense indicating that driving performance suffers when drivers are simultaneously looking at various in-vehicle displays. In this case, drivers tend not to react to situations on the road (McKnight & McKnight, 1992), deviate from their course (Zwahlen & DeBald, 1986), and reduce their average driving speed (Walker, Alicandri, Sedney, & Roberts, 1991). The implication, then, is that if visual performance can be degraded, possibly another sensory modality should be used to decrease the demands on the visual modality. Additional research assesses the workload differences associated with presenting information visually or aurally. Labiale (1990) showed that the workload during guided route-following is lower when navigation information is presented aurally than when it is presented visually. This study also revealed that drivers prefer auditory route guidance information because they feel that it provides a safer system.

A study performed by Walker, Alicandri, Sedney, and Roberts (1991) gauged the safety of the drivers' performance while using guidance devices that varied in complexity and mode of presentation. Visual presentation included both turn-by-turn graphic icon and full route-map information. Auditory information in the form of verbal turn-by-turn guidance was also included along with a paper map as a control condition. The results showed that during high-information load situations, subjects using auditory devices did not reduce their speeds as much as those using visual devices. Yet they made fewer navigational errors than their counterparts. In terms of complexity, the subjects using the devices that presented more complex information (the full maps) drove more slowly than those using the device that presented simpler information (the turn-by-turn information). This study indicates that the audio devices were somewhat safer to use than the visual devices, and that simpler display complexity was generally preferable to higher display complexity.

Despite the inherent advantage of efficiently decreasing the load on the visual modality by using auditory displays while driving, there is also research to support the notion that visual displays should be required in navigational tasks. For example, Aretz (1991) made the case that without some sort of map representation, drivers cannot construct an internal cognitive map of their routes. Therefore, a driver might arrive at a destination with no navigational problems, yet have no idea of the destination's location relative to the driver's starting point or to other landmarks. Along these lines, Streeter et al. (1985) noted that drivers prefer navigation systems that keep them informed of their current location.

Considering auditory information, Dingus and Hulse (1993) cite some potentially negative aspects of using speech presentation in lieu of visual presentation for ATIS. These include: improper prioritization and instinctive reaction to voice commands, even when they conflict with regulatory information; the unintelligibility of low-cost speech synthesis devices; and the inability of low-cost digitized speech devices to provide a vocabulary that gives all desired information. However, some of these issues may be moot in the near future because technological advances are being made in speech technology. Furthermore, much of the information that ATIS will provide is too complex for effective aural display. Also, according to Robinson and Eberts (1987) and Williges, Williges, and Elkerton (1987), performance is optimal for a spatial task like navigating when visual stimuli are coupled with manual responses. If the task is a verbal one, an auditory stimulus should be coupled with a verbal response from the person such as a verbal verification for the acceptance of a new route. This line of research suggests that a visually displayed arrow indicating the direction of travel would assist drivers in a navigation task more than would verbal information (e.g., an auditory speech display).

Another factor to consider in selecting display modalities is a phenomenon known as visual dominance. Wickens (1992) described visual dominance as a reaction that counteracts the automatic alerting tendencies of the auditory modality. It is a bias toward processing information in the visual channels concurrently with information of the same general type or importance provided as auditory or proprioceptive stimuli. Behavior in these situations suggests that the person responds appropriately to the visual information and disregards the cues provided by other modalities. Thus, for ATIS design, given that the visual modality will predominate, it is recommended that in-vehicle information be presented to only one modality at a time, and when visual and auditory information are provided concurrently, care must be taken to design the auditory information in a way that it will enhance the visual information.

If visually oriented navigation information is to be delivered in the car, then the safety aspects of that information should be discussed. French (1990) discussed the safety implications of the time it takes to look at in-vehicle displays associated with driver information systems. The major concern of his research is the average glance time considered safe when a driver is looking at an in-vehicle display. He found that a driver's average glance time at a vehicle display is 1.28 s. Dingus et al. (1995) showed that even though the average glance time is relatively constant with displays of differing complexities, the number of glances required to retrieve information increases substantially for more complex displays. In addition, Dingus et al. found that the number of relatively long glances (over 2.0 s) increases even though the number of these glances is still small compared to the number of total glances. Also, the guidelines in French's (1990) paper indicate that glance times greater than 2 s are unsafe and unacceptable, whereas glance times between 1 and 2 s are considered only marginally safe.

Almost all of the literature suggests that operator performance could be improved by selective use of some combination of auditory and visual stimuli. Dingus and Hulse (1993) recommended that the auditory modality be used to provide an auditory prompt for a visual display of changing or upcoming information, thus reducing the need for the driver to constantly scan the visual display in preparation for an upcoming event. Alternatively, an auditory display could be paired with a simple visual display that supplements the auditory message, so that an auditory message that is not fully understood or remembered can be checked or referred to later via the visual display. Labiale (1990) suggested that for driving safety, the optimal perceptual and cognitive solution seems to be a maximum seven to nine information-unit aural message. In this context, an "information unit" is any single item of information that communicates a single meaning; this item may be a word or short phrase. When both visual and auditory information is used, Labiale recommended the use of an aural message to prompt the driver to look at a simple map or other visual display. Labiale also suggested that it would be useful if drivers could request a repeat of the aural message when the information is complex. Robinson and Eberts (1987) also maintain that optimal display design would combine desirable features from speech displays (e.g., warning or alerting capability) with the spatial orientation provided by visual displays.

Wickens (1987) emphasized the importance of display format redundancy when information is presented in more than one modality. For example, stimuli in the auditory or visual modalities are inevitably masked from time to time by visual glare, background noise, or other elements in the operating environment. However, presenting information in both the auditory and visual modality accomodates this masking, ensuring that the driver receives the information even if one modality is unavailable. Display format redundancy also accommodates the strengths of different ability groups in the population (e.g., high spatial ability versus high verbal ability individuals).

Information presented to the auditory modality, if implemented effectively, has great potential as a "pointer" to guide driver attention at strategic moments to the visual display or, for that matter, to important visual events outside the car (e.g., collision avoidance). Furthermore, selected voice messages can be used to elaborate on the information depicted by the visual display, employing an eyes-on-the-road mode of informing the driver. Voice messages can also indicate whether new information is available so the driver need not glance at the visual display frequently to check for updates. Voice functions can be implemented as a supplement to visual displays or as their own stand-alone system. Williges and Williges (1982) proposed that spoken information should be highly reliable and intended for immediate use because, unlike a visual display, it cannot be referred to at a later point in time. Deatherage (1972) stated that auditory displays should use speech rather than nonverbal signals, thus eliminating the driver's need to decode nonverbal signals and thereby minimizing the driver's information-processing requirements.

According to Means et al. (1992), the precepts which can be applied to the design of an auditory interface to make it more palatable to a driver include minimizing voice "chattering" and "nagging"; maximizing voice intelligibility; providing timely, useful information through voice; and allowing significant driver control of voice functions. Drivers may not be receptive to the use of voice for a system warning unless the condition is urgent. For an open door, a nonverbal auditory signal or a telltale sign on the instrument panel is probably sufficient to alert the driver to the problem. The driver may perceive this use of voice as "nagging" if it resembles an interfering passenger more than it does a machine warning. Drivers may have various reasons for wanting to suppress a voice system at times, and these wishes must be accommodated by giving the driver control over volume as well as activation of voice functions (Means et al., 1992). *Auditory clutter* is the term used by Stokes, Wickens, and Kyte (1990) to describe overuse of the auditory channel, which can distract from the driving task. To minimize it, voice feedback should not be used to note correct maneuvers, driving speed, or system status (uses advocated by Davis & Schmandt, 1989).

The cognitive attention required to process auditory messages increases as the intelligibility of messages decrease. Marketing considerations will dictate low-cost automotive navigation systems. This cost constraint has an important implication for in-vehicle auditory displays: Digitized speech that does not provide street name information to the driver (because of database constraints) will probably be used. Although the quality of low-cost synthesized speech is constantly improving, factors such as tonal quality and inflection currently limit its relative effectiveness (Sanders & McCormick, 1987).

Finally, considerable research has studied the use of nonverbal auditory warnings in aircraft cockpits (see Patterson, Farrer, & Sargent, 1988, for a comprehensive discussion). Some of the knowledge that has accrued from aircraft research may pertain to passenger vehicles (e.g., appropriate volumes and temporal characteristics for auditory tones). However, principles guiding the use of auditory systems in aircraft must not be applied indiscriminately to passenger vehicles. It is important to bear in mind the essential differences between highly trained cockpit personnel and automobile drivers, who range widely in age, driving ability, physical condition, and other characteristics.

Significant research is being conducted on the use of information presented visually and appropriate formats for that information. As previously discussed, navigation display information should be limited to only that which is absolutely necessary. Streeter (1985) recommended that information for drivers following a prespecified route should consist of the next turn, the distance to the turn, the street to turn on, and the direction to turn. Streeter found that people familiar with an area prefer to be given the cross street of the next turn, whereas people who are unfamiliar with an area prefer to be given distance information. In addition to proximal (i.e., next turn) route-following information, notice of upcoming obstacles or traffic congestion also would be of great potential benefit. Such information could conceivably make the composite task of driving safer, given that it can be displayed without requiring substantial driver resources (Dingus & Hulse, 1993).

The use of traffic information for route changes was studied by Ayland and Bright (1991). Their most pertinent finding was that drivers desired reasons for suggested route changes. If an in-vehicle system tells a driver to deviate from a normal path or take an unfamiliar turn, the driver desires information about the reason for the change such as "exit left, accident ahead." The same suggestions were made in the report by Bonsall and Joint (1991) with regard to reasoning for route changes. This study also pointed out the need for accurate information. It was found that drivers' perceptions of the "best" route will be followed rather than the system suggestions, especially if the user has experienced a high rate of inaccurate information.

For ATIS design, sufficient time must be allowed for the driver to respond to any delivered information, auditory or visual. The driver needs time to see and hear the information, decide whether it is relevant, and act on it. It is recommended that 95% to 99% of the driving population should have ample time to respond to in-vehicle information under most driving circumstances. The time required by the driver to process information and respond depends on a number of factors, including the task and the type of display format selected. This topic is covered in more detail in chapter 13 of this volume.

In-Vehicle Haptic Information

Haptic information, although important, has historically been used rarely as the primary channel to transmit information while driving, but it is gaining more and more acceptance, particularly in Europe. Haptic feedback constitutes information gained via the touch senses, including both tactile and proprioceptive information. Considering the vehicle, the most common use of haptic information thus far is in the design of manual controls to provide feedback. For example, Godthelp (1991) used an active gas pedal to provide feedback to the driver about automobile following distances. When the following distance between two cars was too close, the force needed to maintain a given speed with the accelerator was increased. It was found that a gas pedal that offers more resistance when the following distance is too short can increase following distances. This demonstrates that haptic information is a viable method of providing selected information to drivers.

Furthermore, preliminary testing has shown that haptic warning information can have a higher degree of driver acceptance than auditory warning information. Anecdotal observations indicate that drivers may be embarrassed by the activation of an auditory alert when passengers are in the car

(Williams, 1995). Williams indicates that haptic displays are gaining popularity among manufacturers. Tactile sensitivity in the hand is high, and increases from the palm to the fingertips (Deatherage, 1972). However, tactile sensitivity is degraded by low temperatures, a situation that occurs when drivers live in areas where severe winter weather is frequent. The most common type of stimuli for tactile displays has been mechanical vibration or electrical stimulation. Information can be transmitted through mechanical vibrators based on vibrators' location, frequency, intensity, or duration. Electric impulses can be encoded in terms of electrode location, pulse rate, duration, intensity, or phase. Displays can be designed as a function of size, shape, and texture. For shape design, controls that vary by ½ in. in diameter and by ⅜ in. in thickness can be identified very accurately by touch.

The tactile channel rarely is used as the primary channel to transmit information, but is used instead as a redundant form of information. As noted earlier, the most common use of the tactile channel is in the design of manual controls to provide feedback. For example, the "F" and "J" keys on a keyboard often have a raised surface to indicate the position of the index fingers on the home row of the keyboard. Many aircraft have a "stick shaker" tactile display connected to the control column. When the plane is in danger of stalling, the control column will vibrate to alert the pilot to take corrective measures (Kantowitz & Sorkin, 1983). Tactile displays also are used on highways. Reflective strips placed on the white line of the emergency lane or on the yellow dividing lines not only enhance the road contours, but make the car vibrate if one crosses over them.

According to this brief review of haptic displays, it only can be stated that an effort should be made to encode manual controls with tactile information. This will enhance feedback and enable drivers to manipulate controls without taking their eyes off the road. Designers of in-vehicle systems should seriously consider the use of tactile feedback displays. Finally, there is a great need to perform research concerning the use of tactile displays to further define their value for ATIS. In summary, the following guidelines relating to sensory modality can be gleaned from the above discussion:

- The presentation of auditory-only information will provide a good allocation of information-processing resources for individuals' use while driving. However, for many ATIS applications, visually displayed information will be necessary.
- Intelligibility of auditory speech information is of concern in the automotive environment. Auditory messages should be simple, with volume adjustments and a repeat capability. Care in message selection is suggested because there is some evidence that voice commands may be followed in lieu of conflicting warning or regulatory information.

- Auditory information should be used to supplement visual information in alerting the driver of a change in visual display status or to provide speech-based instructions. Speech instructions have been shown to be a beneficial supplement for both simple and complex displays.
- Visual information should be as simple as possible while conveying the necessary information. Complex displays, such as moving maps, should be avoided if possible, or used in conjunction with auditory displays to alleviate demand on attention.
- Tactile information displays are potentially useful for ATIS designs either as a means to provide redundant coding of information or as a means to gain driver attention or to help avert visual modality overload.

DISPLAY FORMAT FOR ROUTE SELECTION AND NAVIGATION

Display formats (i.e., the general way that information is presented) can greatly affect a system's safety and usability. When directions are communicated from one person to another, the message generally uses one of two presentation formats, either a map or a verbal list of instructions. A person either uses a map that shows how to get to a destination, or one person informs the other person with a verbal list of instructions. What must be determined is which option is better for an in-vehicle navigation information system. The answer appears to be task specific. Bartram (1980) tested subjects' ability to plan a bus route using either a list or a map. The subjects who used a map made their decisions more quickly than those who used a list. In another study, Wetherell (1979) found that subjects who studied a map of a driving route made more errors when actually driving the route than did subjects who studied a textual list of turns. Wetherell concluded that two factors could have caused these findings: (a) the spatial processing demands of driving, seeing, and orienting interfere with maintaining a mental map in working memory, and (b) subjects had a harder time maintaining a mental model of a map learned in a North-up orientation when they approached an intersection from any direction but North. In a study conducted by Streeter et al. (1985), subjects who used a route list (a series of verbal directions) to follow a route through neighborhoods drove the route faster and more accurately than subjects who used a customized map with the route highlighted. In terms of attentional resources required by a visual display, a well-designed turn-by-turn format will require less attentional resources than will a full-route format. A turn-by-turn format displays to the driver specific information for each individual turn, and it requires very little information—only the direction of the turn, the distance to the turn, and the street name on which to turn. This information is easy to display in a legible format that

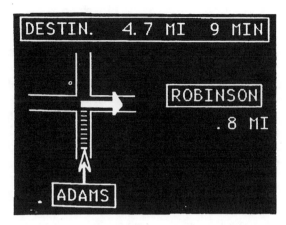

FIG. 12.1. TravTek turn-by-turn guidance map display.

imposes a low attention demand. An example of a TravTek turn-by-turn format is shown in Fig. 12.1. A route map shows the driver the entire route and includes information that may not be pertinent to the specific navigation task. Figure 12.2 shows an example of a TravTek route map display.

McGranaghan, Mark, and Gould (1987) characterized route-following as a series of view–action pairs. For example, information is displayed for an upcoming event (e.g., a turn) and the event is executed; the information for the next event is displayed and the event is executed, and so forth. McGranaghan et al. believed that for route following, only the information for the next view–action pair should be displayed. In their view, any additional information is extraneous and potentially disruptive to the route-following task. However, there are advantages, such as route preview, to displaying an entire route. When the driving task requires relatively little

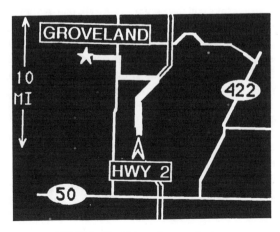

FIG. 12.2. TravTek route map display.

attention, the driver can plan upcoming maneuvers. For complex routes, preplanning could alleviate much of the need for in-transit preview. Drivers may prefer to recall information and review it at their own pace. A second advantage to providing route information arises during maneuvers that happen one right after another, as when two (or more) quick turns are required. The information for the second turn may come up too soon for the driver to comfortably execute the second maneuver. If the route map is displayed, planning for such an event can be done in advance. When selecting a turn-by-turn or route-map visual display format, designers must make sure that the information is displayed in a usable and safe manner. If they choose a turn-by-turn configuration, designers must consider close proximity and preview in general. If a route map is used, it is essential that the designer minimize the information presented so that drivers' attention resources are not overloaded. Even when full-route map information is minimized, it is not clear from the literature that driver resources will not become overloaded in circumstances that require high attention to driving.

Dingus et al. (1995) showed that providing turn-by-turn information in either a well-designed textual or graphic format resulted in effective navigation performance and attention demand level. Some of the studies described previously indicate that textual lists are easier to use than full maps when drivers are navigating to unknown destinations. Note, however, that graphic depictions can provide additional information that textual lists cannot, such as spatial relationships to other streets, parking, hospitals, and other landmarks. Therefore, the choice of a map or a list must depend on the desired task and required information. Depending on the requirements of the system under design, the inclusion of both display formats (displayed in different situations) may provide the most usable overall system.

A study performed by Mitchell (1993) investigated performance differences that resulted from using different display formats to present navigation information. Mitchell investigated a variety of visual display formats, including both textual and graphic versions of turn-by-turn and full-route information displays. The results indicate that a pictorial route map is the worst way to present navigation information for in-car moving-map navigation systems. An example of the pictorial route map used by Mitchell is shown in Fig. 12.3. While using this map configuration, subject performance was consistently poor across all dependent measures. This deficiency in performance could be due to a number of problems inherent to pictorial route maps. First, because most navigation systems will be constrained by size limitations of the display screen, designers are likely to employ some type of algorithm that will determine which information is displayed to the driver. Before displaying the pertinent navigation information, these algorithms consider such factors as proximity, item criticality, and the amount of space available on the display. Because route configurations are highly variant, the algorithm will be unable to display certain information items consis-

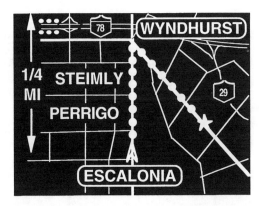

FIG. 12.3. Pictorial full-route map used by Mitchell (1993).

tently. For example, the name of the next cross street may be displayed to the driver in one instance because there is room on the screen, whereas in another situation, perhaps even after the next turn, the next cross street name may not be displayed because of a space limitation. In actual system use, this could become a common occurrence, and would likely confuse drivers by not consistently meeting their expectations of what information will be presented.

Mitchell's research results also indicate that verbal full-route lists are not an optimal means of presentation. However, subjects who used verbal lists tended to perform as well as those who used turn-by-turn maps and far better than those who used the pictorial route counterpart. This increased performance suggests that the inadequacies found with the use of pictorial route maps might result from the inconsistencies in including and positioning the information mentioned previously. The findings indicate that other configurations that do not display full routes might prove more effective in terms of attention and information retrieval. However, if the entire route is to be presented while individuals are driving, it may be better to display it in a verbal instead of a pictorial format (Mitchell, 1993). An example of a verbal turn-by-turn map is shown in Fig. 12.4. The full-route format would include multiple instructions without any landmark information.

It is important to note that in the Mitchell study, the test and evaluation involved only static navigation screens. Therefore, several other factors must be considered in extending this study's findings to real-world driving and navigation. For example, turn-by-turn configurations acquire certain disadvantages when they become dynamic. The dynamic system may not be able to present a series of close-proximity turns quickly enough for a driver to perceive, process, and react to the information about the successive turns. The presentation of entire-route information may be overwhelming in actual driving situations. Therefore, it should be avoided when the

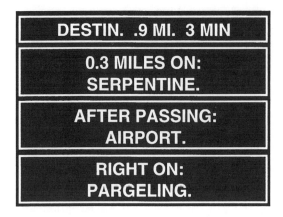

FIG. 12.4. Verbal turn-by-turn format with landmarks.

vehicle is in motion. However, because pictorial full-route maps can offer drivers a valuable preview of their route, full-route maps should be employed as a pretrip planning tool but be made available only when the vehicle is stationary.

Results of Mitchell's study also indicate that a pictorial turn-by-turn map displaying landmark information was most preferred overall. Two pictorial turn-by-turn configurations with lane information presented separately from the intersection (see Fig. 12.5) and integrated into the intersection icon (see Fig. 12.6) were preferred to the remaining verbal lists and graphic maps. The same preferences existed for the perceived ease-of-use results, although not all were found to be significantly different. The pictorial turn-by-turn configurations received higher ratings than either the pictorial route configuration or the verbal configurations (Mitchell, 1993).

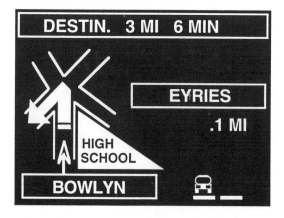

FIG. 12.5. Pictorial turn-by-turn map with landmarks and separated lane icons.

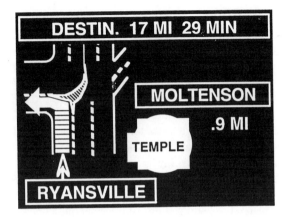

FIG. 12.6. Pictorial turn-by-turn map with integrated lane information.

Most of the results on navigation systems seem to support the recommendation outlined by Streeter (1985): Drivers should be presented with information that is most proximal to their location. However, previous research also suggests that studying paper maps of a given route either substitutes for a cognitive map or aids in developing one. This cognitive map provides an orienting schema that helps people to organize information about an unfamiliar area (Antin et al., 1990). Therefore, presenting drivers with full-route information might aid in the overall navigational task by helping them develop a cognitive map and survey knowledge of the area surrounding their route.

VISUAL DISPLAY CONSIDERATIONS

A major concern with the design of a visual display in the automobile is legibility of information, regardless of whether the information is textual or graphic. A delineation of all appropriate display parameter options (including resolution, luminance, contrast, color, and glare protection) is a complex topic beyond the scope of this chapter. Actual guidelines for determining the proper color, contrast, and luminance levels to be used in cathode ray tube (CRT) displays within vehicles can be found in Kimura, Sugiura, Shinkai, and Nagai (1988). In addition, a number of legibility standards are in existence for visual displays, including those developed for aircraft applications (Boff & Lincoln, 1988). Because the automobile is fraught with many of the same difficulties as those in the aircraft environment, many of these aircraft standards apply as well with respect to display legibility. Note, however, that the selection of a display in the automotive domain will be much more highly constrained by cost and perhaps have limitations well below the state of the art.

Legibility parameters have minimum acceptable standards below which the display is unusable. It is therefore critical to ensure that these minimum standards have been met in spite of the constraints for a given application. In addition, the problem of selecting display parameters is compounded because viewing distance is limited owing to the configuration constraints present for an automotive instrument panel. Adding to the difficulties found in the automobile are those introduced by the user population. Older drivers with poorer visual acuity or bifocal lenses must be carefully considered during the specification of the display parameters. To overcome this combination of limitations, the display parameters must be optimized within practical limits. For example, Carpenter et al. (1991) used a special high-legibility font in a color CRT application to ensure that drivers could glance at the display and grasp the required information as quickly as possible. Other design aspects that aid in legibility include presenting text information always in an upright orientation (even in map applications), maximizing luminance or color contrast under all circumstances, maximizing line widths (particularly on map displays) to increase luminance, and minimizing the amount of displayed information in general to reduce search time (Dingus & Hulse, 1993).

Additional legibility design considerations include contrast, brightness, and character size. As the degree of contrast increases, the detection and identification time becomes quicker for any target (up to a point). The brightness of the instrument panel has been shown to affect reading performance when character size is relatively small (1.5 and 2.5 mm) (Imbeau, Wierwille, Wolf, & Chun, 1989). Character size also plays an important role in response time. Imbeau et al. found that smaller character sizes yielded significant performance decrements for older drivers.

The Use of Color

Another basic visual display concern stems from the presence of color deficiency and color blindness in the population. Approximately 8% of the male population and 4% to 5% of the female population have some degree of color deficiency or color blindness. It is therefore important in consumer product applications to avoid reliance on color coding of critical information. Additional color issues include avoiding certain color combinations (Boff & Lincoln, 1988). For example, when blue lines are placed on a white background they appear to "swim." Boff and Lincoln recommended a sparing use of color coding because too many colors create more information density and an increase in search time.

Brown (1991) found that highlighting techniques using color resulted in quicker and more accurate recognition of targets in a visual display. Although instrument panel color has been shown to have no significant effect on reading and driving performance (Imbeau et al., 1989), Brockman (1991) found that color on a computer display screen can be distracting if used improperly. Brockman recommended several guidelines for averting confusion when color is used to code information. First, color codes should be used consistently. Colors from extreme ends of the color spectrum (red and blue) should not be put next to each other, because doing so makes it difficult for the reader's eye to perceive a straight line. Second, familiar color coding such as red for hot should be used. Third, color alone should not be used to discriminate between items. Brockman recommended designing applications first in black and white, then adding color to provide additional information.

Locating Visual Display

A major component of the driving task is scanning the environment and responding appropriately to unexpected events. Fortunately, humans are very sensitive to peripheral movement. An object moving in the periphery often gains attention instantly. In fact, some human-factors professionals believe that peripheral vision is as important as foveal vision for the task of driving (Dingus & Hulse, 1993).

Given the above considerations, the placement of an information display becomes critical. The information contained on even a well-designed display system will require a relatively large amount of visual attention. Therefore, if the display is placed far from the normal driving forward field of view, none of the driver's peripheral vision can be effectively utilized to detect unexpected movement ahead of the vehicle. Another disadvantage of placing a display far away from the forward field of view is increased switching time. Typical driver visual monitoring behavior involves switching back and forth between the roadway and the display in question. Dingus, Antin, Hulse, and Wierwille (1990) found that switching occurs every 1.0 to 1.5 s while drivers are performing most automotive tasks. The farther the display is away from the roadway, the longer the switching time, and the less time the driver can devote to the roadway or the display (Weintraub, Haines, & Randle, 1985).

The position of an in-vehicle visual display was also studied by Popp and Farber (1991). They found that a display positioned directly in front of the driver resulted in better driving performance, including lane tracking and obstacle detection, than one mounted in a peripheral location. However, performance on a symbolic navigation presentation format was hardly affected because of the change in position, and the results for peripheral location were still quite good. Tarrière, Hartemann, Sfez, Chaput, and Petit-Poilvert (1988), in their review of some ergonomic principles in the designing of in-vehicle environments, echoed the opinion that a CRT display an individual uses while driving should be near the center of the dashboard and not too far below horizontal. They suggest that the screen be mounted 15° below horizontal, but that for optimal driver comfort it should not exceed 30°.

According to the research cited, the display should be placed as close to the forward field of view as is practical. Desirable display locations are high on the instrument panel and near the area directly in front of the driver. There is, however, another automotive option that is currently just beginning to be explored: the head-up display (HUD). Briziarelli and Allen (1989) tested the effect of a HUD speedometer on speeding behavior. Although no significant difference was found between a conventional speedometer and the HUD speedometer, most subjects (70%) felt that the HUD speedometer was easier to use and more comfortable to read than a conventional speedometer. Subjects also reported being more aware of their speed when using the HUD speedometer. Campbell and Hershberger (1988) compare HUDs to conventional displays in a simulator under differing levels of workload. Under both low and high workload conditions, steering variability was less for drivers using a HUD display than for those using a conventional display. Also, steering variability was minimized when the HUD was low and centered in the driver's horizontal field of view. In another simulator study, Green and Williams (1992) found that drivers had faster recognition times between a navigation display and the "true environment" outside the vehicle when the display was a HUD rather than a dash-mounted CRT.

Some design criteria for HUDs in relation to the driving environment have been proposed. Wood and Thomas (1991) found that the windshield installation angle for the HUDs should be 69°. In their study, speed, fuel, temperature, turn signal status, and warning symbols (such as for brake, oil, seat belt) were chosen for testing. The authors also note that there is a trade-off between display field of view and head motion without a loss of display information. In addition, experience with the demonstration unit indicated that an instantaneous field of view of about 6° vertically by 10° horizontally provides a comfortably large field of view.

Greenland and Groves (1991) found that the best location for the HUD is centered in front of the driver and at a slight downward angle (within 15°). This keeps it close to the driver's normal line of sight. HUDs may help older drivers who are farsighted because the ideal HUD focus distance is considered to be a little more than 2 m from the driver. The authors recommended using a separate combiner element as part of the HUD package, rather than the windshield. A separate combiner element would help prevent glare interference and low brightness. This design also would allow the HUD to be a completely self-contained unit, which in turn would allow placement of the HUD in any convenient location in the vehicle. Weintraub, Haines, and Randle (1984) compared the HUD placed at different optical distances with a traditional 10° downward instrument panel. They found that drivers using the HUD display had better decision times, reaction times, and eye reaccommodation than drivers using the 10° downward panel. In their study, Sojourner and Antin (1990) compare speed monitoring, navigation, and salient cue detection under two conditions: a simulated HUD and a dashboard display. For salient cue detection, the subject response time was significantly less for the simulated HUD ($M = 0.57$ s) than for the dashboard ($M = 1.01$ s). However, the number of missed cues did not significantly differ between these conditions. Additional findings included a greater accuracy in detection of speed violation, but no differences in navigation errors or perceived effectiveness. Also, several subjects suggested that the HUD be moved out of their direct line of sight, even though they were not exposed to the HUD in any other location.

Given the above arguments, a HUD providing navigation information on the windshield could be a good choice because it is *in* the forward field of view. Besides the arguments described previously, another advantage of most HUD designs is that they are focused at (or near) infinity, thus reducing the time required for the driver's eyes to accommodate themselves between the display and the roadway. A number of concerns have been raised by Dingus and Hulse (1993) about the use of HUDs. The luminance may be a severely limiting factor in the automobile because of glare and stringent cost constraints. Certainly, a HUD too dim and hard to read could be much worse than an in-dash display. Issues regarding display information density and distraction also must be carefully addressed for HUDs and could result in their own set of problems. Also, an issue exists regarding the division of cognitive attention with HUDs. The fact that a driver is looking forward does not mean that roadway–traffic information is being processed. The importance of this division of attention to driving task performance has yet to be determined.

Human Factors Design Principles Applicable to Display Location

Some general human factors guidelines can be derived from the preceding discussion:

- For optimal driving performance, the in-vehicle displays should be located as close to the center of the front windshield field of view as is practical. Note that the center of the windshield field of view, or central viewing axis, is defined from the driver's eye point and is the center of the lane of travel near, but just below, the horizon.
- Good display location does not alleviate requirements for minimization of visual display complexity. That is, displays must be simple, and those that cannot be made simple should be accessible only while the vehicle is not in motion.

- A display should not be placed too far below the horizontal. The optimal position for mounting the CRT screen is 15° below the horizontal and should not exceed 30°.

- The horizontal comfortable viewing range is between 15° and 30° left or right of the central viewing axis.

For HUDs

- The HUDs should be located in the center of driver's horizontal field of view.

- The holographic combiner can be located in one of two places. One is on the periphery of the driver's field of view. The other is in the lower middle part of the windshield between the driver and the passenger.

- An instantaneous field of view with dimensions of about 6° vertically by 10° horizontally provides a comfortably large field of view.

- An ideal HUD focus range is considered to be a little more than 2 m from the driver.

- Because truck windshields are nearly vertical, the HUD cannot be placed directly on them. Therefore, the combiner should be a separate part of the HUD package. This makes the HUD a completely self-contained unit that can be located in any usable and convenient position in the vehicle.

RESEARCH ON MANUAL CONTROL OF DISPLAYS

Many control-related technological advancements are available for use with computers, and therefore are available for potential use as part of ATIS. The trade-off between hard buttons and soft CRT touchscreen push buttons has become a concern of the ITS human factors community (Dingus & Hulse, 1993). With the use of CRTs and flat-panel displays in the automobile, there has been a strong temptation to use touchscreen overlays for control activation. Although this can be a very good method of control in the automobile for predrive or zero-speed cases, research has shown that this is not true for in-transit circumstances. Zwahlen, Adams, and DeBald (1987) looked at safety aspects of CRT touch-panel controls in automobiles as a function of lateral displacement from the centerline. This study found an unacceptable increase in lateral lane deviation with the use of touch-panel controls. This study found the touchscreen control panels were visually demanding, as demonstrated by the relatively high probabilities of lane deviations. Zwahlen, Adams, and DeBald (1987) suggested that use of touch-panel controls in automobiles should be reconsidered and delayed until more research with regard to driver information acquisition, information processing, eye-hand-finger coordination, touch accuracy, control actions, and safety aspects has been conducted and the designs and applications have been improved to allow individuals to operate them safely while driving.

Monty (1984) found that the use of touchscreen keys while driving required greater visual glance time and resulted in greater driving and system task errors than conventional hard buttons. The reasons for this performance decrement are twofold: (a) the controls are nondedicated (i.e., they change depending on the screen), and (b) soft keys do not provide tactile feedback. For a hard button, the driver must (depending on the control and its proximity) glance briefly at the control and then find it using tactile information to accomplish location fine-tuning. For the soft keys, the driver must glance once to determine the location, and glance again to perform the location fine-tuning (Dingus & Hulse, 1993).

Clearly, one way to minimize control use while in transit is to limit control access severely. Therefore, as with the display information previously discussed, it is important to assess the necessity of every control in terms of both in-transit requirements and frequency of in-transit use to minimize driver control access. Controls that are not absolutely necessary for the in-transit environment can then be allocated to predrive or zero-speed circumstances (Dingus & Hulse, 1993).

The importance of control location has been proved in automotive research. As a control is located farther from the driver, the greater resources are needed to activate the control. This has been demonstrated by Bhise, Forbes, and Farber (1986) and Mourant, Herman, and Moussa-Hamouda (1980), who found that the probability of looking at a control increased with increased distance. Therefore, controls present on the steering wheel, or otherwise in close proximity to the driver, are easier to use. Complexity of the control activation and the potential for interference with steering also have been shown to be important issues. Monty (1984) has shown that continuous controls or controls requiring multiple activations are significantly more difficult to operate. Therefore, limiting controls to single, discrete activations will provide fewer demands on processing resources.

WORKLOAD

Workload, the extent of the resources required for a task, is a complex, multivariate construct that is an important consideration for ATIS. As stated by Kantowitz (1992), the practical benefit of measuring driver workload is that it acts as a means for assessing safety. The importance of workload is clear when one considers that tasks requiring excessive attentional resources will overload, resulting in unsafe behavior. To investigate workload in driving performance, Noy (1989) used the secondary-task method of workload measurement to determine what effects added tasks and and task complexity had on performance. This method requires that the driver perform two tasks concurrently: a primary task and a secondary task. Performance on the secondary task is an indicator of spare capacity. Thus as the primary task requires more resources, performance on the secondary task should decrease. Noy's primary task was driving a car on a winding road. The secondary task consisted of a perception task in which the driver had to detect a short, vertical line among vertical distractor lines on a CRT. In addition, a memory task based on the Sternberg paradigm was used. Noy's study showed that as visual tasks in automobiles increased, headway (the distance to the vehicle in front of the driver's vehicle) and speed control degraded while lane deviation increased. Obviously, neither of these elements can be compromised because the safety costs are too great. Therefore, Noy recommended that before allowing systems to be produced and used by the general public, workload testing must be conducted on all components of the system.

Because safety cannot be proactively and directly measured in driving (i.e., without installing a system and then measuring accidents), human-factors professionals must rely on indirect measures such as workload assessment. Kantowitz (1992) discussed the application of workload techniques traditionally used for aircraft applications to heavy vehicle driving. A summary of existing workload research for driving can be found in a report by Smiley (1989). According to Smiley, systems could be designed so that they automatically avoid overload. For example, if the cellular phone is in use, then the map details could be reduced on a navigation display. Workload could also be reduced by designing into the system smart cards that can determine the user's characteristics in terms of reaction time, age, experience, and so on. In this way a support system could be tuned to the particular needs of each driver. Information could also be prioritized within the system to match the environment. For example, map information could be reduced when the driver is actually driving through an intersection. This would allow the driver to devote full attention to the road at the appropriate time and not to the display.

Predrive Tasks Versus In-Transit Tasks

Most of the ATIS under development (or planned for the future) will be sufficiently demanding to warrant the designation of driver tasks as "predrive" or "in-transit" operations. That is, the functionality of the system available to the driver while the vehicle is in motion should, in many cases, be less than the full system available while the vehicle is not moving. Furthermore, the delineation between predrive and in-transit tasks is necessary because of the attention and information-processing constraints present in the driving environment. For example, Dingus and Hulse (1993) maintained that the only functions required for navigation while the vehicle is in motion are those associated with point-by-point decisions while a driver is traveling from the current location to a destination. Proper selection and design of in-transit functions can allow successful navigation to destinations without substantial driving task interference (Dingus & Hulse, 1993).

A primary requirement of complex in-vehicle systems is that they do not interfere substantially with the driving task. One way to help reduce this interference is by limiting the functions available to drivers while they are driving. Unfortunately, some designers are of the opinion that drivers want or need full ATIS functionality while driving or that drivers will always exercise good judgment about the use of such functionality. Although drivers often will avoid the use of complex ATIS functions in inappropriate circumstances, this is not true in every case (Dingus et al., 1995). As discussed by Dingus et al. (1995), the TravTek operational field test showed that driver acceptance of limited functionality in a moving vehicle was quite good. It is recommended that system functions be prioritized on their necessity and desirability while vehicles are in motion. Based on this scheme, predrive functions would consist of complex planning such as the route to take, the temporal aspects of the trip, the mode to use, and the resources required for the trip. In-transit functions would comprise a large subset of tasks that are necessary for efficient system usage while the vehicle is in motion (Lunenfeld, 1990). These tasks would include selection of an alternative route, secondary stops to make, driving speed, and so forth.

The functionality of the in-transit mode of in-vehicle systems should be limited to those tasks (a) that do not significantly interfere with the driving task, (b) that have benefits outweighing the cost of including the function (i.e., driver resources required for using the function), and (c) that will be used relatively often. Some functions that meet these criteria include providing tense information on navigation (distance to next turn street, direction of next turn, and name of next turn street), proximal traffic congestion, or obstacles (Dingus & Hulse, 1993). An option available to increase in-transit functionality without compromising driving safety is the allocation of functions to a "zero-speed" category (Carpenter et al., 1991). Zero speed describes a stopped vehicle that is still in Drive. The navigation system could allow a certain subset of functions (such as orientation information) to be accessed under these circumstances without concern for overload. However, once the vehicle started to move again, the display and control configuration would return to its in-transit state.

Another argument for minimizing in-transit information is the problem of "out-of-the-loop" loss of familiarity (Dingus & Hulse, 1993). Presently, the driver is required to obtain most information from the outside driving environment (such as street signs or stop lights). As more information is presented within the car in a readily accessible manner, the driver is less likely to obtain the same information from the driving environment. Thus, any problem, deficiency, or inconsistency that requires the driver to shift attention to the driving environment will potentially result in a delay and increased effort because the driver will have become accustomed to having the information provided within the vehicle. Thus, there is a trade-off: the more powerful and informative the system, the more the driver will rely on it to provide information rather than search the driving environment for it (Dingus & Hulse, 1993).

OLDER DRIVERS

Many abilities, including cognitive and visual abilities, decrease with age. In fact, one of the most prevalent issues with regard to user demographics is that of the aging driver. Parviainen, Atkinson, and Young (1991) took an extensive look at both the aging population and the handicapped with regard to in-vehicle systems development. Parviainen et al. (1991) stated that the number of aging drivers will double by the year 2030, and that systems must be designed to accommodate these special populations. According to Franzen and Ilhage (1990) the population of drivers over 65 years of age will soon make up one out of every seven drivers on the road. However, there continue to be problems associated with system design for an aging driver population, including age discrimination and lack of highway traffic engineering. Discrimination is present because of licensure. Waller (1991) stated that the basic information necessary for evaluating driving ability has not been developed by the research community. When highways were first designed, the engineers did not take the older driver into account. Highway design is usually based on measurements derived from young, male drivers (Waller, 1991).

Research has shown that older drivers spend significantly more time looking at navigation displays than do younger drivers. Several experiments have been conducted that compared the visual glance frequencies of both elderly and young drivers directed toward a CRT screen displaying navigation information (Dingus et al., 1989; Pauzie, Marin-Lamellet, & Trauchessec, 1991). In summary, it was found that younger drivers spent 3.5% of their driving time looking at the display, whereas elderly drivers spent 6.3%. Consequently, when navigation systems are involved, the older group directs less time toward the roadway. This greater amount of time required to look at displays dictates that special consideration must be given to this segment of the population, and that minimization of glance time in design of a navigation information display is critically important.

The perception of risk by older drivers is another important consideration for ATIS. Winter (1988) suggested that more older drivers are "running scared," frightened away from traffic situations they can probably handle along with those they cannot. Psychologically, some of them experience fear and anxiety about their vulnerability in a fast, complex traffic world, especially in relation to citations, insurance, and licensing examinations. Winter suggested that older drivers may develop compensatory attitudes and behaviors, some that are positive and contribute to safety and others that are negative and promote unsafe practices. On the positive side, older drivers become more responsible and law-abiding. However, older drivers may deny that their skills are decreasing and continue to drive under conditions highly unsafe for them. Winter reasoned that a prime factor in the immoderate attitudes of the elderly toward driving is the fear of an accident or a violation that would lead to re-examination for licensure and end in a possible loss of both the license and of the independence it affords. Another threat is the cancellation of insurance or the rise in premiums that would make driving too costly.

Many traffic engineering controls designed for older drivers have been suggested in recent research. Lerner and Ratté (1991) suggested that older drivers have a need for advanced signing of upcoming exits. According to Bishu, Foster, and McCoy (1991), older drivers favor larger signs, lower speed limits, more stop signs, and more traffic signals. Garber and Srinivasan (1991) suggested that an increase in amber-colored stoplight time or the protected phase for left-turn lanes will help reduce the large percentage of elderly drivers involved in left-turn lane accidents.

Research has indicated the existence of many age-related performance degradations. Walker et al. (1991) found that older drivers (55 years and older) drove slower, had larger variability in lateral placement, had longer reaction times to instruments, were more likely to be in another lane after a turn, and were more likely to make navigational errors compared to middle aged drivers (35 to 40 years of age) and younger drivers (20 to 25 years of age). Women appear to be at risk from age-related changes in cognitive functioning, particularly those with less driving experience and those over 75 years of age. Men appear to be at risk for both accidents and citations as a result of sensory and psychomotor function degradation. However, there are no significant differences in the rates at which males and females reported either accidents or citations (Bishu et al., 1991; Laux, 1991). For drivers over age 74, slowing of reaction time has a strong association with overall driving performance and with specific driving measures, especially those related to vehicle control (Ranney, 1989). However, Olson and Sivak (1986) found that older drivers, ages 50 to 84, had approximately the same perception–response time to unexpected roadway hazards as younger drivers, ages 18 to 40. Further research in this area has been reported by Chang (1991), Greatorex (1991), Ranney and Pulling (1990), Reynolds (1991), Stelmach and Nahom (1992), and Vercruyssen, Carlton, and Diggles-Buckles (1989).

It is generally acknowledged that vision plays a vital role in safe and proficient driving (Kosnik, Sekuler, & Kline, 1990), and the aging process does bring about a variety of changes in drivers' visual functions and their cognition. Although the process happens gradually, it may affect the interaction of older drivers with the vehicle environment. Older drivers report problems with visual processing speed, visual search, light sensitivity, and near vision (Kosnik et al., 1990). Research has shown that older drivers' visual performance improves through the use of specific display characteristics. Babbitt-Kline, Ghali, and Kline (1990) reported that the use of icons improves user visibility in both distance and lighting conditions (daylight vs. dusk). Hayes, Kurokawa, and Wierwille (1989) reported that many performance decrements in the viewing of visual displays can be countermeasured by increasing the character size of textual labels. Yanik (1989) observed that for color displays, drivers had better visual responses to yellows, oranges, yellow-greens, and whites on contrasting backgrounds. Yanik also found that analog displays (with moving pointers) were preferred over digital or numerical displays. Other studies involving the visual abilities of older drivers are reported by Mortimer (1989), Ranney and Simmons (1992), and Staplin and Lyles (1991).

Obviously, designers of transportation systems will have to consider the needs of the older driver. Mast (1991) stressed a greater need for research and development in transportation systems for older drivers in the areas of traffic control devices, changeable message signs, symbol signing, hazard markers, night driving, sign visibility, intersection design, traffic maneuvers, left turns against traffic, and merging–weaving maneuvers. Lerner and Ratté (1991) reported that focus groups identified the following needs and generated ideas and countermeasures for the safer use of freeways by older drivers:

- Lane restrictions, time restrictions, separate truck roadways, and other methods to reduce interaction with heavy trucks
- Greater police enforcement, new enforcement technologies such as photo radar, new traffic-control technologies, and other methods to reduce speed variability in the traffic stream
- Better graphics, greater use of sign panels listing several upcoming exits, and other methods to improve advance signing so that it better meets the visual and information needs of the elderly
- Wide, high-quality shoulders, increased patrol, brightly lit roadside emergency phones, promotion of citizens band radio use, better night lighting, more frequent path confirmation, and other methods to overcome the frequently expressed concerns about personal security
- More legible maps, map-use training in older-driver education courses, in-vehicle guidance systems, and other pretrip planning aids designed for use by older drivers

- Appreciation for the safety benefits, on-road refresher training for those who have not used high-speed roads recently, training in recovery from navigational errors, and other older-driver education specific to freeway use
- Elimination of short-merge areas along with other methods to improve the interchange geometries that the focus groups identified as contributing to anxiety

DRIVER ACCEPTANCE AND BEHAVIOR

In developing components of ITS it is important to consider the attitudes of the system users. Is the system acceptable, usable, and affordable? A good review of the issues associated with ITS acceptance is presented in a literature review by Sheridan (1991). A survey by Marans and Yoakam (1991) found that most commuters believed ITS was a plausible solution to traffic congestion. The highest approval of ITS (48%) came from commuters whose work commute was from suburban to suburban localities. The survey also reported that 86% of the commuters drove their own cars to work.

A University of Michigan Transportation Research Institute (UMTRI) focus group study (Green & Brand, 1992) elicited attitudes on in-vehicle electronics. Areas discussed were general driver attitudes toward and use of sophisticated in-vehicle display systems, the ways in which people learn to use these systems, automotive gauges and warning systems, entertainment systems, CRT touch screens, trip computers, head-up displays, cellular phones, navigation systems, and road hazard monitoring systems. Drivers regarded the navigation systems with caution. Most indicated that they were good for someone else to use in specific situations. It was noted that men prefer maps and women prefer directions. Left–right rather than compass north–south turning directions were preferred. In a survey by Barfield, Haselkorn, Spyridakis, and Conquest (1989), commuters rated commercial radio as the most useful and preferred medium from which to receive traffic information both before and while driving, as compared to variable message signs, highway advisory radio, commercial TV, and telephone hotline systems. Departure time and route choice were the most flexible commuter decisions. Few commuters indicated that they would be influenced to change their transportation modes.

Driver acceptance of technology was assessed by McGehee, Dingus, and Horowitz (1992) while they were studying a front-to-rear-end collision warning system. They reported that drivers often follow at distances closer than the brake-reaction time permits for accident avoidance. This close driving behavior may result from the rarity of consequences from previous experiences. A front-to-rear-end collision warning system (whether it is visual, aural, or a combination of both) has the potential to provide added driver safety and situation awareness and was generally accepted as a worthwhile device by the subjects tested. Preference for text and voice warning message systems was demonstrated in a study on safety advisory and warning system design by Erlichman (1992). The results of these studies have applicability to In-vehicle Signal and Warning Systems (IVSAWS).

Route Selection and Route Diversion Driver Behavior

A major component of ITS will include navigation aids. These devices can provide the user with current traffic conditions, alternate routes, and guidance over the alternate routes. The effectiveness of these systems in alleviating traffic congestion has been reported in studies by Dingus and Hulse (1993), Halati and Boyce (1991), Hamerslag and van Berkum (1991), and Knapp, Peters, and Gordon (1973).

It has been shown that drivers resist diversion from their present route to avoid congestion (Dudek, 1979). Dudek showed that 50% of drivers were willing to avoid a 20-min delay, but only 8% were willing to avoid a 5-min delay. Thus, congestion would have to be moderately severe before people would take the effort to divert. This unwillingness to detour may depend in large part on the amount and accuracy of the detour knowledge held by the driver. Certainly one of the keys for persuading drivers to use an alternate route is appropriate and timely information. For example, if the driver knows there will be some time savings on an alternate route but does not know how much, the driver may be more reluctant to change routes. In addition, drivers often are not willing to take the shortest or fastest route if it is less pleasant. Cross and McGrath (1977) studied a variety of drivers to determine what they felt was important in selecting a route. The contributing factors used in selecting routes and the percentage of drivers for which a contributing factor was applied were as follows: efficiency (fastest route 54%, shortest route 53%), problem avoidance (safest route 43%, more familiar with route 36%), more miles of multiple lanes 31%, roads in better condition 30%, less chance of getting lost 24%, less traffic 18%, and pleasure and personal convenience (most scenic route 25%). Factors such as road quality, number of junctions and average speed also will be important in route selection.

Allen, Stein, Rosenthal, Ziedman, Torres, and Halati (1991) looked at the route diversion and alternate route selection of drivers using In-vehicle Routing and Navigation Systems (IRANS). Overall, the study showed that navigation system characteristics can have a significant effect on driver diversion behavior, with better systems allowing more anticipation of traffic congestion. Older drivers (55 years of age or older) were more reluctant to choose alternate routes. Navigation systems equipped with congestion monitors effectively changed drivers' behavior, but not all drivers exited to the alternate route suggested by the computer.

Research has shown that a driver faces considerable difficulties in achieving optimum routes from origin to destination (King & Rathi, 1987). These travel inefficiencies have been shown to generate a considerable amount of excessive travel. King and Rathi asked subjects to plan three relatively long trips in unfamiliar areas by using only a road atlas. The routes selected by the subjects were compared with the routes recommended by the American Automobile Association (AAA) for both distance and approximate driving time. Analyses of the data indicated that the excess distance of the routes selected by the subjects increased trip length by an average of 12.1%.

Shirazi, Anderson, and Stesney (1988) looked at commuters' attitudes toward traffic information systems and route diversion. Overall, commuters wanted timely and accurate information, more frequent reporting, and better uses of electronic freeway message signs. Most were in favor of continuous radio traffic reporting (68%) and traffic information phone numbers (53%). In general males were more likely to change routes than females, whereas females (38%) were less likely to know alternative routes than males (62%). Drivers commuting for short periods (less than 45 min) were more likely to change routes to work than those with longer travel times. Nearly 70% of commuters said they would leave the freeway if more accurate information regarding their commute were available, and if they knew that surface streets offered a shorter route to work. The most frequently cited factor for route change was radio traffic reports (30%), followed by personal experience (20%).

Khattak, Schofer, and Koppelman (1991) found also that most commuters indicated that they had changed their trip decisions based on radio traffic reports. Khattak et al. (1991) found that commuters used traffic information more when they were en route than when they were planning their trips. Drivers reported a preference for information on near-term prediction of traffic conditions for congested and unreliable routes as opposed to current traffic conditions. Current conditions were seen as unstable with the potential for rapid change. Overall, radio traffic reports were attributed to reducing en route anxiety and frustration of drivers even if drivers did not modify their trip decisions. Commuters who are more likely to change their departure time in comparison to their route selection wanted long-term prediction of traffic conditions. In related studies, Barfield et al. (1989; see also Haselkorn, Barfield, Spyridakis, & Conquest, 1989; Haselkorn & Barfield, 1990; Haselkorn, Spyridakis, Conquest, & Barfield, 1990; Spyridakis, Barfield, Conquest, Haselkorn, & Isakson, 1991) used cluster analysis to show that four commuter subgroups exist with respect to their willingness to respond to the delivery of real-time traffic information: (a) route changers, those willing to change routes on or before entering Interstate 5 (20.6%); (b) nonchangers, those unwilling to change time, route, or mode (23.4%); (c) time and route changers (40.1%); and (d) pretrip changers, those willing to change time, mode, or route before leaving the house (15.9%).

DINGUS, HULSE, BARFIELD

Bonsall and Parry (1991) tested drivers' compliance with route guidance advice by asking subjects to make several "journeys" from specified origins to specified destinations using an IBM computer-based simulator, Interactive Guidance On Routes (IGOR). The quality of the IGOR advice varied. Overall, about 70% of the advice was accepted. The acceptance of advice declined as the quality of advice decreased. Participants' perception of the usefulness of the advice was strongly conditioned by the physical layout of the network. Generally, if previous advice had been very good, then even poor advice was likely to be accepted. If previous advice, though, had been very poor, then a poor item of advice was almost certain to be rejected. Acceptance of advice generally decreased as familiarity with the network increased. Visible presence of traffic congestion or a road sign apparently confirming advice increased acceptance of advice. Women were found to be significantly more likely to accept advice, particularly nonoptimal advice, than were men. Subjects who normally had high commuting distances were less likely to accept advice.

Finally, to analyze the effectiveness of route guidance systems, Abu-Eisheh and Mannering (1987) developed a route and departure time-choice modeling system that treated departure time as a continuous variable. The models were estimated with a sample of work trip commuters, and the resulting coefficient estimates were of plausible sign and reasonable significance. This study also found that males have a tendency to drive faster than females, that safety belt users tend to drive faster, and that younger commuters drive faster than older commuters.

SUMMARY AND CONCLUSIONS

This chapter discussed many of the important issues for the human-factors design of ATIS. It is not, however, inclusive of all issues of interest, nor should it be utilized as the only source of ATIS design information. There are two reasons for this caveat: Adequate treatment of this complex topic is beyond the scope of this chapter, and there are a number of research issues that must be addressed in order to develop a comprehensive set of human-factors guidelines for ATIS. A preliminary and partial listing of some of these research needs appears in the following discussion.

The ATIS research to date has tended to be oriented toward describing systems, along with details of the research being conducted or that needs to be conducted. A number of research issues have been resolved for ATIS and need not be readdressed. It is apparent that as the development of hardware progresses, the next few years will see a marked growth in the literature available from both U.S. demonstration projects and foreign sources. It is anticipated that the data from initial U.S. operational tests and additional

European and Japanese projects will serve to fill some of the largest gaps in the current human-factors knowledge base. TravTek, as previously described, has a number of IRANS and In-vehicle Motorist Services Information System (IMSIS) features and will provide data from many studies, with well over 1,000,000 miles of data collected. The UMTRI project also will provide data from many subjects, as well as preliminary models that may help in establishing guidelines. The ADVANCE and Guidestar projects promise data from even larger groups of users in the near future. To develop comprehensive and generalizable guidelines that will be usable and useful for years to come, models of the performance of drivers using ATIS systems must be developed. This model development research will likely include application or modification of existing models (such as those provided by Mannering, chapter 14, this volume; UMTRI; or model driver processing proposed by Sheridan, 1991, or Kantowitz, 1990), as well as creation of new models or model parameters associated with ATIS specific applications.

In general, ATIS research is lacking for IVSAWS and In-vehicle Signing and Information Systems (ISIS) applications. Although these ATIS systems probably will not have particularly complex user interfaces, they present unique human-factors and safety issues. Because IVSAWS will be an alarm type of display, issues of timing, modality false alarms, and potential operator reactions must be addressed. Many ISIS issues are similar to those of IRANS and IMSIS. However, presentation of regulatory information carries with it critical issues of message reliability, priority, understanding, and interpretation. In general, further research needs to be performed to investigate the information needs of automobile drivers. The information needed only by urban drivers as opposed to rural drivers, and the information needs of local drivers as opposed to those unfamiliar with the area must be determined. The amount of information available to the driver upon request and the amount normally provided to the driver must be determined. Research is also needed to determine how much information should be provided to the driver in the case of incidents and alternate route selections.

A key knowledge gap requiring both application of existing guidelines and the creation of new guidelines is driver capacity. The human-factors community is currently divided on the issue of what is safe and what is unsafe in the driving environment. The primary cause of this debate has centered around IRANS applications and will require additional research to resolve. Carefully planned and executed experiments that provide generalizable principles instead of system-specific "do's and don'ts" are needed. In addition, a careful understanding of potential safety benefits and costs of utilizing ATIS are needed for meaningful guideline development. It is easy to dismiss a display that provides complex information as requiring too many driver resources. However, until a comparison is made with current techniques for retrieving necessary information, and an assessment is performed to determine the benefit of having the information, a proper assessment cannot be made.

A final general area of necessary research involves driver acceptance of ATIS technology. Even a very safe and efficient system design will not achieve the goals of ITS if the needs and desires of the user are not met. Poor market penetration will result. Although a number of surveys have been conducted that describe desirable ATIS features, ongoing research will be necessary to establish the information and control requirements for system (and product) success. Once information requirements have been established, the process of establishing human factors design guidelines can progress.

REFERENCES

Abu-Eisheh, S., & Mannering, F. (1987). Discrete/continuous analysis of commuters' route and departure time choices. *Transportation Research Record, 1138*, 27–34.

Allen, R. W., Stein, A. C., Rosenthal, T. J., Ziedman, D., Torres J. F., & Halati, A. (1991). A human factors simulation investigation of driver route diversion and alternate route selection using in-vehicle navigation systems. In *Vehicle Navigation and Information Systems Conference Proceedings* (pp. 9–26). Warrendale, PA: Society of Automotive Engineers.

Antin, J. F., Dingus, T. A., Hulse, M. C., & Wierwille, W. W. (1990). An evaluation of the effectiveness and efficiency of an automobile moving-map navigational display. *International Journal of Man-Machine Studies, 33*, 581–594.

Aretz, A. J. (1991). The design of electronic map displays. *Human Factors, 33*(1), 85–101.

Ayland, N., & Bright, J. (1991). Real-time responses to in-vehicle intelligent vehicle-highway system technologies: A European evaluation. *Transportation Research Record, 1318*, 111–117. Washington, DC: National Research Council.

Babbitt-Kline, T. J., Ghali, L. M., & Kline, D. W. (1990). Visibility distance of highway signs among young, middle-aged, and older observers: Icons are better than text. *Human Factors, 32*(5), 609–619.

Barfield, W., Haselkorn, M., Spyridakis, J., & Conquest, L. (1989). Commuter behavior and decision making: Designing motorist information system. *Proceedings of the Human Factors Society 33rd Annual Meeting* (pp. 611–614). Santa Monica, CA: Human Factors Society.

Bartram, D. J. (1980). Comprehending spatial information: The relative efficiency of different methods of presenting information about bus routes. *Journal of Applied Psychology, 65*, 103–110.

Bhise, V. D., Forbes, L. M., & Farber, E. I. (1986, January). *Driver behavioral data and considerations in evaluating in-vehicle controls and displays*. Presented at the Transportation Research Board 65th Annual Meeting, Washington, DC.

Bishu, R. R., Foster B., & McCoy, P. T. (1991). Driving habits of the elderly: A survey. *Proceedings of the Human Factors Society 35th Annual Meeting* (pp. 1134–1138). Santa Monica, CA: Human Factors Society.

Boff, K. R., & Lincoln, J. E., (1988). Guidelines for alerting signals. *Engineering Data Compendium: Human Perception and Performance, 3*, 2388–2389.

Bonsall, P. W., & Joint, M. (1991). Driver compliance with route guidance advice: The evidence and its implications. *Vehicle Navigation and Information Systems Conference Proceedings* (pp. 47–59). Warrendale, PA: Society of Automotive Engineers.

Bonsall, P. W., & Parry, T. (1991). Using an interactive route-choice simulator to investigate drivers' compliance with route guidance advice. *Transportation Research Record, 1306*, 59–68.

Briziarelli, G., & Allen, R. W. (1989). The effect of a head-up speedometer on speeding behavior. *Perceptual and Motor Skills, 69,* 1171–1176.

Brockman, R. J. (1991). The unbearable distraction of color. *IEEE Transactions of Professional Communication, 34*(3), 153–159.

Brown, T. J., (1991). Visual display highlighting and information extraction. *Proceedings of the Human Factors Society 35th Annual Meeting* (pp. 1427–1431). Santa Monica, CA: Human Factors Society.

Campbell, J., & Hershberger, J. (1988). *Automobile head-up display simulation study: Effects of image location and display density on driving performance.* Hughes Aircraft Company, unpublished manuscript.

Carpenter, J. T., Fleischman, R. N., Dingus, T. A., Szczublewski, F. E., Krage, M. K., & Means, L. G. (1991). Human factors engineering the TravTek driver interface. *Vehicle Navigation and Information Systems Conference Proceedings* (pp. 749–756). Warrendale, PA: Society of Automotive Engineers.

Chang, J. (1991). Cross-sectional and longitudinal analyses of set in relation to age. *Proceedings of the Human Factors Society 35th Annual Meeting* (pp. 203–207). Santa Monica, CA: Human Factors Society.

Cross, K. D., & McGrath, J. J. (1977). *A study of trip planning and map use by American motorists* (Tech. Rep. No. FHWA-WV-77-10). Charleston, VA: West Virginia Department of Highways.

Davis, J. R., & Schmandt, C. M. (1989). The back seat driver: Real time spoken driving instructions. In *Vehicle Navigation and Information Systems Conference Proceedings* (No. CH2789, pp. 146–150). Warrendale, PA: Society of Automotive Engineers.

Deatherage, B. (1972). Auditory and other sensory forms of information presentation. In H. Van Cott & R. Kinkade (Eds.), *Human engineering guide to equipment design* (pp. 123–160). Washington, DC: Government Printing Office.

Dingus, T. A., Antin, J. F., Hulse, M. C., & Wierwille, W. W. (1989). Attentional demand requirements of an automobile moving-map navigation system. *Transportation Research, 23A*(4), 301–315.

Dingus, T. A., & Hulse, M. C. (1993). Some human factors design issues and recommendations for automobile navigation information systems. *Transportation Research, 1C*(2), 119–131.

Dingus, T. A., McGehee, D., Hulse, M., Jahns, S., Manakkal, N., Mollenhauer, M., & Fleischman, R. (1995). *Travtek Evaluation Task C₃-Camera Car Study* (Tech. Rep. FHWA-RD-94-076). Washington, DC: Offices of Research and Development, Department of Transportation.

Dudek, C. L. (1979). Human factors considerations for in-vehicle route guidance. *Transportation Research Record, 737,* 104–107.

Erlichman, J. (1992). A pilot study of the in-vehicle safety advisory and warning system (IVSAWS) driver-alert warning system design (DAWS). *Proceedings of the Human Factors Society 36th Annual Meeting* (pp. 480–484). Santa Monica, CA: Human Factors Society.

Franzen, S., & Ilhage, B. (1990). Active safety research on intelligent driver support systems. *Proceedings of the 12th International Technical Conference on Experimental Safety Vehicles (ESV)* (pp. 1–15).

French, R. L. (1990). In-vehicle navigation—status and safety impacts. *Technical Papers from ITE's 1990, 1989, and 1988 Conferences* (pp. 226–235). Institute of Transportation Engineers.

Garber, N., & Srinivasan, R. (1991). Characteristics of accidents involving elderly drivers at intersections. *Transportation Research Record, 1325,* 8–16.

Godthelp, H. (1991). Driving with GIDS: Behavioral interaction with the GIDS architecture. In Commission of the European Communities (Eds.), *Advanced telematics in road transport.* (pp. 351–370). Amsterdam: Elsevier.

Greatorex, G. L. (1991). Aging and speed of behavior: CNS arousal and reaction time distribution analyses. *Proceedings of the Human Factors Society 35th Annual Meeting* (pp. 193–197). Santa Monica, CA: Human Factors Society.

Green, P., & Brand, J. (1992). *Future in-car information systems: Input from focus groups.* SAE Technical Paper Series (SAE No. 920614, pp. 1–9). Warrendale, PA: Society of Automotive Engineers.

Green, P., & Williams, M., (1992). Perspective in orientation / navigation displays: A human factors test. In *Vehicle Navigation and Information Systems Conference Proceedings* (pp. 221–226). Warrendale, PA: Society of Automotive Engineers.

Greenland, A. R., & Groves, D. J. (1991). *Head up display concepts for commercial trucks.* In SAE Technical Paper Series (SAE No. 911681, pp. 1–5). Warrendale, PA: Society of Automotive Engineers.

Halati, A., & Boyce, D. E. (1991). Effectiveness of in-vehicle navigation systems in alleviating non-recurring congestion. In *Vehicle Navigation and Information Systems Conference Proceedings* (pp. 871–889). Warrendale, PA: Society of Automotive Engineers.

Hamerslag, R., & van Berkum, E. C. (1991). Effectiveness of information systems in networks with and without congestion. *Transportation Research Record, 1306,* 14–21.

Haselkorn, M., Barfield, W., Spyridakis, J., & Conquest, L. (1989). Understanding commuter behavior for the design of motorist information systems. *Proceedings of Transportation Research Board 69th Annual Meeting* (pp. 1–11). Warrendale, PA: Society of Automotive Engineers.

Haselkorn, M., Spryridakis, J., Conquest, L., & Barfield, W. (1990). Surveying commuter behavior as a basis for designing motorist information systems. *Transportation Research Record, 128,* 159–167.

Haselkorn, M. P., & Barfield, W. (1990). *Improving motorist information system: Toward a user-based motorist information system for the Puget Sound area* (Tech. Rep. No. WA-RD 187.1). Washington State Transportation Center.

Hayes, B. C., Kurokawa, K., & Wierwille, W. W. (1989). Age-related decrements in automobile instrument panel task performance. *Proceedings of the Human Factors Society 33rd Annual Meeting* (pp. 159–163). Santa Monica, CA: Human Factors Society.

Imbeau, D., Wierwille, W. W., Wolf, L. D., & Chun, G. A. (1989). Effects of instrument panel luminance and chromaticity on reading performance and preference in simulated driving. *Human Factors, 31*(2), 147–160.

Kantowitz, B. H. (1990). Can cognitive theory guide human factors measurement? *Proceedings of the Human Factors Society 34th Annual Meeting* (pp. 1258–1262). Santa Monica, CA: Human Factors Society.

Kantowitz, B. H. (1992). Heavy vehicle driver workload assessment: Lessons from aviation. *Proceedings of the Human Factors Society 35th Annual Meeting* (pp. 1113–1117). Santa Monica, CA: Human Factors Society.

Kantowitz, B. H., & Sorkin, R. D. (1983). Workspace design. *Human factors: Understanding people-system relationships* (pp. 454–493). New York: Wiley.

Khattak, A. J., Schofer, J. L., & Koppelman, F. S. (1991). Effect of traffic reports on commuters' route and departure time changes. In *Vehicle Navigation and Information Systems Conference Proceedings* (pp. 669–679). Warrendale, PA: Society of Automotive Engineers.

Kimura, K., Sugiura, S., Shinkai, H., & Nagai, Y. (1988). *Visibility requirements for automobile CRT displays—color, contrast, and luminance.* SAE Technical Paper Series (SAE No. 880218, pp. 25–31). Warrendale, PA: Society of Automotive Engineers.

King, G. F., & Rathi, A. K. (1987). A study of route selection from highway maps. *Transportation Research Record, 1111,* 134–137.

Knapp, B. G., Peters, J. I., & Gordon, D. A. (1973). *Human factor review of traffic control and diversion projects* (Tech. Rep. FHWA-RD-74-22). Washington, DC: Offices of Research and Development, Department of Transportation.

Kosnik, W. D., Sekuler, R., & Kline, D. W. (1990). Self-reported visual problems of older drivers. *Human Factors, 32*(5), 597–608.

Labiale, G. (1990). In-car road information: Comparison of auditory and visual presentation. *Proceedings of the Human Factors Society 34th Annual Meeting* (pp. 623–627). Santa Monica, CA: Human Factors Society.

Laux, L. F. (1991). A follow-up of the mature driver study: Another look at age and sex effects. *Proceedings of the Human Factors Society 35th Annual Meeting* (pp. 164–166). Santa Monica, CA: Human Factors Society.

Lerner, N. D., & Ratté, D. J. (1991). Problems in freeway use as seen by older drivers. *Transportation Research Record, 1325,* 3–7.

Lunenfeld, H. (1990). Human factors considerations of motorist navigation and information systems. *Proceedings of the First Annual Vehicle Navigation and Information Systems Conference* (pp. 35–42). Warrendale, MI: Society of Automotive Engineers.

Marans, R. W., & Yoakam, C. (1991). *Assessing the acceptability of IVHS: Some preliminary results.* In SAE Technical Paper Series (SAE No. 912811, pp. 657–668). Warrendale, PA: Society of Automotive Engineers.

Mast, T. (1991). Designing and operating safer highways for older drivers: Present and future research issues. *Proceedings of the Human Factors Society 35th Annual Meeting* (pp. 167–171). Santa Monica, CA: Human Factors Society.

McGehee, D. V., Dingus, T. A., & Horowitz, A. D. (1992). The potential value of a front-to-rear-end collision warning system based on factors of driver behavior, visual perception and brake reaction time. *Proceedings of the Human Factors Society 36th Annual Meeting* (pp. 1011–1013). Santa Monica, CA: Human Factors Society.

McGranaghan, M., Mark, D. M., & Gould, M. D. (1987). Automated provision of navigation assistance to drivers. *The American Cartographer, 14*(2), 121–138.

McKnight, J. A., & McKnight, S. A. (1992). *The effect of in-vehicle navigation information systems upon driver attention.* Landover, MD: National Public Services Research, 1–15.

Means, L. G., Carpenter, J. T., Szczublewski, F. E., Fleishman, R. N., Dingus, T. A., & Krage, M. K. (1992). *Design of the TravTek auditory interface* (Tech. Rep. GMR-7664). Warren, MI: General Motors Research and Environmental Staff.

Mitchell, M. (1993). *A comparison of automotive navigation system visual display formats.* Unpublished master's thesis. Moscow, ID: University of Idaho.

Monty, R. W. (1984). *Eye movements and driver performance with electronic navigation displays.* Unpublished master's thesis, Virginia Polytechnic Institute and State University, Blacksburg, VA.

Mortimer, R. G. (1989). Older drivers' visibility and comfort in night driving: Vehicle design factors. *Proceedings of the Human Factors Society 33rd Annual Meeting* (pp. 154–158). Santa Monica, CA: Human Factors Society.

Mourant, R. R., Herman, M., & Moussa-Hamouda, E. (1980). Direct looks and control location in automobiles. *Human Factors, 22*(4), 417–425.

Noy, Y. I. (1989). Intelligent route guidance: Will the new horse be as good as the old? In *Vehicle Navigation and Information Systems Conference Proceedings* (pp. 49–55). Warrendale, PA: Society of Automotive Engineers.

Olson, P. L., & Sivak, M. (1986). Perception-response time to unexpected roadway hazards. *Human Factors, 28*(1), 91–96.

Parkes, A. M., Ashby, M. C., & Fairclough, S. H. (1991). The effect of different in-vehicle route information displays on driver behavior. In *Vehicle Navigation and Information Systems Conference Proceedings* (pp. 61–70). Warrendale, PA: Society of Automotive Engineering.

Parviainen, J. A., Atkinson, W. A. G., & Young, M. L. (1991) *Application of micro-electronic technology to assist elderly and disabled travellers* (Tech. Rep. TP-10890E). Montreal, Quebec, Canada.: Transportation Development Centre.

Patterson, S., Farrer, J., & Sargent, R. (1988). Automotive head-up display. *Automotive Displays and Industrial Illumination, 958,* 114–123.

Pauzie A., Marin-Lamellet C., & Trauchessec, R. (1991). Analysis of aging drivers behaviors navigating with in-vehicle visual display systems. In *Vehicle Navigation and Information Systems Converence Proceedings* (pp. 61–67), Warrendale, PA: Society of Automotive Engineering.

Popp, M. M., & Farber, B. (1991). Advanced display technologies, route guidance systems and the position of displays in cars. In A. G. Gale (Ed.), *Vision in vehicles-III* (pp. 219–225). North-Holland: Elsevier Science Publishers.

Ranney, R. A., & Simmons, L. A. S. (1992). The effects of age and target location uncertainty on decision making in a simulated driving task. *Proceedings of the Human Factors 36th Annual Meeting* (pp. 166–170). Santa Monica, CA: Human Factors Society.

Ranney, T., & Pulling, N. (1990). Performance differences on driving and laboratory tasks between drivers of different ages. *Transportation Research Record, 1281,* 3–10.

Ranney, T. A. (1989). Relation of individual differences in information-processing ability to driver performance. *Proceedings of the Human Factors Society 33rd Annual Meeting* (pp. 965–969). Santa Monica, CA: Human Factors Society.

Reynolds, S. L. (1991). Longitudinal analysis of age changes in speed of behavior. *Proceedings of the Human Factors Society 35th Annual Meeting* (pp. 198–202). Santa Monica, CA: Human Factors Society.

Robinson, C. P., & Eberts, R. E. (1987). Comparison of speech and pictorial displays in a cockpit environment. *Human Factors, 29*(1), 31–44.

Rockwell, T. (1972). Skills, judgment, and information acquisition in driving. In T. Forbes (Ed.), *Human factors in highway traffic safety research* (pp. 133–164). New York: Wiley Interscience.

Sanders, M. S., & McCormick, E. J. (Ed.). (1987). *Human factors in engineering and design.* New York: McGraw-Hill.

Sheridan, T. B. (1991). *Human factors of driver-vehicle interaction in the IVHS environment* (Tech. Rep. DOT-HS-807-837). Washington, DC: National Highway Traffic Safety Administration.

Shirazi, E., Anderson, S., & Stesney, J. (1988). Commuters' attitudes toward traffic information systems and route diversion. *Transportation Research Record, 1168,* 9–15.

Smiley, A. (1989). Mental workload and information management. In *Vehicle Navigation and Information Systems Conference Proceedings* (pp. 435–438). Warrendale, PA: Society of Automotive Engineers.

Sojourner, R. J., & Antin, J. F. (1990). The effects of simulated head-up display speedometer on perceptual task performance. *Human Factors, 32*(3), 329–339.

Sorkin, R. D. (1987). Design of auditory and tactical displays. In G. Salvendy (Ed.), *Handbook of human factors* (pp. 549–574). New York: Wiley.

Spyridakis, J., Barfield, W., Conquest, L., Haselkorn, M., & Isakson, C. (1991). Surveying commuter behavior: Designing motorist information systems. *Transportation Research, 25A*(1), 17–30.

Staplin, L., & Lyles, R. W. (1991). Age differences in motion perception and specific traffic maneuver problems. *Transportation Research Record, 1325,* 23–33.

Stelmach, G. E., & Nahom, A. (1992). Cognitive-motor abilities of the elderly driver. *Human Factors, 34*(1), 53–65.

Stokes, A., Wickens, C., & Kyte, K. (1990). *Display technology: Human factors concepts.* Warrendale, PA: Society of Automotive Engineers.

Streeter, L. A. (1985). Interface considerations in the design of an electronic navigator. *Auto Carta.*

Streeter, L. A., Vitello, D., & Wonsiewicz, S. A. (1985). How to tell people where to go: Comparing navigational aids. *International Journal of Man-Machine Studies, 22,* 549–562.

Tarrière, C., Hartemann, F., Sfez, E., Chaput, D., & Petit-Poilvert, C. (1988). *Some ergonomic features of the driver-vehicle-environment interface.* In SAE Technical Paper Series (SAE No. 885051, pp. 405–427). Warrendale, PA: Society of Automotive Engineers.

Vercruyssen, M., Carlton, B. L., & Diggles-Buckles, V. (1989). Aging, reaction time, and stages of information processing. *Proceedings of the Human Factors Society 33rd Annual Meeting* (pp. 174–178). Santa Monica, CA: Human Factors Society.

Walker, J., Alicandri, E., Sedney, C., & Roberts, K. (1990). *In-vehicle navigation devices: Effects on the safety of driver performance* (Tech. Rep. FHWA-RD-90-053). Washington, DC: Federal Highway Administration.

Walker, J., Alicandri, E., Sedney, C., & Roberts, K. (1991). In-vehicle navigation devices: Effects on the safety of driver preformance. In *Vehicle Navigation and Information Systems Conference Proceedings* (pp. 499–525). Warrendale, PA: Society of Automotive Engineers.

Waller, P. F. (1991). The older driver. *Human Factors, 33*(5), 499–505.

Weintraub, D. J., Haines, R. F., & Randle, R. J. (1984). The utility of head-up-displays: Eye focus vs. decision times. *Proceedings of the of the Human Factors Society 28th Annual Meeting* (pp. 529–533). Santa Monica, CA: Human Factors Society.

Weintraub, D. J., Haines, R. F., & Randle, R. J. (1985). Head-up display (HUD) utility. II. Runway to HUD transition monitoring eye focus and decision times. *Proceedings of the of the Human Factors Society 29th Annual Meeting* (pp. 615–619). Santa Monica, CA: Human Factors Society.

Wetherell, A. (1979). Short term memory for verbal and graphic route information. *Proceedings of the Human Factors Society 23rd Annual Meeting* (pp. 464–469). Santa Monica, CA: Human Factors Society.

Wickens, C. (1987). *Engineering psychology and human performance*. Columbus, OH: Merrill.

Wickens, C. (1992). *Engineering psychology and human performance*. Columbus, OH: Merrill.

Williams, M. (1995). Personal communication.

Williges, R. C., & Williges, B. H. (1982). Structuring human/computer dialogue using speech technology. *Proceedings of the workshop on standardization for speech I/O technology*. Gaithersburg, MD: National Bureau of Standards.

Williges, R. C., Williges, B. H., & Elkerton, J. (1987). Software interface design. In G. Salvendy (Ed.), *Handbook of human factors* (pp. 1416–1448). New York: Wiley.

Winter, D. J. (1988). Older drivers: Their perception of risk. *The Engineering Society for Advancing Mobility Land Sea Air and Space*, 19–29.

Wood, R., & Thomas, M. (1991). A head-up display for automotive applications. *Automotive Displays and Industrial Illumination, 958*, 30–48.

Yanik, A. J. (1989). Factors to consider when designing vehicles for older drivers. *Proceedings of the Human Factors Society 33rd Annual Meeting* (pp. 164–168). Santa Monica, CA: Human Factors Society.

Zwahlen, H. T., Adams, C. C., & DeBald, D. P. (1987). *Safety aspects of CRT touch panel controls in automobiles*. Presented at the Second International Conference on Vision in Vehicles, Nottingham, England.

Zwahlen, H. T., & DeBald, D. P. (1986). Safety aspects of sophisticated in-vehicle information displays and controls. *Proceedings of the Human Factors Society 30th Annual Meeting* (pp. 256–260). Santa Monica, CA: Human Factors Society.

13

APPLICATION OF EXISTING HUMAN FACTORS GUIDELINES TO ATIS

Francine H. Landau
Martha N. Hanley
Cheryl M. Hein
Hughes Aircraft Company

This chapter evaluates the applicability of current human factors guidelines to an Advanced Traveler Information System (ATIS). The system goals are to acquire, analyze, communicate, and present information to assist both commercial and private drivers in moving from one location to a desired destination. This navigation information is provided in the vehicle and includes emergency, congestion, hazard, and "yellow page" categories of information. Although there are currently many worldwide efforts to research, develop, and implement in-car navigation systems, there are no clearly stated design requirements to help the system designer optimize this driver–vehicle interface. Without utilizing good human factors practice and judgment in the design of such systems, the result would surely tax the abilities of most drivers and possibly cause unsafe situations where none exist today. As an initial step in the new system design activity, human factors practitioners consult existing guidelines as an aid in defining the functionality and form of the new system. At the same time, they must determine if the guidelines are appropriate for the new implementation, and what, if any, caveats or complementary research must be considered. To this end, this chapter will describe ATIS functionality, driver–vehicle interface components, and existing guidelines with discussion and recommendations concerning their limitations and research needs.

Nature of Guidelines and Supporting Data

Design guidelines are a critical source of information for human factors and system designers to access when developing and specifying the architecture, functionality, and physical interface of the system. By their nature, guidelines are generally worded so they can be applied to many related system forms. In this respect they are considered an aid and not the specific solution to a design question. Human factors practitioners and system designers must determine which guidelines are relevant, then proceed to develop design rules appropriate for the specific system under development. An example of a design guideline would be: "Displays should be consistently formatted." A design rule resulting from this guideline could be: "Display title, prompts, error messages, and user guidance should always appear in the same area of each screen." The transformation of a guideline to a design rule is not a straightforward nor standardized process. There are, however, types of information, as discussed by Campbell (1992), that can aid the designer in utilizing guidelines more efficiently and appropriately:

1. *The application domain.* Was the purpose or end use of the guideline source similar in nature and functionality as the application under study? For example, are military fixed wing operations similar in functionality or tasking to navigating while driving a car?

2. *The kind of data support for the guideline.* Is the data from military specifications, integrative review of empirical research, or other design standards?

3. *The amount of empirical support and correlated references.* If the guidelines are based on relevant empirical research that is well referenced, the process of determining applicability is made much easier, as greater generalizability will have been investigated.

4. *The operational definitions of independent and dependent variables, and use of a variety of study settings (laboratory, simulation, field) for empirical research.* If these are similar to the characteristics of the intended system, then applicability of the guideline will be greater than if they are dissimilar. As these variables become more dissimilar to those of the system under design, the applicability of the guideline diminishes.

5. *Specification of a range of acceptable values for clearly described users.* For example, are the users young pilots with excellent eyesight, or are they older drivers with degraded visual function?

6. *Specification of factors that interact with the parameter of interest.* For example, specifying colors for a navigation display must take into consideration several shifts in color perception under daytime versus nighttime viewing conditions.

7. *Specification of known constraints in applying the guideline.* For example, would performance be different if tasking is self-paced versus externally paced?

8. *If multiple guideline sources are considered, determination of consistency of recommendations among them.* Consistency is the degree to which the design recommendations specified are in agreement with those from comparable sources.

Expert Judgment and the Automotive Environment

In summary, the goal of a guideline is to be based on highly valid, very applicable, often replicated consistent data. Unhappily, this is almost never the case. In fact, the supporting research for current human factors guidelines has come almost exclusively from the military for applications such as commercial piloting, nuclear power system monitoring, and computer system design. As mentioned previously, there is currently no guideline document developed for the design of automotive navigation systems. Components of the driver–vehicle interface have been studied, and there are some limited standards related to visibility and reach. However, the global issues of navigation functionality and the design of the interface have yet to be specified.

How then is one to proceed in generating a design rule or in estimating the applicability of the design guide when appropriate supporting data are lacking? Expert judgment and knowledge of the automotive environment and driving tasks must be used to bridge the gap. Analysis of the guideline topic areas reveals that there are many common processes, elements, and physical arrangements between driving cars and other operations that originated the guidelines. The sensory modes, legibility issues, and response modes are very similar. However, driving a car places unique demands on the driver. It is this knowledge of the perceptual, cognitive, and motor requirements of the driving task that will enable the human factors professional to filter the assortment of data applicability issues and to formulate a recommendation about the applicability of a specific guideline.

Special Considerations of the Driving Task

Because the process of formulating a guideline recommendation depends on characterizing the driving environment and its tasks, several criteria will be used as standards against which the guideline data will be compared. Does the guideline take the following into consideration:

1. The range of user groups and their varying capabilities
2. The need to guarantee legibility of information during nighttime and daytime ambient illumination conditions

3. The need to assimilate complex graphic data during a task that is externally paced

4. The need to minimize the number of scans and glances to assimilate navigation information

5. The need to minimize additional workload during tasks characterized by several shifts in attention and concomitant interference of memory

6. The manual control requirements of driving a car in addition to manually transacting with the navigation system

7. The visual requirements of driving as a context for the additional visual processing of a navigation system

8. That navigation tasks must be accomplished by the driver's spare capacity for perceptual, cognitive, and motor activities while in transit

Selected Guideline Documents

Because there are hundreds of guidelines contained in journal articles, government technical reports, society and academic publications, it was necessary to limit this review in some way. Analytical activities were therefore focused on large compendia of guidelines related to human–computer interaction as well as military and commercial air and ground operations standards. The elements of in-car navigation are all contained within these systems. A further restriction involves publications after 1985 because so much new and relevant information has become available since then.

Table 13.1 contains a short description of each guideline document reviewed for this effort, which includes chapters considered relevant, the specific authors, the year of publication, the application domain, and an estimate of the number of references.

The Guideline Applicability Rating System

Given the previous discussion about the variability of data supporting guidelines and the necessity to use expert judgment in assessing their applicability to an ATIS system, a rating system is used to determine the level of confidence associated with a set of guidelines. This will have four categories. Each guideline will have 1, 2, 3, or 4 stars, and the ratings will be based on the following criteria:

****　　Based on empirical data for similar functionality from the transportation domain with little or no expert judgment required to apply guideline.

***　　Based on empirical data from other application domains with great consistency across sources. Interacting factors and limita-

TABLE 13.1
Summary of Guideline Sources Reviewed

Title	Authors	Relevant Guideline Topic	References to Empirical Data	Application Domain
1. MIL STD 1472D Human Engineering Design Criteria For Military Systems, Equipment and Facilities (1989)	No individual authors	Input methods, Data display, user-computer interface	None	Military systems and personnel
2. Handbook of Human-Computer Interaction, M. Helander (Ed.), 1988	Czaja, S.; Tullis, S.; Paap, K., & Roske-Hofstrand, R.; Snyder, H.; Streeter, L.	Computing & the elderly, screen layout, menu dialogue, visual display, auditory display	Many within chapters	Military and commerical computer systems
3. Handbook of Human Factors, G. Salvendy (Ed.), 1987	Helander, M.; Sorkin, R.; Sorkin, R. & Kantowitz, B.; Simpson, C. et al.; Williges, R., Williges, B., & Elkerton, J.	Visual display, auditory display, computer interface	Many within chapters	Military and commercial systems design
4. American National Standard for Human Factors Engineering of Visual Display Terminal Workstations (1988)	19 members of the Human Factors Society	Visual display	Some within sections	Visual display workstation environment
5. Engineering Data Compendium, K. Boff & J. Lincoln (Eds.), 1986	No individual authors	Controls, visual & auditory display, computer interface, navigation display	Extensive for each guideline or experiment	Large military and commerical systems
6. Guidelines for Designing User Interface Software (1986)	Smith, S., & Mosier, J.	Human-computer interface, map display	Extensive, for most guidelines	Computer-based information systems

tions have been noted. Some expert judgment is necessary to transform guidelines to new application specifications.

** Based on empirical data from other application domains with some consistency across sources. Sparse or no data on interacting factors and limitations necessitate further research, significant expert judgment, or both.

* Based on primarily expert judgment or design convention. The amount of supporting empirical data is nonexistent or unknown.

NA If the parameter under discussion refers to tasks, procedures, or methods considered to be inappropriate for the driver–vehicle interface because of drivers' physical or cognitive limitations, they will be rated not applicable (NA).

Organization of This Chapter

The remainder of this chapter is divided into two sections. Section 2 describes the postulated functions of an in-vehicle navigation system. Section 3 reviews the guidelines by topic area, rating according to the metric discussed above. If there are several sources of guidelines, the commonalities will be noted, and a summary rating will be applied.

SECTION 2: AN ADVANCED TRAVELER INFORMATION SYSTEM DESCRIPTION

ATIS

The overriding goal of an ATIS is to improve information provided to travelers in both urban and rural settings under normal, congested, poor-weather, and emergency conditions. Several systems are currently being developed jointly by various private and public organizations, each of which contains a subset of ATIS functions. These systems are In-Vehicle Routing and Navigation Systems (IRANS), In-Vehicle Motorist Services Information Systems (IMSIS), In-Vehicle Signing Information Systems (ISIS), and In-Vehicle Safety Advisory and Warning Systems (IVSAWS). The functions of each system as discussed by Lee, Morgan, Wheeler, Hulse, and Dingus (1993) are described in the following sections and collectively represent a comprehensive set of ATIS capabilities.

IRANS. An IRANS provides drivers with information about how to get from one place to another as well as information on traffic operations and on recurrent and nonrecurrent urban traffic congestion. At this time, seven functional components have been identified: trip planning, multimode travel coordination and planning, predrive route and destination selection, dynamic route selection, route guidance, route navigation, and automated toll collection. These are described in greater detail as follows:

Trip Planning

Trip planning is the route-planning of long, multiple-destination journeys that may involve identifying scenic routes and historical sites, as well as coordinating hotel accommodations and restaurant and vehicle service information.

Multimode Travel Coordination and Planning

This element provides coordination information to the driver on different modes of transportation (e.g., buses, trains, and subways) in conjunction with driving a vehicle. This information might include real-time updates of actual bus arrival times and anticipated travel times.

Predrive Route and Destination Selection

This feature allows the driver to select any destination or route while the vehicle is in Park. The characteristics for the predrive selections include inputting and selecting the destination, then selecting a departure time and route to the destination. System information might include real-time or historical congestion information, estimated travel time, and routes that optimize a variety of parameters.

Dynamic Route Selection

Dynamic route selection encompasses any route selection system while the vehicle is not in Park. This capability includes the presentation of updated traffic and incident information that might affect the driver's route selection. In addition, the system would alert the driver if he or she makes an incorrect turn and leaves the planned route. Dynamic route selection can generate a new route that accommodates the driver's new position.

Route Guidance

This capability includes turn-by-turn and directional information. This can be in the form of a highlighted route on an electronic map, icons indicating turn directions on a head-up display (HUD), or a voice commanding turns.

Route Navigation

Route navigation provides information to help the driver reach a selected destination, but it does not include route guidance. It might include presenting an in-vehicle electronic map with streets, direction orientation, current location of vehicle, destination location, and services or attractions locations. Route navigation provides information typically found on paper maps.

Automated Toll Collection

This system would allow a vehicle to travel through a toll roadway without needing to stop and pay tolls; tolls would be deducted from the driver's account automatically as the vehicle drove past toll collection areas.

IMSIS. An IMSIS provides broadcast information on services–attractions, access to a services–attractions directory, destination coordination, and message transfer capability.

Broadcast Services–Attractions Information

Broadcast services–attractions information provides travelers with information that might otherwise be found on roadside signs. It may be very similar to the directory of services and attractions information, but the driver does not need to look for this information; it is presented as the vehicle travels down the road.

Services–Attraction Directory

This capability provides information about motels, hotels, and automobile fuel or repair, as well as entertainment, recreational, and emergency medical services. Using this capability, drivers can access and assess the information concerning businesses or attractions that satisfy their current driving needs.

Destination Coordination

This function enables the driver to communicate and make arrangements with the final destination that may include restaurant and hotel reservations.

Message Transfer

This feature provides the capability for drivers to communicate with others. Currently, this function is implemented by cellular telephones and CB radios. However, future ATIS systems may improve on this technology by automatically generating preset messages at the touch of a button and by receiving messages for future use. Message transfer might involve both text and voice messages.

ISIS. An ISIS provides noncommercial routing as well as warning, regulatory, and advisory information inside the vehicle that is currently depicted on external roadway signs.

Roadway Sign Guidance Information

ISIS guidance information includes street signs, interchange graphics, route markers, and mile posts. Traditionally found on signs outside the vehicle, this information is brought into the vehicle.

Roadway Sign Notification Information

Roadside signs such as merge signs, advisory speed limits, chevrons, and curve arrows that currently notify drivers of potential hazards or changes in the roadway are displayed in the vehicle.

Roadway Sign Regulatory Information
Regulation information such as speed limits, stop signs, yields signs, turn prohibitions, and lane use are provided in the vehicle.

IVSAWS. An IVSAWS provides warnings on immediate hazard and road conditions or situations affecting the roadway ahead of the driver. It provides sufficient advanced warning to permit the driver to take remedial action (e.g., to slow down). This system does not encompass in-vehicle safety warning devices that signal imminent danger requiring immediate action (e.g., lane change–blind spot warning devices, imminent collision warning devices, etc.). In addition, it provides for the capability of both automatic and manual aid requests.

Immediate Hazard Warning
IVSAWS may provide hazard proximity information to the driver by indicating the relative location of a hazard, the type of hazard, and the status of emergency vehicles in the area. Specifically, this might include notifying the driver of an approaching emergency vehicle or warning the driver of an accident immediately ahead.

Road Condition Information
This function provides information on traction, congestion, construction, and so forth within some predefined proximity to the vehicle or the driver's route.

Automatic Aid Request
This function provides a "Mayday" signal in circumstances requiring an emergency response for which a manual aid request is not feasible and in which immediate response is essential (e.g., severe accidents). The signal provides location information and, potentially, severity information to the emergency response personnel.

Manual Aid Request
Manual aid request encompasses those services needed in an emergency (police, ambulance, towing, and fire department). It allows the driver to access these services from the vehicle without the need to locate a phone, know the appropriate phone number, or even know their current location. This function might also include feedback to notify the driver of the status of the response such as the expected arrival time of service.

SECTION 3: GUIDELINE APPLICABILITY

The topic areas listed below are reviewed for guideline availability and applicability to an ATIS. Each guideline topic is introduced by a description of the issue. The six reference sources listed in Table 13.1 are reviewed, and guidelines are integrated and presented in one table of selected guidelines for that topic. The table title has the reference source numbers in superscript that contributed to the recommendations. If there is great consistency among the reference sources, only the title has the superscript reference numbers. If there are original or diverging guidelines for the topic, the reference numbers are noted within the body of the table. After each table, there is a discussion of the deficits in empirical support or constraints in the application of the guidelines to an in-vehicle navigation system. A further distinction must be made in evaluating the applicability of the guidelines: Most of the empirical data were collected for performance with a single task. None of this data has been collected in a dual-task paradigm for which there is a primary task (e.g., tracking) and a secondary task (e.g., recognition of alphanumerics or accessing a computer). For that reason, guideline applicability is rated for predrive and in-transit implementation. Guidelines are evaluated for the following components of an ATIS driver–vehicle interface: At the end of each component section, the relevant guideline applicability ratings are highlighted in a box.

Input Methodology
> Touch
> Pushbutton, Switch Controls

CRT Display and Information Characteristics
> Contrast
> Color
> Character Height
> Font Style
> Strokewidth
> Symbol Width

Auditory Display Characteristics
> Alerting Signals
> Voice Displays
> Voice Messages

Human Computer Interaction
> Dialogue for the Casual User
> Sequence Control
> Menu Dialogue
> Error Message Design
> Display of Text Data
> Design of Abbreviations
> Screen Layout

Navigation Information Format
> Map Displays

INPUT METHODOLOGY

The design of the input mechanisms for an in-vehicle system must consider the accuracy and speed required for transactions. Ill-placed controls with little feedback are certain to cause driver accidents when vehicles are in motion, and to generate errors in requested subsystem functions whether the vehicle is in transit or not. The design and arrangement of two input types—touch and push button—switch mechanisms—are reviewed with respect to current guidelines.

Touch Input. A touch-screen device produces an input signal in response to a touch or movement of the finger at a place on the display. There are two basic principles by which touch-screen devices work. One method uses an overlay that responds to pressure. The other method is activated when the finger interrupts a signal. Touch-screen input is particularly suited to those tasks using data already displayed on the screen and those using menu selection. Some research indicates that workload may be decreased with touch input in plane cockpits for navigation purposes (Beringer, 1979). They are potentially useful in high-stress environments because the number of possible inputs is limited. There are a number of advantages and disadvantages of touch-screen devices: [2, 3, 5, 6]

Touch-Screen Input

Advantages	*Disadvantages*
Direct eye-hand coordination	Arm fatigue
No command memorization needed	Limited resolution
Operator may be led through correct command sequence	Hard-to-select small items
Minimal training necessary	Dirt, smudge on screen reduces contrast
	Slow data entry
High user acceptance	Finger or arm may obscure screen
	Overlays may lead to parallax

Table 13.2. contains the guidelines for touch-screen input.

Discussion

The availability of data on touch input for navigation tasks and the enhancements of eye—hand coordination for data input are advantages for this method of data entry. However, some disadvantages such as reduced contrast and having to look at the display for data entry may make this input mechanism inappropriate for data entry while a vehicle is in motion.

TABLE 13.2
Guidelines for Touch-Screen Input[1,5]

1. Touch-screen control may be used to provide a display overlaying control function where direct visual reference access and optimum direct control access are desired.

2. When used, touch-screen displays shall have sufficient luminance transmission to allow the display with touch-screen installed to be clearly readable in the intended environment and meet the display luminance requirements herein.

3. A positive indication of touch-screen actuation shall be provided to acknowledge the system response to the control action.

4. The dimensions and separation of responsive areas of the touch-screen shall be:

Dimensions		Separation	
Minimum	Maximum	Minimum	Maximum
3/4 in.	1 1/2 in.	1/8 in.	1/4 in.

The only quantitative specifications for touch input come from MIL STD 1472D, and they are based on switch activation areas. Again, methods for measuring luminance from a display with a touch overlay are not presented in the guideline source documents reviewed and must be considered when this data input method is specified.

Touch-Screen Applicability Rating:	Predrive:	**
(See Table 13.2)	In Transit:	NA

Mechanical Controls.   This section describes the guidelines relating to push buttons and switches. These guidelines are extensively reviewed in MIL STD 1472D. Because it seems likely that fast transaction times will be required of the driver while in motion, push-button or switch activation on the steering wheel may provide the driver with the quickest and most accurate solution. However, the number of buttons must be kept to a minimum, and they should be tactually coded for "blind" operation. Table 13.3 contains the guidelines for push buttons or switch controls.

Discussion

Because the motivation for much of the research supporting these control guidelines was workstation console and cockpit design of large military systems, the recommendation that controls should be located with the

TABLE 13.3
Selected Guidelines for Control Inputs[1]

1. Controls shall be selected and distributed so that none of the operator's limbs are overburdened.
2. Where applicable, control selection shall include consideration of operation under variable g-loading on the operator.
3. Detent controls shall be selected when the operational mode requires control operation in discrete steps.
4. Direction of control movement shall be consistent with the related movement of an associated display, equipment component, or vehicle. In general, movement of a control forward, clockwise to the right, or up, or pressing a control, shall turn the equipment or component on, cause the quantity to increase, or cause the equipment or component to move forward, clockwise, to the right, or up.
5. Controls shall be oriented with respect to the operator. Where the operator may use two or more vehicle operator stations, the controls shall cause movement oriented to the operator at the effecting station unless remote visual reference is used.
6. All controls that function in sequential operation necessary to a particular task, or which operate together, shall be grouped together along with their associated displays.
7. The most important and frequently used controls shall have the most favorable position with respect to ease of reaching and grasping.
8. Where controls are operated at a position remote from the display, equipment, or controlled vehicle, they shall be arranged to facilitate direction-of-movement consistency.
9. Minimum spacing between controls shall be specified as described on page 69 of reference 1.
10. The use of coding mode (e.g., size and color) for a particular application shall be governed by the relative advantages and disadvantages of each coding type. Where coding is used to differentiate among controls, application of the code shall be uniform throughout the system.
11. Controls shall be compatible with handwear to be utilized in the anticipated environment. Unless otherwise specified, all dimensions cited herein are for barehand operation and should be revised where necessary for use with gloves or mittens.
12. Where "blind" operation is necessary, hand controls shall be shape-coded or separated from adjacent controls by at least 5 in.
13. Primary use of shape coding for controls is for identifying the control by feel; however, shapes shall be identifiable both visually and tactually. When shape coding is used,

   • The coded feature shall not interfere with ease of control manipulation.
   • Shapes shall be identifiable by the hand regardless of the position and orientation of the control knob or handle.
   • Shapes shall be tactually identifiable when gloves need to be worn.
   • A sufficient number of identifiable shapes shall be provided to cover the expected number of controls that require tactual identification.
   • Shape-coded knobs shall be positively and nonreversably attached to their shafts to preclude incorrect attachment when replacement is required.
   • Shapes shall be associated with or resemble control function, and not alternate functions.

14. Controls shall be designed and located so that they are not susceptible to being moved accidentally.
15. Any method of protecting a control from inadvertent operation shall not preclude operation within the time required. For situations in which controls must be protected from accidental actuation, one or more of the following methods shall be used:

   • Locate and orient the controls so that the operator is not likely to strike or move them accidentally in the normal sequence of control movements.
   • Recess, shield, or otherwise surround the controls by physical barriers. The control shall be entirely contained within the envelop described by the recess or barrier.
   • Provide the controls with resistance (i.e., viscous or coulomb friction, spring-loading, or inertia) so that definite or sustained effort is required for actuation.

display they operate makes good sense. However, this guideline may be violated for good reason in an in-vehicle navigation system. In-transit transactions should be minimized, but when they are necessary, single actions of controls placed on the steering wheel may be the optimal solution. Shape coding of these controls will add much to their usability and Fig. 13.1 shows 25 shape-coded push buttons tested for blind discrimination. Codes 1, 4, 21, 22, 23, and 24 had the highest accuracy and least confusions. The only caveat that must be considered is inadvertent activation and its consequences. If this is considered during design, it should not prove to be too significant a problem. Locating the controls where the driver can activate them without reaching and looking at them is very desirable for in-transit procedures. However, the location of controls is confounded with reach questions and the type of display used for information presentation. With traditional CRT and direct-view displays drivers can move about somewhat and still see the

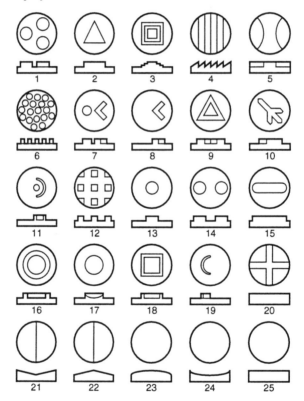

FIG. 13.1. Shapes (front and side views) tested for tactile discrimination. From K. R. Boff and J. E. Lincoln (1988). *Engineering Data Compendium. Human Perception and Performance.* AAMRL, Wright-Patterson Air Force Base, Ohio. Originally from T. G. Moore (1974). Tactile and kinesthetic aspects of push buttons. *Applied Ergonomics, 52,* 66–71.

display from many positions. The same is not true of HUDs because the location of the driver's head will constrain the ability to see the information. This may dictate a location for controls. In general, the applicability of these guidelines is high, but further verification of appropriate design should be undertaken for gloved operation and older drivers who have limitations in flexibility and touch discrimination.

Controls Applicability Rating:	Predrive:	***
(See Table 13.3)	In Transit:	**

CRT DISPLAY AND INFORMATION CHARACTERISTICS

This section covers guidelines related to both legibility and readability of a CRT display. Legibility is the rapid identification of single characters that may be presented in a noncontextual format. Readability is the ability to recognize the form of a word or group of words for contextual purposes. Because legibility is the more stringent requirement, all guidelines will be evaluated with that standard. If there are any exceptions, they will be noted.

Contrast. The term "contrast" refers to the difference, usually expressed as a ratio or a percentage, between the luminance of a symbol and the luminance of the background. Contrast has been expressed as modulation, contrast ratio, dynamic range, brightness contrast, and relative contrast (Farrell & Booth, 1984).

If L_{max} is used to symbolize the brighter of two contrasting areas and L_{min} is used to symbolize the darker of two contrasting areas, contrast can be specified as a ratio (i.e., $L_{max}:L_{min}$, Eq. 1). The same values can be used to specify luminance contrast as a percentage, as shown in Equation 2, or as modulation, as in Equation 3.

$$\text{Contrast (as a ratio)} \quad = \frac{L_{max}}{L_{min}} \; :1 \tag{1}$$

$$\text{Contrast (as a percentage)} = \frac{L_{max} - L_{min}}{L_{max}} \times 100 \tag{2}$$

$$\text{Contrast (as a modulation)} = \frac{L_{max} - L_{min}}{L_{max} + L_{min}} \tag{3}$$

Table 13.4 lists guidelines for contrast utilizing monochrome displays.

TABLE 13.4
Contrast Guidelines for Displays[2,3,4]

---

1. Character luminance modulation shall be equal to or greater than 0.5 (contrast ratio of 3:1). A luminance modulation of at least 0.75 (contrast ratio of 7:1) is preferred. [2,4]

2. Small characters between 10 and 17 arcminutes shall have a minimum luminance modulation calculated + rom the following formula:

$$\text{Luminance modulation } M = 0.3 + 0.07(20 - S)$$

where S is the vertical height of the character set in arcminutes.[4]

3. Helander and Snyder discuss contrast in these terms but relate contrast to resolution and image quality. The metric Modulation Transfer Function Area (MTFA) is used. The required values of MTFA depend on whether the display is used for graphics or for alphanumerics. For most office applications in which the equipment is used for displaying alphanumeric information, the MTFA should be greater than 5. [2,3]

Discussion

There are several issues associated with display contrast in the automobile environment and for ATIS in particular. Contrast will vary as a function of changes in symbol luminance and, more important, changes in the ambient illumination reflected from a display. Sources 2, 3, and 4 recommend that luminance, and therefore contrast, be measured in the ambient conditions representative of the conditions in which the driver will view the display. Because a glass or plastic screen is interposed between the display surface and the driver, under certain conditions, this screen acts as a glare source by reflecting ambient illumination into the eyes of the driver. This reduces display contrast. Sources 2 and 3 recommended similar methods to reduce glare: The display should be positioned to avoid glare, or an antiglare treatment such as an antireflection coating, a mesh, or faceplate-filter geometry should be applied to the display screen.

The applicability of these guidelines is high when applied to monochrome CRTs because the supporting justification for these guidelines is extended, and description of measurement techniques is provided in the sources discussed. However, it seems unlikely that a monochrome display will be used in a navigation system in which maps must convey large information sets quickly. Color-coded symbology could help drivers identify types of information more quickly. Therefore, it seems that color contrast for color displays is the parameter of interest (see Table 13.5 on color guidelines).

The empirical basis for most of the recommendations does not include performance with subjects 60 years of age and older. Unfortunately, both contrast sensitivity and tolerance of glare degrade as the population ages. Older drivers may need as much as three times the luminance (Czaja, 1988) as younger drivers to see small symbols on a map display. At the same time, the requirements for contrast increase substantially as a person ages. Changes in the pupil, lens, and vitreous body increase intraocular light scattering and reduce retinal illumination. Because of this, older people have difficulty tolerating glare and also need more contrast in order to resolve alphanumerics. In addition, numerous studies of age and dark adaptation have shown that the visibility threshold increases with age even after final adaptation (McFarland, Domey, Warren, & Ward; cited in Czaja, 1988). Czaja in the same review described research that has found perception and comprehension rates to be slower for older than for younger subjects. Contrast specifications, as a consequence, must consider the perceptual deficits of the older driver.

Contrast Applicability Rating:	Predrive:	***
(See Table 13.4)	In Transit:	**

Color. Our perception of color is derived from variations in the wavelength (or spectral composition) of emitted light. Color perception can be described in terms of three psychological dimensions; hue, saturation, and brightness. Hue is related to the dominant wavelength of the stimulus; saturation is somewhat more loosely related to its spectral bandwidth; and brightness is related to its luminance. Saturation, apparent brightness, and color contrast (all of them related) can make a greater contribution to legibility than hue itself (Campbell, 1988). When more than one color is displayed on a screen, estimates of color contrast predict legibility to some extent. Color contrast refers to perceived differences between hue and saturation of two colors. Brightness difference is luminance contrast (although there is empirical evidence that colors of equal luminance may not appear equally bright). Color contrast research has been aimed at developing a uniform color space (i.e., one in which equal distances on a color diagram correspond to equal perceptions regarding color differences). The distance between two colors within a uniform color space is used to indicate the magnitude of color contrast. The measure of color contrast is then correlated with human performance. Several metrics have been proposed using the 1976 CIE Uniform Color Space (UCS) color diagram as discussed in Table 13.5.

Choosing colors for use on displays depends on the application. Color coding may be helpful if the display is unformatted, the symbol density is high, and the operator must search for specified information. Guidelines for the specification of symbology colors are listed in Table 13.5.

Discussion

Several factors must be considered in determining a suitable color-coding set for a particular application: ambient illumination, symbol luminance, and

TABLE 13.5
Guidelines for Symbology Color Specification[1,2,3,4,5]

1. Color coding may be employed to differentiate between classes of information in complex, dense, or critical displays. For simple coding of information, no more than 10 colors should be used.

2. Information shall not be coded solely by color if the data must be accessed from monochromatic as well as color terminals or printed in hardcopy versions[1,2]

3. Task analysis (breaking down the task into primary and secondary units of information) should be used to analyze the number of necessary colors.

4. The selection of color codes should take into consideration established meanings that are already associated with colors.[1,3,5]

> Red--no go, failure, error, malfunction.
> Yellow--caution, marginal situation, check.
> Green--safe, fully operational, all conditions satisfactory.
> White--used to indicate action in progress, all conditions satisfactory.
> Flashing red--emergency conditions, immediate action required.

5. Because the eye's peak sensitivity to color is approximately .5 cycles/deg. colored characters must be much larger than black and white characters whose recommended minimum spatial frequency is about 2 cycles/deg.[3]

6. Saturated blue should not be used for fine detail.

7. Avoid using low luminance (no level specified) colors for older viewers because of the decline of transparency of the lens and vitreous humor in the eye.[2]

8. Increasing ambient illuminance decreases color purity and, consequently, color discriminability. Accordingly, color measurements should be made under the presumed ambient lighting conditions in which the display will be used.

9. Simultaneous presentation of both pure red and pure blue (or to a lesser extent red and green, or blue and green) on a dark background may result in chromostereopsis (a three-dimensional effect) and should therefore be avoided unless chromostereopsis is acceptable or intentional.

10. Dominant wavelengths above 650 nm in displays should be avoided because protanopes are noticeably less sensitive to these wavelengths.

11. Because approximately 8% of the male population has some form of color blindness, saturated reds and greens should be avoided. Colors should be specified so that they have different amounts of red, green, and blue in them.[2,3,4,5]

12. When color coding is used for *discriminability or conspicuousness* of displayed information, all colors in the set should differ from one another by a minimum of 40 ΔE(CIE L*u* v*) distances. This approach will make available at least 7 to 10 simultaneous colors. The difference formula is:[4]

$$\Delta E \text{ units (CIE } L^*u^* v^*) = [(L_1^* - L_2^*)^2 + (u_1^* - u_2^*)^2 + (v_1^* - v_2^*)^2]^{1/2}$$
(see Note 1 at end)

TABLE 13.5
(continued)

---

13. For adequate *legibility*, color symbols should differ from their colored background by a minimum of 100 ΔE(CIE Yu'v') distances. The elements required for the calculation are the luminance in cd/m²(Y) and the UCS coordinates (u',v') of the text and background. The formula is:

$$\Delta E(CIE\ Yu'v') = [(155\ (\Delta Y/Ym))^2 + (367\ \Delta u')^2 + (167\ \Delta v')^2]^{1/2}$$
(see Note 2 at end)

This formula should be used with caution in assessing legibility for characters in colors having small luminance differences. Unusually large or small characters also may lead to erroneous estimates of legibility.

Note 1	*Note 2*
$L^* = 116(Y/Y_0)^{1/3} - 16;\ 1.0 > Y/Y_0 > .01$	ΔE (CIE Yu'v') = the color contrast metric
$u^* = 13L^*(u' - u'_0)$	ΔY = difference in luminance between text (symbology) and background
$v^* = 13L^*(v' - v'_0)$	Ym = the maximum luminance between text (symbology) and background
$u' = 4X / (X + 15Y + 3Z)$	Δu' = difference between u' coordinates of text (symbology) and background (per the 1976 CIE UCS)
$v' = 9Y / (X + 15Y + 3Z)$	Δv' = difference between v' coordinates of text (symbology) and background (per the 1976 CIE UCS)
u'_0 and v'_0 are the UCS coordinates for the reference white derived from the 1976 UCS.	
For reference white, Δ6500 K° u'_0 = .198 and v'_0 = .468	
For reference white, 9300 K° + 27 Minimum Perceptible Color Difference u'_0 = .181 and v'_0 = .454	
Y = luminance in cd/m². Y_0 is the luminance of the reference white.	

---

symbol color interact. At low levels of ambient illumination, symbol luminance must be reduced to maintain dark adaptation, but if reduced to less than 3 candela/meter², colors cannot be reliably differentiated.

At high levels of ambient illumination, symbol luminance must be significantly increased to compensate for the apparent color fading or washout that can occur with the reduction of the symbol-to-background contrast.

This occurs because CRT displays are not capable of reproducing highly saturated colors. They can produce enough saturation for most tasks, but high levels of ambient illumination (6,000 lux) produce veiling luminance on the screen, washing out screen contrast and reducing the number of colors that can be perceived. The ambient light mixes with the emitted light to create an additive mixture based on the proportions of each. The primary effect of veiling luminance is a desaturation of colors, and in some instances, it makes them indiscriminable from white. Depending on the color of the light, the desaturation may be different. Because most ambient light sources are broad band, there is a desaturation of all three phosphors.

Performance with specific wavelengths is not uniformly affected by ambient illumination, the light falling on a surface. At lower illuminance levels, response times are shorter to wavelengths in the blue-to-green region of the spectrum. At high levels of illuminance, response to red targets is faster than for either green or yellow targets at all levels of symbol luminance. It is apparent that the entire range of ambient lighting conditions must be taken into consideration in determining a suitable color-coding set and the required luminance levels for a specific application. If designers follow guideline instructions on empirically determining appropriate color sets with measured luminances for the intended lighting environment, then applicability of these guidelines is high. Research must also confirm discriminability of color pairs for the older driver who has difficulty seeing blue-green under both nighttime and daytime illumination.

Symbol Color Applicability Rating:	Predrive:	***
(See Table 13.5)	In Transit:	**

Character Height. Character height refers to the vertical distance between the top and the bottom edges of a nonaccented uppercase letter, such as an M. Because the distance between the observer and the display can vary across environments and applications, character height is specified as the visual angle subtended by the symbology in minutes of arc (arcmin). This value may be approximated by using equation 4, for which the character height and viewing distance must use the same linear unit:

$$\text{visual angle (arcmin)} = \frac{\text{character height} \times 3437.7}{\text{viewing distance}} \quad (4)$$

Character height is difficult to specify in isolation because it interacts with symbol contrast, luminance, and color. As symbol luminance, contrast, or character height decreases, one or both of the other factors must be increased. In addition, adapting luminance level, task criticality, and dynamic versus static presentation will influence whether a minimum or maximum value is necessary.

TABLE 13.6
Guidelines for Character Height[2,3,4]

1. The size of characters is dependent on the task and the display parameters (resolution, contrast, glare treatment, etc.). Characters that are too small can make reading as difficult as can characters that are too large. Characters larger than 24' are not appropriate for reading.

2. Character size that is 40% to 50% larger than threshold character size results in comfortable reading of text.

3. A minimum character height shall be 16 min of arc, but the preferred character height is 20 to 22 min of arc.

4. Characters should not be larger than 45 min of arc when groups of characters are displayed.[4]

Table 13.6 lists the guidelines for character height.

Discussion

MIL STD 1472D makes character height recommendations for critical and noncritical printed labels, which are not considered appropriate for electronic displays. The guidelines listed previously are appropriate for office environment CRT presentation. There are no character heights specified by task criticality or for older drivers. For this specification, an especially relevant visual parameter is the static and dynamic visual acuity performance of the older population. Acuity refers to the ability of the observer to see fine detail. Kline and Schieber have reviewed numerous studies that show the degradation of static acuity (cited in Czaja, 1988). After age 20 there is a moderate decrease in visual acuity through age 50 and a steady decline thereafter. Presbyopia starts to degrade near visibility in the 40s so that users may need bifocals to see symbology at instrument panel distances. Dynamic visual acuity is the ability to resolve fine spatial detail for moving objects. This ability declines more rapidly with age than static acuity (Czaja, 1988). The admonition that character height cannot be specified unless other variable levels are known for contrast, color, and glare is good advice, but there are no character height values specified for each level of the interacting factor. As for contrast, there seems to be a need for research to determine proper character height for symbology presented peripherally to an older driver.

| Character Height Applicability Rating: | Predrive: | *** |
| (See Table 13.6) | In Transit: | * |

TABLE 13.7
Guidelines for Dot-Matrix Fonts[1,2,3,4]

---

1. Where alphanumeric characters appear on CRT-like displays, the font style shall allow discrimination of similar characters such as letter l and number 1, letter Z and number 2.[1,2,3]

2. Font design may have an impact on display legibility and readability. Examples of methods for comparing font designs may be found in Shurtleff (1980) and Snyder and Maddox (cited in Helander, 1987). The font chosen for the specific application must be tested for legibility and readability.

3. A 5 x 7 (width to height) character matrix shall be the minimum matrix used for numeric and uppercase-only presentations. The vertical height should be increased upward by two dot positions if diacritical marks are used.

4. A 7 x 9 character matrix shall be the minimum matrix for tasks that require continuous reading for context, or when individual alphabetic character legibility is important. The vertical height should be increased upward by two dot (pixel) positions if diacritical marks are used. If lower case is used, the vertical height should be increased downward by at least one dot (pixel) position, preferably two or more, to accommodate descenders of lowercase letters.

Font. Font refers to the geometric characteristics or style of alphanumeric symbology. The style of symbology is related to the technique used to generate the symbology. For CRTs, it is the dot-matrix technique. This technique uses individual points of light to generate symbols. Dot-matrix characters are typically described by the size of the matrix (e.g., 5 × 7, 7 × 9, or 9 × 11). Table 13.7 lists guidelines for dot-matrix characters.

Discussion

These guidelines are appropriate for the office environment where most tasks are self-paced, illumination can be controlled, and direct-view displays are used. Unless tested for older drivers under some level of time pressure, a chosen font may not be adequate for a navigation system that presents messages while the vehicle is in transit. In addition, if a HUD is being considered for text presentation, font design needs to be evaluated because, unlike direct-view displays, font serifs and ornamentation degrade legibility. The supporting research was done with Lincoln/Mitre and Huddleston fonts. If those are not used, additional development and research should be considered.

| Font Applicability Rating: | Predrive: | ** |
| (See Table 13.7) | In Transit: | * |

TABLE 13.8
Guideline for Dot-Matrix Strokewidth[2,4]

---

1.   Miminum strokewidth should be greater than 0.08 times the character height.
Strokewidth should not exceed .20 times the character height.

Strokewidth. The strokewidth-to-character-height ratio of alphanumeric symbology refers to the ratio of the thickness of the stroke of the symbol to the height of the symbol. Once a character is well above the minimum size, and contrast and luminance level are adequate, strokewidth is not critical for performance. There is only one guideline on strokewidth in the references reviewed. It is listed in Table 13.8.

Discussion

MIL STD 1472D has guidelines for black printed characters, which are considered inappropriate for electronic display. This guideline is based on empirical data using the Lincoln/Mitre font in self-paced single tasks with younger subjects. Although the variable itself has not been shown to be critical, the use of this guideline for navigation with complex map data requires testing. If a HUD is the presentation display, testing is essential because strokewidth has a pronounced effect on contrast and legibility for lower luminance image sources.

Strokewidth Applicability Rating: (See Table 13.8)	Predrive: ** In Transit: *

Symbol Width. Symbol width refers to the width of the alphanumeric character, which is typically expressed as a ratio of symbol width to the symbol height. Guidelines for dot-matrix symbol width are listed in Table 13.9.

TABLE 13.9
Guidelines for Dot-Matrix Symbol Width[2,4]

---

1.   For fixed (as opposed to proportionally spaced) column presentations, the width-to-height ratio shall be between 0.7:1 and 0.9:1. For display formats requiring more than 80 characters on a line, ratios as low as 0.5:1 are permitted.

2.   For proportionally spaced presentations, a width-to-height ratio closer to 1:1 shall be permitted for some characters such as the capital letters M and W.

Discussion

The same guideline issues that were relevant for character height, font, and strokewidth apply also for symbol width. Data are needed on legibility of character symbol widths for time-critical tasks in which contrast may be low and the user is an older driver. In addition, intercharacter spacing interacts strongly with this variable. If a HUD presentation is being considered, testing is essential because unpublished research conducted by the authors has shown symbol widths of 0.5:1 as difficult to read.

Symbol Width Applicability Rating:	Predrive:	**
(See Table 13.9)	In Transit:	*

AUDITORY DISPLAY CHARACTERISTICS

Auditory displays include both nonverbal and verbal aural displays. Nonverbal displays use auditory alerting signals to signify events. Verbal displays use voice signals or messages to signify events and to provide more complex information. Auditory displays can supplement visual systems, particularly in situations of high visual attentional demand. If properly designed, they have the potential for making the task of driving easier and safer (Dingus & Hulse, 1993). Within an ATIS application, auditory displays can become a means for imparting information to the driver. Voice messages can be used to present route guidance, navigation, and traffic information without creating a visual distraction (Means et al., 1993). The desirability of using vocal messages for navigation tasks has been shown experimentally (Streeter, Vitello, & Wonsiewicz, 1985). Voice messages can transmit a much higher rate of information than auditory signals. However, there are problems complicating the use of voice messages in an ATIS application. Numerous forms of noise can interfere with the auditory channel. Among these are road noise, noise from one's own vehicle, noise from other vehicles, competing speech communication from passengers in the vehicle, vehicle entertainment, communication and information systems (radio, CD player, cellular phone, CB radios, computer, fax and copier machines), and vehicle heating and air-conditioning systems. Older drivers have an attenuated ability to decipher speech through noise (Bergman, 1971). The intelligibility of the most likely form of voice message display, computer synthesized speech, has limitations that can potentially add to driver workload (Sanders & McCormick, 1987).

Alerting Signals. Auditory alerting signals are characterized by the parameters of frequency and amplitude. Frequency, which is equivalent to the perception of pitch, is measured in cycles per second or hertz (Hz). Males speak at an average frequency of 200 Hz, females at 400 Hz. Musical instruments are in the range of 50 to 5,000 Hz. Alerting signals generally are in the range of 150 to 4,000 Hz. Amplitude is loosely perceived as loudness or intensity. Amplitude for ordinary conversation is 70 dB. Sound at 108 dB causes humans discomfort. The amplitude required for alerting signals is not simply a fundamental dB range. It depends on the background noise the signal must overcome, which is referred to as the masking threshold. The amplitude of alerting signals generally should be at least 15 dB above the amplitude of the masking threshold. Auditory alerting signals can be either high-priority signals used for immediate warning or low-priority signals used to communicate noncritical information. High-priority alerting signals should be used for time-critical event detection and response. The priority of a signal should be based on the time an operator has to respond before the point where a response will not change the outcome of the situation. High-priority alerting signals should be used along with visual signals and voice signals for the fastest response time. Noncritical auditory signals can be used to supplement visual displays. They may be used for situations in which the visual task loading on the operator is high, or when visually presented messages may be overlooked or misinterpreted. Applications include getting attention, announcing changes in system state, and declaring system or input–output failure. Examples are alerting tone preceding voice messages, prompting tone for screen glances when new data is presented, and error tone for inappropriate button or touch-screen input. Table 13.10 contains the guidelines for auditory alerting signals.

Discussion

There is consistent agreement among Table 13.10 references 1, 3, and 5, on the guidelines for alerting signals, with a few exceptions. The values for frequency range and amplitude were slightly lower in reference 3 than in the other references, and only reference 3 specified a maximum dB over masking threshold value. Reference 5 specified that the frequency and amplitude of the signal should be varied over time. In discussing the uses of a voice display, reference 5 recommended that an alerting signal should be preceded by an identifiable spoken word. References 1 and 5 conflict in the area of onset of the signal, with reference 1 recommending sudden onset of the alerting signal, and reference 5 recommending slower onset. Reference 1 is a military standard that does not discuss the basis for its guidelines. Reference 5 guidelines were developed from experimental work and from an expert handbook developed for the military. It is unclear whether the guidelines from the latter source were empirically validated. Reference 3 guidelines are based on experimental work. Particularly relevant to the ATIS application are reference 3 guidelines on the use of standard signals to convey noncritical information and the limitation of four distinct signals.

TABLE 13.10
Guidelines for Auditory Alerting Signals[1,3,5]

High-Priority Auditory Alerting Signals

---

1. Present high-priority alerting signals visually as well as aurally.
2. The detectability of auditory alerting signals should be maximized as follows:

   - Auditory alerts should be multiple frequency with more than one frequency in the range of 250 to 4000 Hz.
   - The pitch of alerting signals should be between 150 and 1000 Hz.[3]
   - The amplitude of an auditory signal should be at least 15 dB above the amplitude of the masking threshold. The signal should be less than 30 dB above masking threshold to minimize disruption of communication. [3]
   - An auditory alerting signal should be intermittent or changing over time.[5]
   - Auditory alerting signals should be dichotically separated from auditory distracters and noise.
   - An attention-intruding signal (e.g., the person's name) should be given at the beginning of an alerting signal.[5]
   - Signal burst duration should be approximately 100 ms to ensure detection but not disrupt communication with other personnel.[3]

3. A warning signal should not be capable of being confused with any other signal.
4. Alerting signal onset and offset rates should be limited to 1 dB/ms to prevent a dangerous startle response from the operator.[5]
5. The onset of critical alerting signals should be sudden.[1]
6. The design of the audio display devices and circuits should preclude false alarms.
7. The first 0.5 sec of a critical signal should be discriminable from the first 0.5 sec on any other signal that might occur.[1]

Noncritical Auditory Alerting Signals

---

1. The auditory signals should be used to alert and direct the user's attention to the appropriate visual display or other subsequent additional response.
2. The optimum type of signal should be carefully evaluated for ready notice by users while not startling them. Because of variable background noises, the intensity should be adjustable.
3. The intensity, duration, and source location of the signal should be compatible with the acoustical environment of the intended receiver.
4. Noncritical auditory signals should be capable of being turned off at the discretion of the user.
5. Standard signals should not be used to convey new meanings.
6. When several different audio signals are to be used to alert an operator to different types of conditions, a discriminable difference in intensity, pitch, or use of beats and harmonics should be provided. If absolute discrimination is required, the number of signals to be identified should not exceed four.

Drivers operating multiple vehicles may become confused or need constant retraining if signals are not standardized or if too many distinct signals are used. There are no guidelines on the repeat cycle for nonverbal auditory signals. Research needs to be done on the optimized use of auditory signals in cars, so that drivers are cued to respond and not irritated to the point of turning off the auditory display.

Alerting Signals Applicability Rating:	Predrive:	**
(See Table 13.10)	In Transit:	*

Voice Displays. Voice displays present information aurally using speech. Voice displays for ATIS applications can choose from either computer-generated synthesized or digitized speech. Digitized speech is human speech recorded and stored in computer memory. Synthesized speech is the real-time generation of humanlike speech from computer algorithms. Digitized speech has the advantage of naturalness, but it has the disadvantage in ATIS applications of requiring a prohibitive amount of prerecording and computer storage of all possible street names, route guidance, navigation, and traffic information. Synthesized speech provides the capability of generating route guidance based on computer selection and providing flexible text-to-speech traffic information from keyboard input at a central location.

One of the significant parameters for a voice display is its intelligibility. When applied to voice displays, intelligibility is meant to cover both the understandability of the speech as well as the comprehension of the information being presented. The intelligibility of synthesized speech depends on many variables (e.g., background noise, rate of speech, prosodics, pronunciation, pitch, loudness, message length, message content, message complexity, and the predictability of the message). In synthesized speech the prosodics (intonation pattern or rhythm of the output of speech), pronunciation, and pitch of individual syllables and phonemes (smallest unit of speech) are subject to programmable manipulation for improved intelligibility. Comprehension can be enhanced through careful attention to voice message design, presentation, and content. However, the acceptability of synthesized speech depends not only on its intelligibility, but also on the quality of the speech as perceived by the user. Quality in this sense is a function of user preference for a particular sounding voice. It is possible, though, for a user to understand and comprehend spoken words that have an unacceptable voice quality. Table 13.11 contains the guidelines for voice displays.

Discussion

Because many voice display topics were covered by only one reference, consistency in guidelines was precluded. There was broad agreement on guidelines for speech being used as a warning signal. Only reference 1

TABLE 13.11
Guidelines for Voice Displays[1,2,3,5]

Background Noise

1. Speech intelligibility increases as the ratio of signal (speech) power to noise power increases.
2. Speech intelligibility is reduced more by voice masks than by noise masks.
3. Older listeners have more difficulty attending to distorted, masked, or clipped speech than younger ones.
4. Noise created by mixing a number of voices interferes more with speech intelligibility than does noise consisting of a single voice.
5. If speech signal and masking speech noise are on different topics, the signal is more intelligible.

Speech Signals

1. Clipping the peak amplitude of the voice signal by as much as 90% and then amplifying the remainder to original levels before mixing with noise improves intelligibility.
2. Satisfactory speech communication is obtainable by filtering out all sounds below 800 Hz and all those above 2500 Hz.[3]

Sound Source Location

1. Presenting the speech signal to one ear and the noise to both ears, or the signal and noise to opposite ears reduces the masking effect of the noise.
2. When speech signal and masking speech are presented from different speaker locations, speech signal intelligibility increases as the distance between signal and masking speech increases.

Voice Type

1. Presenting a speech message and distractor in different voices reduces interference.
2. Select a voice type according to the source of speech messages: For machine messages to the operator, use machine-sounding voice quality; when simulating human speech, use human-sounding voice quality.[3]

Quality of Speech

1. If an application consists of only a few messages and if hearers will be exposed to these messages fairly frequently, then low-quality speech synthesis may suffice.[2]
2. Synthesized speech should be assessed in terms of quality and intelligibility.[2]

Speech Characteristics

1. Regardless of voice type, use the best approximation to natural prosodics.
2. Rate of speech should be approximately 150 words per minute.[3]

TABLE 13.11
(continued)

Speech Used As a Warning Signal

1. Speech signals should be used along with visual signals.
2. The effectiveness of speech signals may be maximized as follows:

- The language and phraseology should be familiar to the operator.
- The message should be preceded by an alerting tone, word, or phrase.
- Synthesized speech systems may be used if every effort is made to simplify the communication task.
- The warning system should have the capability of attenuating other speech systems while the warning is activated.

3. The voice should be distinctive and mature, and the warning should be presented in a formal, impersonal manner.[1]
4. The loudness of an audio warning system should be designed to be controlled by the operator.
5. Critical verbal warning signals should be repeated with not more than a 3-sec pause between messages until the condition is corrected or overridden by the operator.[1]

---

discusses voice type to be used for warning signals, and suggested a guideline of "distinctive, mature, formal and impersonal" that was clearly aimed at prerecorded human voice warnings rather than synthesized voice. Reference 3 suggests using voice types that are either machinelike or humanlike, depending on the source of the message: For machine messages to the operator, machine-sounding voice quality should be used; in simulating human speech, human-sounding voice quality is recommended. Reference 2 discusses the importance of the synthesized speech quality as well as intelligibility. This guideline is based on experimental work that found users' ratings of the usefulness of a synthetic speech application related to the evaluation of the particular synthetic voice that was used and not to the users' performance with the system (Rosson & Mellon, 1985). It is recommended in reference 2 that more sophisticated quality measures be found. There is a gap in the guidelines on the choice of gender for voice, and on the apparent body size of the voice (e.g., large, medium, or small person). The repeat cycle for critical verbal warning signals of at least every 3 s, from reference 1 guidelines, may not be applicable to ATIS. It was intended for military settings.

| Voice Display Applicability Rating: | Predrive: | ** |
| (See Table 13.11) | In Transit: | * |

Voice Messages. In an ATIS application, voice messages can be used for the delivery of simple warnings ("You have left your planned route.") or complex information ("There is a sigalert on the northbound 405 freeway just before the Crenshaw off-ramp. It will add 30 min to your drive. It will be in effect until 9:00 a.m."). Messages may be used for route guidance, navigation assistance, and traffic information. Voice messages require definition of design (e.g., number of syllables, words, sentences; which words, grammar, context), definition of presentation (e.g., timing of message in relation to acceptable response times; timing of message in relation to other messages; alerting preface to message; mechanical or more natural sounding voice, gender, and body size of voice), and definition of content (e.g., words to use for each maneuver to be specified, type of units to use for specifying distance, type of information to present, reliability of information). Table 13.12 contains the guidelines for voice message design, voice message presentation, and voice message content.

TABLE 13.12
Guidelines for Voice Messages[1,2,3,5]

Voice Message Design

1. Redundancy improves intelligibility. Noise interference is less if speech signal is composed of words with high frequency of occurrence, contains grammatical sentences, is selected from a small set of alternatives, and consists of meaningful words.
2. Message length should be a minimum of four syllables to provide sufficient linguistic context for warning comprehension after first enunciation of the message.
3. Messages should use sentences instead of isolated words for increased intelligibility.
4. Messages should not be used where an auditory signal (tone) will suffice.[2]
5. In selecting words to be used, priority should be given to intelligibility, aptness, and conciseness in that order.

Voice Message Presentation

1. When time-critical information such as warnings are delivered by voice, a priority system should be incorporated to concurrently trigger voice messages so that the most critical is presented first.
2. Conflicts between multiple voice messages should be considered.
3. For warning messages, the message after an appropriate time interval should be repeated only if the condition that triggered the warning message is still true.
4. For warning messages, the length of time allowed before the same message is presented again should depend on the severity of the consequences of the user not correcting the problem.
5. For spoken menus, without concurrent visual display, the number of menu items should be limited to three.

Voice Message Content

1. Message content should be appropriate for the task and use terminology familiar to the users.
2. Spoken information should be highly reliable.[3]
3. Spoken information should be intended for use in the immediate future because of its poor retention in short-term memory.[3]

Discussion

Voice display guidelines are oriented to verbal warning messages and not to imparting more complex information. There is a gap in the guidelines concerning what type of voice information to provide in an ATIS-type application. There are no guidelines applicable to the optimal information content of route guidance and traffic messages. In addition, there are no guidelines to the timing of route maneuver messages or to the kind of distance units that are meaningful to drivers. Further research is needed to optimize comprehension of voice display information for ATIS applications.

Reference 2 discusses the problem of speech as an interface that can become annoying. The basis for this discussion is the experience of "talking automobiles" in the 1980s. These automobile models used computer speech chips to give warnings about common problems and were perceived by consumers as annoying and not particularly informative. The "talking" vehicle models were discontinued after two years. The guideline for not using a voice message when a simple tone will suffice is based on this experience.

Voice Message Applicability Rating:	Predrive:	**
(See Table 13.12)	In Transit:	*

HUMAN–COMPUTER INTERACTION

The interaction between a driver and an ATIS system will be modeled to a great degree on human–computer systems because the nature and complexity of the transactions are so similar to current computer interfaces. Therefore, the applicability of human–computer interface guidelines will be reviewed.

Casual User Dialogue Design. The design of human–computer dialogue for casual, novice, or infrequent users differs greatly from the design of systems specifically intended for experienced or professional users. Designers must consider the consequences of infrequent use, including poor retention of system and training details. Casual users employ imprecise logic in dealing with computer systems and often make assumptions that the system knows contextual references without the user needing to state them. Failure to specifically consider the abilities and needs of the casual user can result in poor user performance because of extended time in training, increased error frequency, error recovery time, and for an ATIS system, user frustration and low acceptance. Table 13.13 presents the selected guidelines for casual user systems that were found in reference 5.

TABLE 13.13
Selected Guidelines for Designing Person-Computer Dialogue for the Casual User[5]

Design for a Forgetful User

1. Train users in principles, not details.
    - Emphasize conceptualization of systems as a whole.
    - Provide concise groundings in the principles of the interface.
    - Do not rely on existing skills in users.

2. Provide explicit, constrained choices for user inputs.
    - Use menu selection or prompting messages.
    - Make available choices apparent.

3. Use natural language interface for communication on user's terms and provide system guidance.
    - Restrict queries within system competence.
    - Continue progression of a dialogue.
    - Supplement information presented in display, preferably online.

Minimize System-Provided Opportunity for User Errors

1. Limit number of things user must consider at one time.
    - Menu selection: 10 or fewer items, consistent selection method.
    - Prompting systems: Restrict responses to well-defined range of values; utilize query-in-depth for system-initiated remediation of inappropriate responses.

2. Include corrective features in system-detected errors.
    - Attempt prediction of user's intended entry.
    - Initiate subdialogues for clarification.
    - Describe corrective actions in error messages.

3. Word error message for user acceptance.
    - Use humble wording to retain user's goodwill.
    - Provide at least two alternatively phrased messages.

Provide Feedback for User Guidance and Reassurance

1. Include unambiguous system responses for all user entries.
    - Reinforce correct responses.
    - Provide error detection.
    - Supply user identification of system actions.

2. Use easily understandable dialogue.
    - Avoid computer jargon.
    - Avoid unusual terms or abbreviations.

3. Maintain a natural flow of dialogue and a natural ordering of questions and inputs.

Match Database Query System to Infrequent User's Abilities

1. Reduce data structure, content, or semantic knowledge requirements.
    - Data should be requested by descriptive terms.
    - Data should guide user through valid choices.
    - Requests by field (attribute) or record (relational) names should be minimal.
    - Names and descriptive phrases should be displayable upon user request.
    - When multiple logic paths occur, choices should be explained in user-oriented terms.

TABLE 13.13
(continued)

---

2. Design for deviations in query precision.
   - Anticipate vague, exploratory queries before precise questions.
   - Guard against excessive output from broad or erroneous requests, even if query appears legitimate.
   - If specific attributes of an entity are requested, consider supplying others to facilitate query formulation and data retrieval.

3. Match dialogue language to the needs and abilities of users.
   - Reduced language formality can facilitate casual users.
   - System-aided syntax-error detection and correction is desirable.
   - Implicit logic specification is less error-prone than explicit use of logical connectives and quantifiers.
   - Plan for logical errors in syntactically correct queries if explicit logic is required.

Discussion

The guidelines for designing computer dialogue for the casual user are based on a single reference (Cuff, 1980), which in turn is based loosely on some empirical findings but no experimental studies. None of the other guidelines reviewed have referenced this article. Even so, Cuff has indicated a great understanding of the peculiar and unique interaction strategies that characterize novice system users as well as reinforcement and guidance consequently required from the system. In addition, there are a number of guidelines in the table that are recommended independently by the other reference sources. These guidelines are more relevant for large data retrieval, entry, or text editing systems than for an in-vehicle navigation system. However, the authors recommend using these guidelines as a departure for design guidance, subject to verification, especially for the predrive case. They are not especially relevant for the in-transit case.

Novice User Computer Interaction	Predrive:	**
Applicability Rating:	In transit:	NA
(See Table 13.13)		

Sequence Control. Sequence control refers to user actions and computer logic that initiate, interrupt, or terminate transactions. Sequence control governs the transition from one transaction to the next. Methods of sequence control require explicit attention in interface design. General design objectives include consistency of control actions, minimized memory load on the user, minimized need for control actions, and flexibility of sequence control to adapt to user needs. The guidelines for Table 13.14 are based almost wholly on references 5 and 6 with some support from reference 1.

TABLE 13.14

User Considerations

---

1. Minimize user actions
   - Simplify control actions to maximum extent.
   - Design for minimum number of required control entries consistent with user abilities.

2. Match ease of sequence control with desired ends.
   - Provide easy and quick control for frequent and urgent actions.
   - Supply distinctive actions under explicit user control for potentially destructive actions.

3. Match control to level of user skill, which may require mixed dialogue or entry-stacking option.

4. Require explicit user actions.
   - Computer should not interrupt user entries until conclusion.
   - Routine actions can benefit from computer control.

5. Permit initiative and control by user.
   - Anticipate all possible user actions and consequences.
   - Provide appropriate options for each potential user action.
   - Avoid "dead ends" in dialogues.
   - Allow user to interrupt, defer, or abort transaction sequences.

6. Design user pace of sequence control for user needs, attention span, and time available.

7. Prevent interference between simultaneous users.

Logic Considerations

---

1. Design consistent control actions throughout a system.
2. Make control actions independent of prior actions.
3. Base linked transaction sequence on logical unit of user task analysis not on logical unit of the computer system.
4. Establish consistent terminology for instructional materials, online messages, and command terms.
5. Offer active options only.

Language Considerations

---

1. Base choice of dialogue and design of sequence control on user and task characteristics.
   - *Question-and-answering dialogue* should be used for routine data-entry tasks when data items are known and their ordering can be constrained, when user has little or no training, and when computer response is expected to be moderately fast.

TABLE 13.14
(continued)

- *Form-filling dialogue* should be provided when some flexibility in data entry is needed such as the inclusion of optional as well as required items, when users have moderate training, and when computer response may be slow.
- *Menu selection* should be designed for tasks such as scheduling and monitoring that involve little entry of arbitrary data, when users have relatively little training, and when computer response is expected to be fast.
- *Function keys* should be made available for tasks requiring only a limited number of control entries that must be made quickly without syntax error.
- *Command language* should be used for tasks involving a wide range of user control entries, when users are highly trained in the interests of achieving efficient performance, and when computer response is expected to be relatively fast.
- *Query language*, which is a specialized subcategory of general command language, should be employed for tasks emphasizing unpredictable information retrieval (as in many analysis and planning tasks), with moderately trained users and fast computer response expected.
- *Graphic interaction* should supplement other forms of human-machine dialogue when special task requirements exist; effective implementation of graphic capabilities requires very fast computer response.

Input-Output Considerations

---

1. Match computer response time to transactions, using a faster response for those perceived to be simpler.
2. Do not pace entries by computer response delays.
   - When delays are unavoidable, keyboard should automatically lock.
   - Following lockout, computer readiness should be signaled to user.
   - User should be provided with means of aborting transaction during lockout.

3. Provide unambiguous feedback for control entries.
   - Signal completion of processing
   - Design unambiguous feedback to require immediate execution, change in state, or acceptance-rejection message.
   - Provide a standby message if system response is longer than 2 sec.[1]

4. Design sequence control features to be distinctive in position or format.
5. Provide a means to prevent accidental actuation of potentially destructive control actions, including the possibility of accidental erasure or memory dump.[1]

Discussion

In general, although all these guidelines do not apply to in-vehicle navigation systems, there are elements in that application that are similar to human–computer sequence control processes. By and large, there is a paucity of empirical data supporting these guidelines. In the experimental studies that were conducted the subjects were typically programmers, system operators, or designers. The definition of a novice user was frequently a programmer who was not familiar with the system under design (or occasionally a secretary). This strata of subjects obviously has different capabilities than the broad range of drivers. They are younger, better educated, computer fluent, and tend to like advanced technology. Their visual, motor, and cognitive abilities are not representative of the older driver at all. This situation also describes most of the supporting data for the design of menus and screens, error messages, text data display, and abbreviations. Most of the recommendations are relevant for the predrive case. In-transit sequence control should be kept to a minimum except for those transactions that aid the drivers in finding their way while their vehicles are in motion.

| Sequence Control Applicability Rating: | Predrive: | ** |
| (See Table 13.14) | In Transit: | * |

Menu Dialogue. In Table 13.14 guidelines are provided for all types of dialogues so that the designer can make informed decisions about the advantages or disadvantages of each. Of the dialogue types described, only three seem to be appropriate for in-vehicle navigation systems: question and answer, menu, and function keys. Some combination of these three will probably suffice for the range of functions that will be needed and the type of users who will be interacting with the system. A primary dialogue type must be chosen that will consistently structure the interaction between user and system. The skills, abilities, and attitudes required for interaction with menu dialogues match the minimum characteristics required of the driver population. Menus are a set of options displayed on the screen in which the selection or execution of one (or more) of the options results in a change in the state of the interface. Menu systems are good when limited flexibility in control access is desirable for system databases that are not exceedingly large. Because of its expected utility for in-vehicle navigation systems, additional guidelines for menu dialogue are listed in Table 13.15. If designers are interested in other types of dialogues, references 1 and 6 present extensive guidelines.

Discussion

As with the other computer system interaction guidelines reviewed here, the data on which menu design guidelines are based are biased in favor of computer programmers, designers, and operators. Furthermore, Paap and Roske-Hofstrand (1988) described many of the menu design recommendations as based on "lore." When using undergraduate novice and experienced subjects in an evaluation of menu, natural language, and command dialogue types. Hauptman and Green (cited in Paap & Roske-Hofstrand, 1988) found no difference in time or accuracy between the two subject groups and the three dialogue types. Although there is a lack of data supporting the menu guidelines, there is a wealth of experience behind them. This data should not be dismissed just on the basis of experimental deficiencies. These menu

TABLE 13.15
Selected Guidelines for the Design of Menus[1,2,6]

1. Menu selection tasks should be considered when there is choice among a constrained set of alternative actions that require little entry of arbitrary data, when users have little training, and when system response is relatively fast.
2. Each page of menu options should have a title that clarifies the purpose of that menu.
3. Names for menu titles and options must be precise and have the same contextual references that users will have.[2]
4. Users should have the capability to stack menu selections (i.e., to make several menu selections without having each menu displayed).
5. A menu shall not consist of a long list of multipage options, but shall be logically segmented to allow several sequential selections among a few alternatives.
6. The system shall present menu selections only for actions that are currently available.
7. Menus shall be presented in a consistent format throughout the system and should be readily available at all times.
8. Menu selection shall be listed in a logical order or, if no logical order exists, in the order of frequency of use.
9. When the number of selections can fit on one page in no more than two columns, a simple menu shall be used. If the number of options exceeds two columns, hierarchical menus may be used.
10. It is better to maximize breadth than depth (i.e., two panels of eight options is better than four panels of four options).[2]
11. Selection codes and associated descriptors shall be presented on single lines.
12. If several levels of hierarchical menus are provided, a direct function call capability shall be provided such that the experienced user does not have to step through multiple menu levels.
13. When selections are indicated by coded entry, the code associated with each option shall be included on the display in some consistent manner.
14. When menu traversal can be accomplished by clearly defined hierarchical paths, the user should be given some indication of the displayed menu's current position in the overall or relevant structure, such as having an optional display of "path" information. A menu tree showing the menu hierarchy should be included in the user manual.
15. When using hierarchical menus, the user shall be able to return to the next higher level by using single key action until the initial, top-level menu or display is reached.
16. A function shall be provided to directly recall the initial, top-level menu or display without the user needing to step through the menu or display hierarchy.
17. A menu system hierarchy should be based on the cognitive pathways derived from the user population and not just that suggested by the design team.[2]
18. When providing feedback in menu systems, the system should indicate the following:

- Which options are selectable
- When an option is under the pointer or cursor and can be selected
- Which options have been selected so far
- The end of the selection process

recommendations provide an initial design for the predrive tasks, subject to experimental verification.

Menu Dialogue Applicability Rating:	Predrive: **
(See Table 13.15)	In Transit: NA

Error Detection, Message Design, and Correction. Because users could spend an inordinate amount of time recovering from an erroneous input, it is imperative that a system aid the user in correcting and recovering from system- or driver-detected errors. The utility of these aids may decrease as the driver becomes familiar with the structure and transactions of the navigation system, but the novice user will benefit from a system that takes into consideration feedback, directional guidance, the opportunity for immediate correction, and the availability of easy-to-understand documentation. These guidelines were taken from reference 5 which were, in turn, based on references 1 and 6.

TABLE 13.16

Guidelines for Error Detection, Message Design, and Correction[1,5,6]

Error Detection

1. Users should be able to stop and return to previous levels of a multilevel control process at any point in a sequence as a result of user-detected error.
2. Rejected inputs should result in an error message with highlighting of the erroneous portion.
3. Batched or stacked strings of entries should be processed (executed) to the point of error, and then an error message should be sent.
4. Error messages should be provided as soon as possible after detection by the system.
5. For multiple errors, the number of errors detected and their locations should be displayed until they are corrected.
6. Errors made while correcting other errors should result in new error messages.
7. User-input errors should be minimized through internal software validation of entries, such as detection of numerics entered in alpha fields.

Error Message Design

1. System-detected errors should result in messages providing as much diagnostic information and remedial action as can be inferred reliably from the error condition.
2. Error messages should reflect the user's point of view of what is needed for recovery.
3. All error messages should indicate the following:
    *   Location of error
    *   Nature of error
    *   One or more ways to recover or clues for finding out how to recover
4. Error messages should appear as close as possible to the erroneous entry.
5. Error messages should be understandable and nonthreatening to user (avoid computer jargon and humorous or condemning messages).
6. User should be able to select the amount of detail contained in error messages: Two levels of messages will be sufficient for most cases.

TABLE 13.16
(Continued)

Error Correction

1. An easy means of correcting erroneous entries should be provided.
2. When an error has occurred, the system should allow immediate correction.
3. A user should not have to reenter an entire line because of an omission or misspelling of one word.
4. Lines of input should be alterable during as well as after entry.
5. Users should be able to stop and return, at any point, to previous levels of multilevel control processes.
6. A simple, standard action that is always available should be provided for requesting HELP.[1]

Guidelines for error detection, message design, and correction are listed in Table 13.16.

Discussion

These recommendations refer more to text editing or tasks that require entering large sets of data. An in-vehicle navigation system should be designed when possible to avoid having the driver enter alphanumeric data. However, there can be occasions when databases are not comprehensive enough or updated frequently enough, requiring the driver to enter hotel or street names on an alphanumeric keypad. In that case, error detection and message design guidelines are helpful. There is no reason to believe that office workers need any less help with this than drivers. Unfortunately, most of the error message guidelines are based on an Engel and Granda (1975) report that contains no description of research or subject parameters. Given the proviso that computer operators and programmers are different from older drivers in many respects, it seems likely that these recommendations should be confirmed by literature review and experimental validation.

Error Detection, Message Design, and Correction Applicability Rating: (See Table 13.16)	Predrive:	**
	In Transit:	*

Display of Text Data. The vast majority of human–computer systems include some type of text. The purpose of the following guidelines is to make labels easy to read and data easy to find. The guidelines, taken from reference 5, are based on an article by Williges and Williges (1984). References

TABLE 13.17
Considerations for Display of Text Data[5]

Display of Text Data

1. Small screen should be used with not more than 50 to 55 characters per line of data.
2. Large screen should be used with two or more columns of 30 to 35 characters per line.
3. Mixture of upper- and lowercase should be preferred.
4. When search is most important, then all uppercase letters are recommended.[2]
5. Text should be left-justified.
6. Consistent spacing between words should be used.[2,6]
7. Spacing between lines should be equal to or slightly greater than the height of the characters.[2,4]
8. Paragraphs should be separated by at least one blank line.
9. For reading ease, field width should be 40 characters or less. For extended text, line lengths should be 60 to 80 characters. Line length interacts with character height and interline spacing.[2]

Display of Alphanumeric Data

1. Character types should be grouped rather than interspersed.
2. Strings of five or more alphanumerics should be grouped at natural breaks or should be grouped into three or four characters when no natural split or predefined break occurs.

Multicolumn Displays

1. Columns should be separated by at least eight spaces when text is right-justified.
2. Columns should be separated by three to four spaces when text is left-justified.

Grammatical Style

1. Statements should be made in the affirmative.
2. Active voice should be used when possible (active voice is generally easier to understand than passive voice).
3. If a sentence describes a sequence of events, the word order in the sentence should correspond to the temporal sequence of events.
4. Short simple sentences should be used.
5. Sentences should begin with the main topic.

---

1 and 6 refer to text editing. Editing of any nature was felt to be inappropriate for an in-vehicle navigation system.

Table 13.17 lists the guidelines for display of text data.

Discussion

All the relevant factors for the design of the user–computer interface have already been mentioned previously. Review of the literature and experimental validation are needed to confirm readability in short glances for older drivers while their vehicles are in motion.

| Display of Text Data Applicability Rating: | Predrive: | ** |
| (See Table 13.17) | In Transit: | * |

TABLE 13.18
Guidelines for Design of Abbreviations and Acronyms[5,6]

General Guidelines

1. Option should be provided for use of abbreviations or full command.
2. User should be instructed concerning method used for selecting command abbreviations.
3. Definition of data-entry codes or abbreviations by the user should be allowed.
4. Contractions should not be used on electronic displays.

Abbreviation Design

Abbreviations should have the following characteristics:

- Limited to one per word
- Considerably shorter than the original term
- Mnemonically meaningful
- Distinctive to avoid confusion
- Composed of unrestricted alphabetic sets when alphabetic data entry is required
- Consistent with unabbreviated command input
- Simply truncated when used with command names.

Expansion of Abbreviations

Abbreviations should be permitted in text entry and expanded later by the computer.

Design of Abbreviations. For input displays, unambiguous abbreviations serve to lower input errors and increase productivity and user satisfaction. For output displays abbreviations reduce reading time and use precious display space optimally. Unless thoughtfully designed, abbreviations can become confusing to users. Table 13.18 presents guidelines for the successful design of abbreviations. They are based on references 5 and 6, which in turn were based on an article by Ramsey and Atwood (1979).

Discussion

There is much consistency among the references concerning rules for abbreviations, but again, they refer to text editing and data entry tasks. Because automotive displays typically have smaller display areas than computer displays because of packaging and cost constraints, abbreviations may be used more frequently in ATIS systems. For the predrive tasks it would seem that these recommendations are good advice except that there

TABLE 13.19
Guidelines for Screen Layout and Structuring[1,2,5,6]

Windowing and Partitioning of Display

1.   The display should not be divided into many small windows.
2.   The user should be permitted to divide the screen into windows or functional areas of an appropriate size for the task.
3.   Dashed lines may be used to segment the display.
4.   The unused areas should be used to separate logical groups, rather than having all the unused area on one side of the display.
5.   In data-entry and retrieval tasks, the screen should be functionally partitioned into different areas to discriminate among different classes of information for commands, status messages, and input fields.
6.   To enhance important or infrequent messages and alarms, they should be placed in the central field of vision relative to the display window.

Organization of Fields

1.   The organization of displayed fields should be standardized. Functional areas should remain in the same relative location on all frames. This permits users to develop spatial expectancies. For example, functional areas reserved for a particular kind of data should remain in the same relative display location throughout the dialogue.
2.   For data-entry dialogues, an obvious starting point in the upper left corner of the screen should be provided.
3.   To avoid clutter, data should be presented using spacing, grouping, and columns to produce an orderly and legible display.
4.   Data should be arranged in logical groups: sequentially; functionally; by conventional usage; by importance; by frequency; and by alphabetical order,[2] which is the least desirable.
5.   Logically related data should be clearly grouped and separated from other categories of data. On large, uncluttered screens, the display or functional areas should be separated by blank spaces (3 to 5 rows or columns). On smaller and more cluttered screens, structure can be defined by geometric shapes, color, and so forth.
6.   Data should be arranged on the screen so that the observation of similarities, differences, trends, and relationships is facilitated for the most common uses.
7.   "Chunks" of data assimilated in one fixation usually subtend approximately a 5 degree square. Screen layout should consider this rule of thumb.[2]

Instruction and Supplemental Information

1.   In computer-initiated dialogues, each display page should have a title that indicates the purpose of the page.
2.   Instructions should stand out (e.g., instructions may be preceded by a row of asterisks).
3.   Instructions on how to use a data entry screen should precede the screen or appear at the top of the screen.
4.   Instructions concerning how to process a completed data-entry screen should appear at the bottom of the screen.
5.   Symmetrical balance should be maintained by centering titles and graphics.
6.   In data-entry and retrieval tasks, the last four lines on each display page should be reserved for messages that indicate errors, communication links, or system status.
7.   When command language is used for control input, an appropriate entry area should be provided in a consistent location on every display, preferably at the bottom of the screen if the cursor can be conveniently moved there.
8.   Displays should be designed so that information relevant to sequence control should be distinctive in position, format, or both.
9.   Frequently appearing commands should appear in the same area of the display at all times.

are no procedures given on how to make abbreviations "mnemonically meaningful" or distinctive to avoid confusion. As with the other human–computer interaction guidelines, abbreviations must be tested with the end-user population, especially if abbreviations are presented to the drivers while they are in transit.

Design of Abbreviations Applicability Rating:	Predrive: **
	In Transit: *
(See Table 13.18)	

Screen Layout and Structure. Structuring a screen layout allows presentation of information that reinforces normal scan patterns, enhances reading ease, and makes relationships easy for users to see. Display of too much information can cause confusion and tax the user's memory. The results are high error rates and user dissatisfaction with the system. Table 13.19 provides a series of guidelines focused on functional grouping and display consistency concepts that produce increased user awareness of the perceptual organization of the data. The guidelines taken from reference 5 are based on a Williges and Williges 1984 article. There is substantial corroborating empirical support from references 1, 2, and 6.

Discussion

The aforementioned human–computer interaction factors apply here as well. The critical question of how screen layout aids or degrades readability during driving has not been tested.

Screen Layout and Structuring Applicability Rating:	Predrive: **
	In Transit: *
(See Table 13.19)	

NAVIGATION INFORMATION FORMAT

Navigation information is typically portrayed by maps that provide direction and distance relationships in a plan view presentation. Another type of navigational format is a turn-by-turn sequential list. There is almost no information on the proper design of maps, turn-by-turn sequential navigation formats, or perspective view formats for particular applications in the guidelines documents reviewed. Two sources have provided guidelines, references 5 and 6, although they are derived from three different domains. The former provides guidelines for "you are here" maps (i.e., maps that appear

TABLE 13.20
Selected Guidelines for Navigation Displays[5,6]

1. Provide maps to display geographic data for direction and distance relationships among physical locations.
2. When it is necessary to show the geographic location of changing events, combine event data with a map background.
3. When several different maps will be displayed, adopt a consistent orientation so that the top of each map will always represent the same direction.
4. Label significant features of a map directly in the display when that can be done without cluttering the display. If the display gets too crowded, consider providing a zoom-in and zoom-out feature.
5. Position labels on a map consistently in relation to the displayed features that they designate.
6. When a map exceeds the capacity of a single display frame in terms of the required extent and detail of coverage, consider providing users a capability to pan the display frame over the mapped data in order to examine a different area of current interest.
7. For "you are here" maps, a map should be aligned with the terrain (alignment principle). This means that on a horizontally placed map, the direction between any pair of locations in the terrain should be parallel to the direction connecting the map symbols for those locations.
8. For "you are here" maps, the upward direction of a map in a vertical position should be equivalent to the forward direction of a map in a horizontal position (forward-up equivalence principle).

on building walls, at elevators, and so forth to orient people in or near buildings). These are based on well-documented experimental research. Reference 6 provides very general guidelines on map displays based on two articles: one pertaining to aviation and another for school atlases. The guidelines are listed in Table 13.20.

Discussion

The database is too small and the application domains too different for automatically applying these recommendations to an in-vehicle navigation display. User spatial abilities will probably vary greatly, and it is already known that some people have great difficulty in deriving and maintaining a global cognitive map as it relates to their immediate vicinity. Some people are able to use (and prefer) plan-view route maps. Some people use landmarks extensively to orient themselves in a turn-by-turn cognitive map. The question of orientation of a plan-view map (north-up or heading-up) is a further complication; how many people prefer north-up versus heading-up? Does it depend on the task? For example, is driving to downtown Los Angeles, California, from Beverly Hills, California, the same kind of navigation task as driving from Beverly Hills, California, to Boulder, Colorado? Suffice it to say that the present reference sources are inadequate in providing usable guideline data for designing navigation formats for in-vehicle navigation displays.

Navigation Display Formats	Predrive:	*
Applicability Rating:	In Transit:	*
(See Table 13.20)		

When discussing research needs for an ATIS system, one should keep in mind that a great volume of information pertaining to human-factors issues is available in the form of journal articles and technical reports. These were not reviewed for this analysis. Unfortunately, the guidelines that were reviewed uniformly lacked the appropriate supporting data necessary to specify driver–vehicle interface guidelines. Additional review of the journal literature and public reports may provide the supplementary information that is missing from guideline supporting data, such as reports on driver–display interfaces that are being designed and evaluated for TRAVTEK, DRIVE, PROMETHEUS, Minnesota Guidestar Travlink, the California Smart Traveler, and Ontario's Travelguide programs. Because research related to Advanced Traveler Information Systems is being pursued with energy and by great numbers at universities worldwide, the likelihood of being able to obtain pertinent driver–vehicle interface data is good. However, there are always gaps in supplementary data sources because of the variability in implementation functions and design objectives. In those cases, additional experimentation must be undertaken to specify in-vehicle system guidelines.

A final note concerning the scope of this chapter. Because of space limitations, a minimum set of ATIS driver–vehicle interface components were considered for this evaluation. New technologies such as voice recognition and liquid crystal and head-up displays may be available with enough capability and reliability to be used in ATIS systems in the near future. With the advent of new technologies, more automation, or better designs, functions currently acceptable for the predrive case may become acceptable for in-transit operation. In that case, human-factors assessments will be mandatory.

REFERENCES

Bergman, M. (1971). Hearing and age. *Audiology, 10*, 164–171.

Beringer, D. B. (1979). The design and evaluation of complex systems: Application to a man–machine interface for aerial navigation. *Proceedings of the Human Factors 23rd Annual Meeting* (pp. 75–79). Santa Monica, CA: Human Factors Society.

Boff, K. R., & Lincoln, J. L. (Eds.). (1988). *Engineering data compendium human perception and performance, Vols. I, II, III.* Wright-Patterson Air Force Base, Ohio: Harry G. Armstrong Aerospace Medical Research Laboratory.

Campbell, J. L. (1988). *Alphanumeric symbology analysis* (Tech. Rep. 88-27-76/G3396-002). Culver City, CA: Hughes Aircraft Company.

Campbell, J. L. (1992). Task H. In C. M. Hein, *Human factors design of automated highway systems (AHS)*, Technical Proposal to Federal Highway Administration (Solicitation No. DTFH61-92-R-00100). Culver City, CA: Hughes Aircraft Company.

Cuff, R. (1980). On casual users. *International Journal of Man-Machine Studies, 12*, 163-187.

Czaja, S. J. (1988). Microcomputers and the elderly. In M. Helander (Ed.), *Handbook of human–computer interaction* (pp. 581-598). Amsterdam, The Netherlands: Elsevier.

Department of Defense. (1989). *Military Standard: Human Engineering Design Criteria for Military Systems, Equipment and Facilities* (MIL-STD-1472D). Washington, DC.

Dingus, T. A., & Hulse, M. C. (1993). Some human factors design issues and recommendations for automobile navigation information systems. *Transportation Research, 1*, 119-131.

Engel, S. E., & Granda, R. E. (1975). *Guidelines for man/display interfaces* (Tech. Rep. TR00.2720). Poughkeepsie, NY: IBM.

Farrell, J. J., & Booth, J. M. (1984). *Design handbook for imagery interpretation equipment.* Boeing Aerospace Co.

Helander, M. G. (1987). Design of visual displays. In G. Salvendy (Ed.), *Handbook of human factors* (pp. 507-548). New York: Wiley.

Helander, M. G. (Ed.). (1988). *Design of human–computer interaction.* Amsterdam, The Netherlands: Elsevier.

Henderson, R. L. (Ed.). (1986). *NHTSA driver performance data book.* Santa Monica, CA: Vector Enterprises, Inc.

Lee, J. D., Morgan, J. M., Wheeler, W. A., Hulse, M. C., & Dingus, T. A. (1993). Development of human factors guidelines for Advanced Traveler Information Systems and Commercial Vehicle Operations. *Task C Working Paper: Description of ATIS/CVO Functions.* Seattle, WA: Battelle Seattle Research Center.

Means, L. G., Fleischman, R. N., Carpenter, J. T., Szczublewski, F. E., Dingus, T. A., & Krage, M. K. (1993). Design of TRAVTAK auditory interface. *Transportation Research Record* (1403, pp. 1-6). Washington, DC.

Paap, K. R., & Roske-Hofstrand, R. J. (1988). Design of menus. In M. Helander (Ed.), *Handbook of human–computer interaction* (pp. 205-235). Amsterdam, The Netherlands: Elsevier.

Ramsey, H. R., & Atwood, M. E. (1979). *Human factors in computer systems: A review of the literature* (SAI-79-111-Den). Englewood, CO: Science Applications.

Rosson, M. B., & Mellon, N. M. (1985). Behavioral issues in speech-based remote information retrieval. In L. Lerman (Ed.), *Proceedings of the Voice I/O Systems Applications Conference '85* (pp. 23-24). San Francisco: AVIOS.

Salvendy, G. (Ed.). (1987). *Handbook of human factors.* New York: Wiley.

Sanders, M. S., & McCormick, E. J. (1987). *Human factors in engineering and design.* New York: McGraw-Hill.

Shurtlett, D. A. (1980). *How to make displays legible.* La Mirada, CA: Human Interface Design.

Simpson, C. A., McCauley, M. E., Roland, E. F., Ruth, J. C., & Williges, B. H. (1987). Speech controls and displays. In G. Salvendy (Ed.), *Handbook of human factors* (pp. 1490-1525). New York: Wiley.

Smith, S. L., & Mosier, J. N. (1986). *Guidelines for designing user interface software* (Tech. Rep. TR MTR-100900). Bedford, MA: The MITRE Corporation.

Snyder, H. L. (1980). *Human visual performance and flat panel display image quality.* Blacksburg, VA: Virginia Polytechnic Institute and State University: Human Factors Laboratory.

Snyder, H. L., Lynch, E. F., Abernathy, C. N., Companion, M. A., Green, J. M., Helander, M. G., Hirsch, R. S., Hunt, S. R., Korell, D. D., Kroemer, K. H., Murch, G. M., Palacios, N., Palermo, S. A., Rinalducci, E. J., Rupp, B. A., Smith, W., Wagner, G. N., Williams, R. D., & Zwalen, H. (1988). *American National Standard for Human Factors Engineering of Visual Display Terminal*

Workstations (ANSI/HFS 100-1988). Copyright 1988 by the Human Factors and Ergonomics Society.

Sorkin, R. D. (1987). Design of auditory and tactile displays. In G. Salvendy (Ed.), *Handbook of human factors* (pp. 549–576). New York: Wiley.

Sorkin, R. D., & Kantowitz, B. H. (1987). Speech communications. In G. Salvendy (Ed.), *Handbook of human factors* (pp. 294–309). New York: Wiley.

Streeter, L. A., Vitello, D., & Wonsiewicz, S. A. (1985). How to tell people where to go: Comparing navigational aids. *International Journal of Man-Machine Studies, 22*, 549–562.

Tullis, S. (1988). Screen design. In M. Helander (Ed.), *Handbook of human–computer interaction* (pp. 377–412). Amsterdam, The Netherlands: Elsevier.

Williges, R. C., Williges, B. H., & Elkerton, J. (1987). In G. Salvendy (Ed.), *Handbook of human factors* (pp. 1416–1449). New York: Wiley.

Williges, B. H., & Williges, R. C. (1984). Design considerations for interactive computer systems. In F. A Muckler (Ed.), *Human factors review: 1984* (pp. 167–208). Santa Monica, CA: Human Factors Society.

AUTHOR INDEX

SUBJECT INDEX